Nutritional Value and Health Benefits from Fruits, Vegetables, Nuts and Spices

The Author

Dr. U.D. Chavan obtained his B.Sc. and M.Sc. (Agri. in Biochemistry) degrees from Mahatma Phule Krishi Vidyapeeth, Rahuri in 1985 and 1987 respectively. He received his Ph.D. degree in Food Science from Memorial University of Newfoundland St. John's Canada in 1999. He has done International training on "Global Nutrition 2002" at Uppsala University Uppsala, Sweden in 2002. He also attended follow-up International workshop on "Global Nutrition 2002" at Hanoi, Vietnam in 2002. Dr. Chavan visited Denmark, Finland, Ireland, France, Switzerland, Poland, Spain, Vietnam, Thailand, England, and U.S.A. under "Global Nutrition 2002" Programme sponsored by Swedish International Development Agency (SIDA). During this programme he worked on human nutritional diet and disorders as well as on genetically modified organisms (GMO).

Dr. Chavan worked as Senior Research Assistant in the Department of Biochemistry and Food Science and Technology at Mahatma Phule Krishi Vidyapeeth Rahuri from 1988 to 2000. During his Ph.D., he worked as Technician/Research Associate at Atlantic Cool Climate Crop Research Center and Agriculture and Agri-Food Canada. He received D.Sc. degree in 2006 from USA. He has guided 21 students for M.Sc. (Agri.) in Biochemistry and Food Science and Technology. From 2000 to 2004 he worked as Assistant Professor of Biochemistry at Mahatma Phule Krishi Vidyapeeth Rahuri.

Dr. Chavan received Korgavkar Trust Fellowship, under-graduate and post-graduate merit scholarship as well as Senior Research Fellowship from ICAR, New Delhi and United States Department of Agriculture U. S. A. He received University Graduate Fellowship from Memorial University of Newfoundland St. John's Canada for his Ph. D. programme. Dr. Chavan also received international Scholar Award and Excellentiam Award for his Ph. D. research work from Memorial University of Newfoundland St. John's Canada. He has received a "Certificate of Appreciation" from the U.S. Department of Agriculture in 1992 for his work on processing of groundnut under the guidance of Dr. S. S. Kadam (Principle Investigator). He has written 110 research papers and 105 popular articles. He has authored 18 books in Marathi, 7 books in English and six book chapters in English. He has been awarded "Literary Award" for his best book on "Growth Regular" on agriculture during 1997. Dr. Chavan was selected as a best group leader and best presentation for "Global Nutrition 2002" by SIDA. He has been awarded "Life Time Achievement Award 2004" for his outstanding contribution in post-harvest technology of fruits and vegetables and allied fields by United Writers' Association of India. He has been elected as a Fellow of United Writers' Association of India in 2004. Dr. Chavan received pride of nation and Maharashtra Gunigan Ratna Award for 2006. He also received Jewels of India Award for 2006 for his contribution in the field of Food Science and Technology. He has been awarded "Literary Awards" for his best book on "Sorghum Grain Processing" on agriculture during 2009-2010 and Pulses cultivation to processing and value addition for 2012. He is recipient of "Krishi Goorav Award-2011" from Bhartiya Krishak Samaj, Maharashtra State, Nasik. He also contributed in the development of crop varieties in wheat NIAW-917 and in Sorghum Phule Anuradha (RSV 423), Phule Chitra (RSV 1546), Phule Suchitra (RSV 1098), Phule Vasudha (RSV 423), Phule Revati (RSV 1006), SPV-1626, Phule Panchami (RSPOV-3) for Sorghum pops and SPV-2057 for forage single cut. Now he is working as a Senior Cereal Food Technologist in the Department of Food Science and Technology at Mahatma Phule Krishi Vidyapeeth, Rahuri, Dist. Ahmednagar, Maharashtra.

Nutritional Value and Health Benefits from
Fruits, Vegetables, Nuts and Spices

U.D. Chavan
Professor
Department of Food Science and Technology
Mahatma Phule Krishi Vidyapeeth,
Rahuri – 413 722, Maharashtra, India

2015
Daya Publishing House®
A Division of
Astral International Pvt. Ltd.
New Delhi – 110 002

ISBN 9789351306153

Published by : **Daya Publishing House®**
A Division of
Astral International Pvt. Ltd.
– ISO 9001:2008 Certified Company –
4760-61/23, Ansari Road, Darya Ganj
New Delhi-110 002
Ph. 011-43549197, 23278134
E-mail: info@astralint.com
Website: www.astralint.com

Laser Typesetting : **Classic Computer Services**, Delhi - 110 035

Printed at : **Thomson Press India Limited**

PRINTED IN INDIA

Acknowledgement

Author is thankful to the Hon. Dr. T.A. More, Vice-Chancellor, Mahatma Phule Krishi Vidyapeeth, Rahuri for his critical and fruitful suggestions while the preparation of *"Nutritional Value and Health Benefits from Fruits, Vegetables, Nuts and Spices"* book. I am thankful to those scientists who worked on development of various analyses and documentation of fruits, vegetables, nuts, seeds and spices and their contribution is acknowledged in this book as well their sources are cited in the reference section. Author is also thankful to all scientists and friends those helped directly or indirectly while preparation of this manuscript. I would like to express over profound thanks to our family members for their constant support, patience and devotion, which inspired me to complete this manuscript.

U.D. Chavan

Foreword

The Indian Council of Agricultural Research and the State Agricultural Universities have made concentrated efforts in increasing productivity of fruits and vegetables. In recent years, the Government of Maharashtra has launched an ambitious Horticulture Development Programme to increase area under fruit crops by providing grant-in-aid to the growers. It is envisaged that the area covered under fruit crops in Maharashtra would be 10 lakh hectares in the next few years. However, very little attention has been paid to reduce the post harvest losses that occur due to poor handling, storage and transportation. According to reports of Swaminathan Committee, post harvest losses of fruits and vegetables in India vary from 20-30 per cent of production with estimated value of Rs. 1500-2000 crores annually. If these losses are controlled by using improved post harvest technology, the supplies of fresh fruits and vegetables will be increased to the extent of their existing losses. This will help to increase the per capita availability of fruits and vegetables. If the people know the importance of the fruits, vegetables, nuts and spices by knowing their nutritional values and health benefits they will take more amounts of these food items in their diet.

Recent developments in nutrition have created greater demands for fresh fruits and vegetables. The processing technology that increases the functionality without changing their fresh like properties has significant role in modern fruit and vegetable industry. This book addresses factors that are involved in maintaining the quality of fresh fruits, vegetables, nuts and spices. It is vital that consumers in health-conscious society would be provided with food products which retain their quality characteristics leading to more healthy individuals. From this people will now the nutritional value and their benefits to the health.

Fruits and vegetables, being perishable, are processed into several products such as jams, jellies, marmalade, syrups, squashes and canned, dehydrated and fermented products by employing different methods of processing. This book incorporates information on principles involved in preparation of various products as well as methodology employed in home scale as well as industrial processing of fruits and vegetables. I hope, this information will be useful for students in the field of horticulture, food science and nutrition and professionals in food industries.

The ever-increasing scientific research on the subject has resulted in a large body of knowledge which can be found scattered in various scientific journals and technical reports and is not easily accessible. The available material needs to be assembled and briefed in order that the students/technologists/agriculturists/horticulturists/ engineers and others actively engaged in quantitative studies can have it as a ready reference. This book provides a comprehensive but concise treatise on the subject matter.

J.K. Chavan

Ex. Head, Food Science and Technology

MPKV, Rahuri

Preface

Fruits and vegetables are important constituents of our diet which provide dietary fibers, vitamins, and minerals in addition to proteins, carbohydrates and a small amount of lipids. These contribute more than 50 per cent requirement of vitamin A. The dietary fibers aid in digestion, bowel movement and utilization of more concentrated foods in human diet. Fruits and vegetables are highly perishable commodities. Losses to the extent of 20-30 per cent occur during post-harvest handling of fruits and vegetables. If these losses are controlled by using improved post-harvest technology, the supplies of fresh fruits and vegetables can be increased to the extent of their existing losses. This will certainly help to increase per capita availability of fruits and vegetables.

The consumption of fruits, vegetables, nuts and spices in fresh, unprocessed form is economically cheaper and nutritionally more desirable. Nowadays people are very conscious about their health and therefore, they want to know about their food materials. They also are trying to get information about benefits of the food material to the health. The minimal processing involves treatment to these commodities without Changing wholesomeness and nutritional properties of fresh produce. It includes sorting, grading and packaging of fresh produce so that consumer can get processed product having physical, nutritional and aesthetic properties similar to that of fresh produce. In recent years, there are significant developments in canning, freezing and dehydration of fruits and vegetables. Novel product in tetra pack like Frooti has captured the Indian market as ready-to-serve beverage. Similarly, canned corn and canned mixed beans have become popular in Indian market. Ketchup and sauces are also produced and marketed under various trade names. The processed products like jams and jellies have become integral part of breakfast served in middle class families of our society. Considering the importance of fresh as well as processed products, an attempt has been made to present the fruits, vegetables, nuts and spices nutritional values and their benefits to the human being in this book. I hope, this will be useful book to under-graduate as well as post-graduate students in agriculture, food technology, horticulture, home science, consumers and mostly industrial sectors.

U.D. Chavan

Contents

B. Nutritional Value and Health Benefits of Vegetables

C. Nutritional Value and Health Benefits of Nuts and Seeds

A. Nutritional Value and Health Benefits of Fruits

Introduction

Fruits are nature's wonderful gift to mankind; indeed, medicines packed with vitamins, minerals, anti-oxidants and many phyto-nutrients (Plant derived micro-nutrients). They are an absolute feast to our sight, not just because of their colour and flavour but for their unique nutrition-profile that help the human body free of diseases and stay healthy.

1. Fruits are low in calories and fat and are a source of simple sugars, fiber, and vitamins, which are essential for optimizing our health.
2. Fruits provide plenty of soluble dietary fiber, which helps to ward of cholesterol and fats from the body and to get relief from constipation as well.
3. Fruits contain many anti-oxidants like *poly-phenolic flavonoids, vitamin-C,* and *anthocyanins.* These compounds, firstly, help human body protected from oxidant stress, diseases, and cancers, and secondly; help the body develop capacity to fight against these ailments by boosting our immunity level. Many fruits, when compared to vegetables and cereals, have very high anti-oxidant value, which is something measured by their "Oxygen Radical Absorbent Capacity" or (ORAC).
4. Anthocyanins are flavonoid category of poly-phenolic compounds found in some "blue-fruits" like blue-black grapes, mulberries, acaiberry, chokeberry, blueberries, blackberries, and in many vegetables featuring blue or deep purple colour. Eating fruits rich in blue pigments offers many health benefits. These compounds have potent anti-oxidant properties, remove free radicals from the body, and thus offer protection against cancers, ageing, infections, etc. These pigments tend to concentrate just underneath the skin.
5. Fruit's health benefiting properties are because of their richness in vitamins, minerals, micro-nutrients, anti-oxidants, which helps the body prevent or

at least prolong the natural changes of ageing by protecting and rejuvenating cells, tissues and organs. Their overall benefits are manifold! Fruit nutrition benefits are infinite! You are protecting yourself from minor ailments like wrinkling of skin, hair-fall, and memory loss to major ailments like age-related macular degeneration (AMRD) of the retina in the eyes, Alzheimer's disease, colon cancers, weak bones (osteoporosis) etc., and the list of fruit nutrition benefits never ends!

How much Fruit Nutrition should be Included in our Daily Diet?

Research studies suggest that one may eat any servings of fruits daily. It is recommended to eat at least 2-3 servings of fresh fruits every day. The term one fruit serving is about 250 g of cleaned, edible-portion of the fruit, excluding discards like peel, seeds, etc. Include seasonal fruits in the daily diet. Variety of fruit's type, colour, and flavour should be encouraged to get maximum health benefits. Yellow and orange colour fruits are rich sources of α- and β-carotenes, zeaxanthin and crypto-xanthin, while blue, black coloured like black or blueberries are a good source of poly-phenolic anthocyanin anti-oxidants.

Selection of Fruits

To avail maximum fruit nutrition benefits eat organic produce. Many "wild" varieties of berries and "tropical tree" fruits have still not treated with any kind of fertilizers or chemicals and can be readily purchased from local farm owners. Organic fruits tend to be smaller; however, they feature special flavour and richness in vitamins, minerals and stuffed with numerous anti-oxidants. In the store, however, choose fruits that feature freshness, bright in colour and flavour and feel heavy in your hands. Look carefully for blemishes, spots, molds, and signs of insecticide spray. Buy whole fruits instead of section of them (for example, buy a small size watermelon instead of a section of big size melon).

Dry Fruits

Dry fruits are rich source of nutrients such as vitamins and minerals. Consumption of some dry fruits daily enhances the overall bioavailability of nutrients. Try raisins (dried grapes), apricots, dates, figs, tamarind, which are indeed very good in iron, calcium, zinc, selenium and manganese. In addition, mixing dry fruits with some fresh fruits would help you get vitamin C which in-turn facilitates complete absorption of iron inside the stomach.

How to Use Fruits?

Barring few fruits like berries, try to wash them soon after shopping. Rinse them in salt water for few minutes and then wash in cool running water until you are satisfied with complete cleanliness. This way, you make sure that the fruit you eat is rid-off dust, sand and any residual chemical spray. Eat them while they are fresh because, certain fruits have very short shelf life and the health benefiting properties of fruit declines with time. If you need to store them, place them in plastic wrappings or in zip pouches to keep them fresh for short periods until you use them. Eat fruits as a whole without any additions to get their original flavour, taste, and to get maximum fruit nutrition benefits.

If you need to store them, place them in plastic wrappings or in zip pouches to keep them fresh for short periods until you use them. Eat fruits as a whole without any additions to get their original flavour, taste, and to get maximum fruit nutrition benefits.

Nutritional Significance

Fruits and vegetables play an important role in human nutrition particularly in supplying certain constituents in which other food materials are deficient. Fruits and vegetables contribute vitamins and minerals and the bulk of the dietary fiber to the diets of human beings. They neutralize the acid substances produced in the course of digestion of meat, cheese, and other high-energy foods. The value of fruits and vegetables as dietary fiber has increasingly been realized in recent years. The possible beneficial effects of fiber, vitamin C, β-carotene, and the vitamin B-complex derived from fruits on human health are being reexamined with the objective of minimizing certain important human diseases related to lack of fiber and antioxidant properties of some vitamins and minerals in diets. Ascorbic acid (vitamin C) is the principal vitamin supplied by fruits and vegetables in the diet. About 90 per cent of a person's dietary vitamin C requirement is obtained from fruits and vegetables. In addition, fruits and vegetables are rich source of β-carotene, thiamine, niacin, folic acid and vitamin A.

The β-carotene is essential structural component of the eye, and a prolonged deficiency of this vitamin may lead to blindness. These are also rich in calcium, magnesium, phosphorus, and iron, which are not present in many other food materials in quantities sufficient for the needs of the body.

Fruits and vegetables are particularly rich in pectin and gums. Pectin has been shown to delay gastric emptying, which may lead to favourable changes in glycaemic responses to foods. Pectin may also induce a satiety effect. It acts as a general intestinal regulator and detoxifying agent. It is used as a standard additive in commercial baby food formulations. Pectin has been shown to have antibacterial, antiviral, wound-healing, and metal-binding properties. It is shown to be effective in reducing cholesterol levels.

A substantial proportion of carbohydrates in fruits and vegetables are present as dietary fiber in the form of cellulose, hemicelluloses, pectin substances, and lignin. It is not digested and passed through the human intestinal system, as people are not capable of secreting the digestive enzymes necessary to break down these polymer complexes into simple units in the forms absorbed by the intestinal tract. Human body lacks enzymes such as cellulase, hemicellulase, and pectinase. Fibre was once considered to be an unnecessary component in the human diet although it was thought to relieve constipation. The epidemiological evidence obtained so far shows that dietary fiber can be the panacea to cure several human diseases, especially in the affluent societies.

Fruits and vegetables contain some flavour compounds. These include sugars, amino acids, organic acids, volatiles such as aromatic hydrocarbons, aldehydes, acetals, ketones, alcohols, ester and sulphur compounds when used along with other food items. These flavouring compounds make food more palatable Onions and garlic are used as flavouring agents in variety of dishes, soups, sausage and curries. The pungency of garlic is due to volatile sulphur compound, diallyl disulphide which is produced by the action of allinase enzyme on the amino acid alliin present in garlic. Several volatile sulphur compounds are responsible for characteristic flavour of cole crops. Dimethyl trisulphide has been indicated as a major aroma component in cooked cole vegetables.

Therapeutic Properties

Many therapeutic drugs in use in modern medicine originated as plant extracts. Certain fruit and vegetable components exert pharmacological or therapeutic effects. Limonin and nomilin (and other limonoids) are present in citrus fruits such as ornage, lemon; β-carotene may also play a role in the prevention of some forms of cancer.

Fruits and vegetables are very well known sources of different types of flavonoids which include flavonones, flavones, flavonols, anthocyanins, catechins and biflavans. Apart from fat soluble tocopherols, flavonoids are commonest and most active antioxidant compounds in human food, being active in hydrophillic as well as in lipophillic systems. This has considerable practical impact on nutrition. It protects food stuff from oxidative deterioration, prolongs their shelf life and keeping quality and improves taste, acceptability and wholesomeness of mixed dishes by inhibiting the oxidation of accompanying lipids. Flavonoids increase or stabilize biological activity of ascorbic acid. All fruits and vegetables rich in flavonoids are marked by the unusual stability of their vitamin C. The ascorbic acid stabilizing effect of flavonoids results in increased production of ascorbic acid-2-sulfate which improves membrane stability by synthesis of sulfated mucopolysaccharides and enhances cholesterol excretion. Ellagic acid is a naturally occurring phenolic constituent present in fruits, especially strawberries and other berries and has been shown to be effective as an anti mutagen and anticarcinogen and a potential inhibitor of common inducers of cancer.

Onion and garlic are known to reduce low-density lipoprotein and induce the formation of high-density-lipoprotein. The increase in high-density lipoprotein is more evident from consumption of white and yellow onion and is reduced with cooking. The allicin which is produced in garlic when tissue is injured by cutting or crushing has been reported to have hypocholesterolaemic action.

Cole vegetables contain a group of compounds known as indoles which have been recently linked with the prevention of cancers of colon, rectum and breast. Some of the species of *Cucumis. Luffa, Coccinia, and Momordica* exhibit varying degrees of bitterness in fruits, leaves and twigs. This is caused by terpenes called momordicins. Owing to its bitter principle, the bitter gourd is used in Ayurvedic system of medicine. A hypoglycemic ingredient cheratin isolated from bitter gourd lowers blood sugars. Sapogenin compound called diosgenin present in some species of yam is used for production of cortisome and contraceptive drugs.

1

Acid Berry

Recent revelations establish that it has no extra beneficial nutrients than some of commonly used fruits like mango, blueberry, pomegranate, etc. The berry has been a traditional staple food of Amazon basin tribes for centuries, but somehow recently gained recognition as so-called "super food" in the western world. Nonetheless, this wonderful berry has very good levels of anti-oxidants, minerals, and vitamins that would benefit overall health and fitness.

Acai berry is a small, round, dark purple fruit obtained from acai palm tree. Botanically, acai belongs to palm or *Arecaceae* family of tall trees in the genus *Euterpe* and known as *Euterpe oleracea.*

Acai palm is a tall, slender tree, which grows 15 to 25 m in height. The average mature plant has 3-10 well-developed stems (10-18 cm in diameter) from a single seed and root system. It grows better under waterlogged low-lying plains receiving good annual rainfall. Each stem behaves like individual tree and a bear 3-5 bunches with each bunch holding about hundreds of berries in a way similar to areca palm or coconut palm. Each acai berry measures about the size of small sized grape, 2-3 cm in diameter. They appear dark-green initially, later turn to deep-purple colour on attaining maturity. Technically, the fruit is a drupe which consists of outer-edible pulp surrounding a central large seed. Only the pulp part, comprising about just 10-15 per cent of berry weight, is edible.

Health Benefits of Acai Berry

- ☆ Acai berry has very good levels of anti-oxidants, minerals, and vitamins that have health benefiting and disease preventing properties.
- ☆ Unlike other berries and fruits, acai has high caloric values and fats. 100 g of berries provide about 80-250 calories depending up on the preparation and serving methods. In fact, fresh acai berry has been the staple nutritious

diet of native Amazonian for centuries. There is no solid research evidence suggesting the use of these berries will help lose weight.

- ☆ Acai berry contains many polyphenolic anthocyanin compounds like *resveratrol, cyanidin-3-galactoside, ferulic acid, delphinidin, petunidin* as well as astringent pro-anthocyanidin tannins like *epicatechin, protocatechuic acid* and *ellagic acid*. Scientific studies suggest that these compounds have been claimed to act as anti-ageing, anti-inflammatory, anti-cancer functions by virtue of their anti-free radical fighting actions. In addition, *tannins* are known to have anti-infective, anti-inflammatory and anti-hemorrhagic properties.
- ☆ Primary research studies suggest that *ellagic acid* in acai has anti-proliferative properties by virtue of its ability to directly inhibit DNA binding of certain carcinogens (nitrosamine toxins in the food).
- ☆ Acai berry is also rich in medium chain fatty acids like oleic acid (omega-9) and linoleic acid (omega-6). These compounds help reduce LDL-cholesterol level and raise good HDL-cholesterol levels in the body and thus help prevent heart disease. In addition, the essential fats help prevent skin dryness by maintaining adequate moisture in the skin.
- ☆ Acai pulp has good levels of dietary fiber. Adequate fiber in the diet helps clear cholesterol through the stools.
- ☆ ORAC value (oxygen radical absorbance capacity) of acai berry is thought to be at mid-level range for fruits, higher than that of oranges but less than pomegranate. USDA so far not validated exact ORAC value of acai.
- ☆ Acai berries contain a good amount of minerals like potassium, manganese, copper, iron, and magnesium. Potassium is an important component of cell and body fluids that helps controlling heart rate and blood pressure. Manganese is used by the body as a co-factor for the antioxidant enzyme, *superoxide dismutase*.
- ☆ Further, they are rich in B-complex vitamins and vitamin-K. Contain very good amounts of niacin, vitamin B-6 and riboflavin. These vitamins are function as co-factors and help body in the metabolism of carbohydrates, proteins, and fats.

Acai Uses in Traditional Medicine

- ☆ Acai extracts have been used by Amazon basin tribes as the treatment remedy for diarrhea, parasitic infections, hemorrhages, and ulcer treatment.
- ☆ A kind of decoction obtained from crushed seeds has been used in the treatment of fever.
- ☆ Decoction obtained from the roots has been used in the treatment of menstrual pain, liver diseases, malaria in Peruvian culture.

2

Apple

Delicious and crunchy apple fruit is one of the most popular fruits, favourite of health conscious, fitness lovers who believe in the concept "health is wealth." This wonderful fruit is packed with rich phyto-nutrients that in the true senses indispensable for optimal health. The antioxidants in apple have much health promoting and disease prevention properties; thus truly justifying the adage, *"an apple a day keeps the doctor away."* Scientific name: *Malus domestica.*

Apples are obtained from the medium-sized tree belonging to the *Rosaceae* family. The apple tree is originated in the mineral-rich mountain ranges of Kazakhstan, and now being cultivated in many parts of the world. Apple fruit features oval or pear shape; and the outer skin has different colours depending upon the cultivar type. Internally, the juicy pulp has been off-white to cream in colour and has to mix of mild sweet and tart taste. Its seeds are inedible because of their bitter taste. Hundreds of varieties of apples are either eaten as table fruits or as dessert fruit grown in the US and worldwide. Some of the apples are sought-after in cooking and baking too.

Health Benefits of Apple

- ☆ Delicious and crunchy apple fruit is notable for its impressive list of phtyto-nutrients, and anti-oxidants. Studies suggest that its components are essential for normal growth, development and overall well-being.
- ☆ Apples are low in calories; 100 g of fresh fruit slices provide only 50 calories. They, however, contain no saturated fats or cholesterol. Nonetheless, the fruit is rich in dietary fiber, which helps prevent absorption of dietary-LDL or bad cholesterol in the gut. The fiber also saves the colon mucous membrane from exposure to toxic substances by binding to cancer-causing chemicals inside the colon.
- ☆ Apples are rich in antioxidant phyto-nutrients *flavonoids* and *polyphenolics.* The total measured anti-oxidant strength (ORAC value) of 100 g apple fruit

is 5900 TE. Some of the important flavonoids in apples are quercetin, epicatechin, and procyanidin B2. Additionally, they are also good in tartaric acid that gives tart flavour to them. Altogether, these compounds help the body protect from deleterious effects of free radicals.

Apple Fruit (*Malus domestica*), Fresh, Nutritive Value per 100 g, ORAC Value-5900

Principle	*Nutrient Value*	*Percentage of RDA*
Energy	50 Kcal	2.5 per cent
Carbohydrates	13.81 g	11 per cent
Protein	0.26 g	0.5 per cent
Total Fat	0.17 g	0.5 per cent
Cholesterol	0 mg	0 per cent
Dietary Fiber	2.40 g	6 per cent
Vitamins		
Folates	3 µg	1 per cent
Niacin	0.091 mg	1 per cent
Pantothenic acid	0.061 mg	1 per cent
Pyridoxine	0.041 mg	3 per cent
Riboflavin	0.026 mg	2 per cent
Thiamin	0.017 mg	1 per cent
Vitamin A	54 IU	2 per cent
Vitamin C	4.6 mg	8 per cent
Vitamin E	0.18 mg	1 per cent
Vitamin K	2.2 µg	2 per cent
Electrolytes		
Sodium	1 mg	0 per cent
Potassium	107 mg	2 per cent
Minerals		
Calcium	6 mg	0.6 per cent
Iron	0.12 mg	1 per cent
Magnesium	5 mg	1 per cent
Phosphorus	11 mg	2 per cent
Zinc	0.04 mg	0 per cent
Phyto-nutrients		
Carotene-β	27 µg	—
Crypto-xanthin-β	11 µg	—
Lutein-zeaxanthin	29 µg	—

✰ Apple fruit contains good quantities of *vitamin-C* and *beta-carotene.* Vitamin C is a powerful natural antioxidant. Consumption of foods rich in vitamin

C helps the body develop resistance against infectious agents and scavenge harmful, pro-inflammatory free radicals from the body.

- ✰ Further, apple fruit is a good source of B-complex vitamins such as riboflavin, thiamin, and pyridoxine (vitamin B-6). Together these vitamins help as co-factors for enzymes in metabolism as well as in various synthetic functions inside the body.
- ✰ Apple also contains a small amount of minerals like potassium, phosphorus, and calcium. Potassium is an important component of cell and body fluids helps controlling heart rate and blood pressure; thus, counters the bad influences of sodium.

3

Apricot

Sweet, fragrant rich, golden-orange apricot fruits are another summer delicacy of Asian origin. These much-prized fruits were first brought to Europe by Greeks who called them as "golden eggs of the sun." Today, the most important commercially producing countries are Turkey, Iran, Italy, France, Spain, Syria, Greece, and China. Botanically, the fruit is closely related to peaches and nectarine, sharing with them in the broader *Rosaceae* family of fruit trees in the genus, *Prunus*. Scientific name: *Prunus armenia*.

Apricot is a medium sized deciduous tree growing best in the well-drained mountainous slope soils. During the spring, the plant bears plenty of beautiful pinkish-white flowers, which attract bees. The fruits have almost uniform size, 4-5 cm in diameter, and weigh about 35 g. In structure; the fruit is a drupe, consist of a centrally located single seed surrounded by crunchy, aromatic edible flesh. The seed is enclosed in a hard stony shell, often called a "stone."

Fresh, ripe apricots have a sweet flavour similar to plums. Sun dried organic fruits have concentrated nutrient values than fresh ones, although they are lesser in vitamin-C content. Its seed is also edible and taste like that of almonds. Oil extracted from the seeds has been used in cooking.

Health Benefits of Apricots

- ☆ Fresh fruits are low in calories, composing just 50 calories per 100 g weight. Nonetheless, they are rich source of dietary fiber, antioxidants, vitamins, and minerals. The fruits are enriched with numerous health promoting phyto-chemicals; that helps prevent heart disease, reduce LDL, ("bad cholesterol") levels and offer protection against cancers.
- ☆ Apricots are excellent sources of vitamin-A, and carotenes. 100 g fresh fruits have 1926 IU or 64 per cent of daily-required levels of vitamin A. Both

compounds are known to have antioxidant properties and are essential for vision. Vitamin-A is also required for maintaining healthy mucus membranes and skin. Consumption of natural fruits rich in carotenes helps protect the body from lung and oral cavity cancers.

Apricots (*Prunus armeniaca*), fresh, Nutritive Value per 100 g

Principle	*Nutrient Value*	*Percentage of RDA*
Energy	50 Kcal	2.5 per cent
Carbohydrates	11 g	8.5 per cent
Protein	1.4 g	2.5 per cent
Total Fat	0.4 g	1 per cent
Cholesterol	0 mg	0 per cent
Dietary Fiber	2 g	5 per cent
Vitamins		
Folates	9 µg	2 per cent
Niacin	0.600 mg	4 per cent
Pantothenic acid	0.240 mg	5 per cent
Pyridoxine	0.054 mg	5 per cent
Riboflavin	0.040 mg	3 per cent
Thiamin	0.030 mg	2.5 per cent
Vitamin A	1926 IU	64 per cent
Vitamin C	10 mg	16 per cent
Vitamin E	0 mg	0 per cent
Vitamin K	3.3 µg	3 per cent
Electrolytes		
Sodium	1 mg	0 per cent
Potassium	259 mg	5.5 per cent
Minerals		
Calcium	13 mg	1.3 per cent
Copper		
Iron	0.39 mg	5 per cent
Magnesium	10 mg	2.5 per cent
Manganese	0.077 mg	3 per cent
Phosphorus	23 mg	3 per cent
Zinc	0.2 mg	2 per cent
Phyto-nutrients		
Carotene-α	19 µg	—
Carotene-β	1094 µg	—
Crypto-xanthin-β	104 µg	—
Lutein-zeaxanthin	89 µg	—

- Fresh fruits contain vitamin-C, another natural anti-oxidant. Vitamin-C helps the body develop resistance against infectious agents and scavenge harmful oxygen-free radicals.
- They are an also good source of minerals such as potassium, iron, zinc, calcium and manganese. Potassium is a heart-healthy mineral; an important component of cell and body fluids that help regulate heart rate and blood pressure.
- The total anti-oxidant or ORAC value of raw apricots is 1115 umol TE/100 g. Much of this in these fruits comes from some important health promoting flavonoid poly phenolic anti-oxidants such as *lutein, zeaxanthin and beta cryptoxanthin.* Altogether, these compounds act as protective scavengers against oxygen-derived free radicals and reactive oxygen species (ROS) that play a role in ageing, cancers and various disease processes.
- Further, zeaxanthin, a carotenoid selectively absorbed into the retinal "macula lutea" in the eyes where it is thought to provide anti-oxidant and protective light-filtering functions. Thus, consumption of fruits like apricots rich in zeaxanthin helps eyes protect from age-related macular disease (AMRD), especially in the elderly people.

4

Avocados

Avocados are distinct fruits with high-fat content and calories. Subtly flavourful yet buttery, they are among the most popular fruits having nutrition profile similar to that of some edible nuts and seeds. Some of the common names for this staple fruit are alligator pear, aguacate, butter pear, etc. Botanically, the fruit belongs to the family of *Lauraceae*, the family that also includes some unusual members like bay laurel, cinnamon, etc. Scientific name is *Persea americana*.

Avocados are medium sized evergreen trees of about 20-30 feet in height with large green leaves. The tree prefers fertile soil with high moisture content to flourish. Small light green flowers appear during winter. After about 8-10 months later, hundreds of pear shaped green colour fruits appear on the tree. Avocados, like bananas, mature on the tree but ripen only after their harvest. Once Ripen, they turn dark green or deep purple and yield to gentle pressure. Inside, cream colour flesh has butter-like consistency with bland taste yet pleasant aroma. The fruit features centrally placed single brown colour seed. On an average, each fruit weighs about 300-700 g, although heavier avocados are quite common in the markets.

Health Benefits of Avocado

- ☆ Avocados, like olives, are high in mono-unsaturated fats and calories. However, they are very rich in dietary fiber, vitamins, and minerals and packed with numerous health benefiting plant nutrients.
- ☆ Their creamy pulp is a very good source of mono-unsaturated fatty acids like oleic and palmitoleic acids as well as omega-6 poly-unsaturated fatty acid linoleic acid. Research studies suggest that Mediterranean diet that is rich in mono-unsaturated fatty acids help lower LDL or bad cholesterol and increase HDL or good-cholesterol, thereby, prevent coronary artery disease and strokes by favouring healthy blood lipid profile.

Avocado (*Persea americana*), Nutrition Value per 100 g.

Principle	*Nutrient Value*	*Percentage of RDA*
Energy	160 Kcal	8 per cent
Carbohydrates	8.53 g	6.5 per cent
Protein	2.0 g	3.5 per cent
Total Fat	14.66 g	48 per cent
Cholesterol	0 mg	0 per cent
Dietary Fiber	6.7 g	18 per cent
Vitamins		
Folates	81 µg	20 per cent
Niacin	1.738 mg	11 per cent
Pantothenic acid	1.389 mg	28 per cent
Pyridoxine	0.257 mg	20 per cent
Riboflavin	0.130 mg	10 per cent
Thiamin	0.067 mg	5.5 per cent
Vitamin A	146 IU	5 per cent
Vitamin C	10 mg	17 per cent
Vitamin E	2.07 mg	14 per cent
Vitamin K	21 µg	17.5 per cent
Electrolytes		
Sodium	7 mg	0.5 per cent
Potassium	485 mg	10 per cent
Minerals		
Calcium	12 mg	1 per cent
Copper	0.190 mg	21 per cent
Iron	0.55 mg	7 per cent
Magnesium	29 mg	7 per cent
Manganese	0.142 mg	6 per cent
Phosphorus	52 mg	7 per cent
Selenium	0.4 µg	<1 per cent
Zinc	0.64 mg	6 per cent
Phyto-nutrients		
Carotene-α	24 µg	—
Carotene-β	62 µg	—
Cryptoxanthin-β	28 µg	—
Lutein-zeaxanthin	271 µg	—

☆ They are a very good source of soluble and insoluble dietary fiber. 100 g fruit provides 6.7 g or about 18 per cent of recommended daily intake. Dietary fibers help lower blood cholesterol levels and prevent constipation.

- ☆ In addition, the fruit, like persimmons, contain high concentration of tannin. Tannin, a poly-phenolic compound, which was once labeled as an anti-nutritional agent is in-fact, has beneficial anti-inflammatory, anti-ulcer and anti-oxidant properties.
- ☆ Its flesh contains health promoting flavonoid poly-phenolic antioxidants such as *cryptoxanthin, lutein, zeaxanthin, beta and alpha carotenes* in small amounts. Together, these compounds act as protective scavengers against oxygen-derived free radicals and reactive oxygen species (ROS) that play a role in ageing and various disease processes.
- ☆ Total antioxidant strength (ORAC) of avocados (raw, Hass variety) is 1933 µmol TE/100 g.
- ☆ They are also good in many health-benefiting vitamins. Vitamin A, E, and K are especially concentrated in its creamy pulp.
- ☆ Avocados are also excellent sources of minerals like iron, copper, magnesium, and manganese. Magnesium is essential for bone strengthening and has a cardiac-protective role as well. Manganese is used by the body as a co-factor for the antioxidant enzyme, *superoxide dismutase*. Iron and copper are required in the production of red blood cells.
- ☆ Fresh avocado pear is a very rich source of potassium. 100 g of fruit provides 485 mg or about 10 per cent of daily-required levels. Potassium is an important component of cell and body fluids where it helps regulate heart rate and blood pressure, countering bad effects of sodium.

5

Banana

Go for banana fruit, natures own energy-rich food that comes in a safety envelope! Fresh, delicious bananas are available year around and in fact, one of the cheapest fruits. Botanically, the fruit belongs to the family of *Musaceae.* Commercially, it is one of the widely cultivated crops in the tropical and subtropical zones. Scientific name: *Musa acuminata colla.*

Banana is a perennial herbaceous plant that develops from the underground rhizome. It flourishes well under tropical moisture-rich, humid low-lying farmlands. In fact, the whole plant is a false stem (pseudostem), consisting of broad leaves along with their petioles overlapping around each other in a circular fashion standing up to 2 to 6 meters tall from the ground surface depending upon the cultivar types. At maturity, the rhizome gives rise to flower (inflorescence) that is carried up on a long smooth un-branched stem through the centre of the pseudo-stem emerging out at the top in the centre of the leaf cluster. The flower subsequently develops to hanging bunch consisting of 3 to 20 hands (tiers), each with at least 5-10 fingers (fruits) in each hand (tier). Banana plant or plantain bears hanging clusters of fruits as a bunch. Fruits are arranged in tiers, with 6-20 fruits in each tier. There are several cultivars of banana grown with different size (4"-9"inch), colour (yellow to brown), weight (70-150g) and taste. Structurally, fruit has a protective outer skin and delicious, sweet and tart, creamy-white colour edible flesh inside. Plantains are other cultivar types, more often used as *cooking bananas*. They are closely related to the familiar fruit banana or dessert banana. Plantains are used as a staple diet in Thailand, Laos, and other Southeast Asian as well as in many parts of tropical African and Caribbean regions.

Health Benefits of Banana Fruit

- ✰ Banana fruit is one of the high calorie tropical fruits. 100 g of fruit provides 90 calories. Besides, it contains good amounts of health benefiting anti-oxidants, minerals, and vitamins.

Banana Fruit (*Musa acuminata* colla), Nutritive Value per 100 g

Principle	*Nutrient Value*	*Percentage of RDA*
Energy	90 Kcal	4.5 per cent
Carbohydrates	22.84 g	18 per cent
Protein	1.09g	2 per cent
Total Fat	0.33 g	1 per cent
Cholesterol	0 mg	0 per cent
Dietary Fiber	2.60 g	7 per cent
Vitamins		
Folates	20 μg	5 per cent
Niacin	0.665 mg	4 per cent
Pantothenic acid	0.334 mg	7 per cent
Pyridoxine	0.367 mg	28 per cent
Riboflavin	0.073 mg	5 per cent
Thiamin	0.031 mg	2 per cent
Vitamin A	64 IU	2 per cent
Vitamin C	8.7 mg	15 per cent
Vitamin E	0.10 mg	1 per cent
Vitamin K	0.5 μg	1 per cent
Electrolytes		
Sodium	1 mg	0 per cent
Potassium	358 mg	8 per cent
Minerals		
Calcium	5 mg	0.5 per cent
Copper	0.078 mg	8 per cent
Iron	0.26 mg	2 per cent
Magnesium	27 mg	7 per cent
Manganese	0.270 mg	13 per cent
Phosphorus	22 mg	3 per cent
Selenium	1.0 μg	2 per cent
Zinc	0.15 mg	1 per cent
Phyto-nutrients		
Carotene-α	25 μg	—
Carotene-β	26 μg	—
Lutein-zeaxanthin	22 μg	—

☆ Banana pulp is composed of soft, easily digestible flesh with simple sugars like fructose and sucrose that when eaten replenishes energy and revitalizes the body instantly; thus, for these qualities, bananas are being used by athletes to get instant energy and as supplement food in the treatment plan for underweight children.

- ✰ The fruit contains a good amount of soluble dietary fiber (7 per cent of DRA per 100 g) that helps normal bowel movements; thereby reducing constipation problems.
- ✰ It contains health promoting flavonoid poly-phenolic antioxidants such as*lutein, zeaxanthin,* β *and* α*-carotenes* in small amounts. These compounds help act as protective scavengers against oxygen-derived free radicals and reactive oxygen species (ROS) that play a role in ageing and various disease processes.
- ✰ It is also a very good source of vitamin-B6 (pyridoxine), provides about 28 per cent of daily-recommended allowance. Pyridoxine is an important B-complex vitamin that has a beneficial role for the treatment of neuritis, and anemia. Further, it helps decrease *homocystine* (one of the causative factors in coronary artery disease (CHD) and stroke episodes) levels within the body.
- ✰ The fruit is an also moderate source of vitamin-C (about 8.7 mg per 100g). Consumption of foods rich in vitamin-C helps the body develop resistance against infectious agents and scavenge harmful oxygen-free radicals.
- ✰ Fresh bananas provide adequate levels of minerals like copper, magnesium, and manganese. Magnesium is essential for bone strengthening and has a cardiac-protective role as well. Manganese is used by the body as a co-factor for the antioxidant enzyme, *superoxide dismutase*. Copper is required in the production of red blood cells.
- ✰ Fresh banana is a very rich source of potassium. 100 g fruit provides 358 mg potassium. Potassium is an important component of cell and body fluids that helps control heart rate and blood pressure, countering bad effects of sodium.

6

Black Berry

Sweet, succulent blackberries are summer delicacies in the northern temperate regions. As in raspberries, they grow on shrubs known as brambles in the vast Rosaceae family bush berries. The plant is native to Europe and now grown at a commercial scale from North America, particularly in the USA, to as far as Siberia. Binomially, the plant is a small perennial shrub belonging to the family *Rosaceae*, of the genus: *Rubus*. Botanical name: *Rubus fruticosus.*

Depending up on cultivar type, blackberry bush can be classified into erect, semi-erect and trailing types. Erect type bush generally features cane thorns and spreads by root suckers (which send cane shoots) along hedgerows, whereas, trailing shrubs require a trellis to support growth, and they spread by fresh shoots known as canes or primocanes. In the second year, these primocanes become floricanes. White or pink flowers appear in the floricanes. Technically, the berry is an aggregate fruit consisting of small drupelets arranged in circular fashion. Each drupelet composes of juicy pulp with single tiny seed. The berry measures about 3-4 cm in length containing about 80-100 drupelets. *Loganberries* and *phenomenon-berries* are hybrids between blackberry and red raspberry. Several other bramble berries such as *boysenberry, nessberry, youngberry, marionberry* etc., are hybrids of *dewberry, blackberry,* and *wild raspberry* cultivars.

Health Benefits of Blackberries

- ☆ As in other kinds of bush berries, blackberries too are packed with numerous plant nutrients such as vitamins, minerals, anti-oxidants, and dietary fibers that are essential for optimum health.
- ☆ The berries are very low in calories. 100 g provide just 43 calories. Nonetheless, they are rich in soluble and insoluble fiber (100 g whole berries consist of 5.3 g or 14 per cent RDA of fiber). Xylitol, a low-calorie sugar substitute in the fruit fiber, absorbs more slowly than glucose inside the gut, and thus does not cause rapid fluctuations in blood sugar levels.

Blackberries (*Rubus fruticosus*), Nutritive Value per 100 g

Principle	*Nutrient Value*	*Percentage of RDA*
Energy	43 Kcal	2 per cent
Carbohydrates	9.61 g	7 per cent
Protein	1.39 g	2 per cent
Total Fat	0.49 g	2 per cent
Cholesterol	0 mg	0 per cent
Dietary Fiber	5.3 g	14 per cent
Vitamins		
Folates	25 µg	6 per cent
Niacin	0.646 mg	4 per cent
Pantothenic acid	0.276 mg	5.5 per cent
Pyridoxine	0.030 mg	2 per cent
Thiamin	0.020 IU	2 per cent
Vitamin A	214 IU	7 per cent
Vitamin C	21 mg	35 per cent
Vitamin E	1.17 mg	8 per cent
Vitamin K	19.8 µg	16.5 per cent
Electrolytes		
Sodium	1 mg	0 per cent
Potassium	162 mg	3 per cent
Minerals		
Calcium	29 mg	3 per cent
Copper	165 µg	18 per cent
Iron	0.62 mg	8 per cent
Magnesium	20 mg	5 per cent
Manganese	0.646 mg	3 per cent
Selenium	0.4 µg	1 per cent
Zinc	0.53 mg	5 per cent
Phyto-nutrients		
Carotene-β	128 µg	—
Carotene-α	0 µg	—
Lutein-zeaxanthin	118 µg	—

☆ Blackberries compose significantly high amounts of phenolic flavonoid phytochemicals such as*anthocyanins, ellagic acid, tannin*), *quercetin, gallic acid, cyanidins, pelargonidins, catechins, kaempferol* and *salicylic acid*. Scientific studies show that these antioxidant compounds may have potential health benefits against cancer, ageing, inflammation, and neurological diseases.

- ☆ Fresh berries are an excellent source of vitamin-C (100 g of berries contain 23 mg or 35 per cent of RDA), which is a powerful natural antioxidant. Consumption of fruits rich in vitamin C helps develop resistance against infectious agents, counter inflammation, and scavenge harmful free radicals from the human body.
- ☆ They contain adequate levels of vitamin A, vitamin E, and vitamin K (16 per cent of RDA/100 g) and in addition; they are rich in much other health promoting flavonoid poly-phenolic antioxidants such as *lutein, zeaxanthin, and β-carotene* in small amounts. Altogether, these compounds help act as protective scavengers against oxygen-derived free radicals and reactive oxygen species (ROS) that play a role in ageing and various disease processes.
- ☆ Blackberries have an ORAC value (oxygen radical absorbance capacity, a measure of anti-oxidant strength) of about 5347µmol TE per 100 grams.
- ☆ Further, blackberries contain a good amount of minerals like potassium, manganese, copper, and magnesium. Copper is required in the bone metabolism as well as in production of white and red blood cells.
- ☆ They contain moderate levels of B-complex vitamins. It contains very good amounts of pyridoxine, niacin, pantothenic acid, riboflavin, and folic acid. These vitamins are acting as cofactors help the body metabolize carbohydrates, proteins, and fats.

7

Black Currents

Incredibly rich in numerous health benefiting phyto-nutrients and anti-oxidants, black currants are indeed one of the very popular summer season berries. Currants are native to central and northern Europe and Siberia. Currant is a small shrub belonging to the family of *Grossulariaceae*, of the genus; *Ribes* and has the scientific name: *Ribes nigrum*. In general, currants grow best in regions where summer is humid but winter is severe and chilling. It is a fast growing, deciduous, small shrub reaching about 5-6 ft tall.

During each season, the shrub bears pendulous chain of small berries. The fruit has size of about 1 cm in diameter, very dark purple, almost black, with a glossy skin and a persistent calyx at the apex, and containing 3-10 tiny seeds. Berries of red, white and pink currants are translucent, sweeter in taste, but less nutritional than black varieties.

Health Benefits of Black Currants

☆ Black currants have significantly high amounts of phenolic flavonoid phytochemicals called *anthocyanins*. Scientific studies have shown that consumption of blackcurrants have potential health effects against cancer, ageing, inflammation, and neurological diseases.

☆ Black currants have anti-oxidant value (Oxygen radical absorbance capacity- ORAC) of **7950** Trolex Equivalents per 100g, which is one of the highest values for fruits after chokeberries, elderberry, and cranberries. Red currants, however, possess comparatively less ORAC value at 3387 TE than the black variety.

☆ These berries are an excellent source of antioxidant vitamin, vitamin-C. 100 g of fresh currants provide more than 300 per cent of daily-recommended intake values of vitamin C. Research studies have shown that consumption of fruits rich in vitamin C helps the body develop immunity against infectious agents and also help scavenge harmful oxygen-free radicals from the body.

Black Currants (*Ribus nigrum*), Nutrition Value per 100 g.
ORAC Value (Anti-oxidant level) 7950

Principle	*Nutrient Value*	*Percentage of RDA*
Energy	63 Kcal	3 per cent
Carbohydrates	15.38 g	12 per cent
Protein	1.4 g	2.5 per cent
Total Fat	0.41 g	2 per cent
Cholesterol	0 mg	0 per cent
Dietary Fiber	4.3 g	11 per cent
Vitamins		
Folates	8 µg	2 per cent
Niacin	0.300 mg	2 per cent
Pantothenic acid	0.398 mg	8 per cent
Pyridoxine	0.066 mg	5 per cent
Riboflavin	0.050 mg	4 per cent
Thiamin	0.050 mg	4 per cent
Vitamin A	230 IU	7.5 per cent
Vitamin C	181 mg	301 per cent
Electrolytes		
Sodium	2 mg	0 per cent
Potassium	322 mg	7 per cent
Minerals		
Calcium	55 mg	5.5 per cent
Copper	0.086 mg	9.5 per cent
Iron	1.54 mg	19.5 per cent
Magnesium	24 mg	6 per cent
Manganese	0.256 mg	11 per cent
Phosphorus	59 mg	8.5 per cent
Zinc	0.27 mg	2 per cent

☆ Black currants are very good in vitamin A, and flavonoid anti-oxidants such as *beta-carotene, zeaxanthin and cryptoxanthin* levels. 100 g fresh berries provide 230 IU of vitamin A. These compounds are known to have antioxidant properties. Vitamin A is also required for maintaining integrity of mucus membranes and skin, and essential for healthy eye-sight. Furthermore, consumption of natural fruits rich in flavonoid anti-oxidants helps to protect from lung and oral cavity cancers.

- Fresh blackcurrants are also rich in many essential vitamins such as pantothenic acid (vitamin B5), pyridoxine (vitamin B-6) and thiamin (vitamin B-1). These vitamins are essential in the sense that body requires them from external sources to replenish and required for metabolism.
- They also contain good amounts of mineral **iron**. 100 g currant berries provide about 20 per cent of daily recommended levels. Iron is an important co-factor for cytochrome oxidase guided cellular metabolism. It is also required for red blood cell (RBC) production in the bone marrow.
- Additionally, the berries are also a very good source of other important minerals like copper, calcium, phosphorus, manganese, magnesium, and potassium, which are very essential for body metabolism.

8

Blueberries

Sweet, juicy blueberries are rich in pro-anthocyanin natural pigment anti-oxidants. These tiny, round blue-purple berries have long been attributed to the longevity and wellness of indigenous natives living in the subarctic regions in the Northern hemisphere. Botanically, the plant is a deciduous shrub belonging within the family of *Ericaceae*, of the genus: *Vaccinium.*

Broadly, *vaccinium* species are classified according to their growth habit as high-bush and low-bush berries.

- ✰ High-bush blueberry (*Vaccinium corymbosum*) is highly branched, erect deciduous shrub with rich foliage. It grows up to 10-12 feet tall in cultivated orchards and bears clusters of small, cream-white flowers during spring, which subsequently develop into berries after about two months. In the wild, high bush-blueberry is found on the edges of marshes, lakes, ponds, and streams. *Rabbiteye* blueberry (*Vaccinium virgatum*, also known as *V. ashei.*) is a medium-sized shrub grows naturally in South Eastern parts of USA.
- ✰ Low-bush blueberry (*Vaccinium angustifolium*) is a short, erect plant grows about one-two feet in height and spread by underground rhizomes. In cultivated farms, it is grown as two-year cycle crop, since the whole plant is either mowed down or burnt to allow new shoots that appear only during next season.

Both species require well-drained sandy, acidic soil to flourish. This berry shrub prefers open sunny conditions and intolerant of shade. In general, the berries are ready to harvest when the green- pink colour turns completely blue, juicy and sweeter. Traditionally, they handpicked and therefore, require intense laboring. Soon after the berries separated from the shrub, they are sorted out and transport to cold facility for storage.

Health Benefits of Blueberries

- ✰ Blueberries are very low in calories. 100 g fresh berries provide only 57 calories. However, they possess notable health benefiting plant-nutrients such as soluble dietary fiber, minerals, vitamins, and pigment anti-oxidants that contribute immensely towards optimum health and wellness.
- ✰ Blueberries are among the highest anti-oxidant value fruits. The ORAC value of 100 g fresh blueberry is 5562 TE (Trolex equivalents). Their antioxidant value largely derived from poly-phenolic anthocyanidin compounds such as *chlorogenic acid, tannins, myricetin, quercetin* and *kaempferol*.

Blueberries (*Vaccinium* spp.) ORAC Value 5562, Nutrition Value per 100 g

Principle	*Nutrient Value*	*Percentage of RDA*
Energy	57 Kcal	3 per cent
Carbohydrates	14.49 g	11 per cent
Protein	0.74 g	1 per cent
Total Fat	0.33 g	1 per cent
Cholesterol	0 mg	0 per cent
Dietary Fiber	2.4 g	6 per cent
Vitamins		
Folates	6 µg	1.5 per cent
Niacin	0.418 mg	2.5 per cent
Pantothenic acid	0.124 mg	2.5 per cent
Pyridoxine	0.052 mg	4 per cent
Riboflavin	0.041 mg	3 per cent
Vitamin A	54 IU	2 per cent
Vitamin C	9.7 mg	1.5 per cent
Vitamin E	0.57 mg	4 per cent
Vitamin K	19.3 µg	13 per cent
Electrolytes		
Sodium	1 mg	0 per cent
Potassium	77 mg	2 per cent
Minerals		
Calcium	6 mg	0.5 per cent
Iron	0.28 mg	3.5 per cent
Magnesium	6 mg	1.5 per cent
Manganese	0.336 mg	14 per cent
Zinc	0.16 mg	1.5 per cent
Phyto-nutrients		
Carotene-β	32 µg	—
Lutein-zeaxanthin	80 µg	—

- ☆ In addition, these berries have other flavonoid anti-oxidants such as *carotene-β, lutein and zeaxanthin*.
- ☆ Altogether, the phyto-chemical compounds in the blueberry help rid off harmful oxygen-derived free radicals from the body, and thereby, protect the human body against cancers, ageing, degenerative diseases, and infections.
- ☆ Further, research studies suggest that chlorogenic acid in these berries help lower blood sugar levels and control blood-glucose levels in type-II diabetes mellitus condition.
- ☆ Fresh berries contain a small amount of vitamin C, vitamin A and vitamin E. Altogether these vitamins work as potent anti-oxidants, which help limit free radical mediated injury to the body.
- ☆ The berries also contain a small amount of B-complex group of vitamins such as niacin, pyridoxine, folates and pantothenic acid. It contains very good amounts of vitamin B-6, niacin, riboflavin, pantothenic acid and folic acid. These vitamins are acting as co-factors help the body metabolize carbohydrates, protein, and fats.
- ☆ Furthermore, they contain a good amount of minerals like potassium, manganese, copper, iron and zinc. Potassium is an important component of cell and body fluids that helps controlling heart rate and blood pressure. Manganese is used by the body as a co-factor for the antioxidant enzyme, *superoxide dismutase*. Copper is required for the production of red blood cells. Iron is required for red blood cell formation.

9

Cantaloupe/Muskmelon

Wonderfully delicious and unique flavoured cantaloupe or "muskmelon" is a member in the large *Cucurbitaceae* family. Some of the popular fruits and vegetables in the cucurbita family include squash, pumpkin, cucumber, gourd etc., and like its relatives, melons grow on the ground surface as a trailing vine. Muskmelons thought to be originated either from India or ancient Persia or Africa. They grow best on draining sandy soil with good irrigation facility, and require honeybees for effective pollination. Melons, just as mangoes, watermelon etc., are actually summer season fruits. Their season runs from April through August, when they are at their best.

Varieties of muskmelons grown all over the world; however, *two* common varieties that named after their place of origin have become popular in the western world. The European cantaloupe (*Cucumis melo* cantalupensis) derives its name from the Italian papal village of "Cantalup" and features lightly-ribbed, pale green skin that looks quite different from the North American cantaloupe. Galia melon and charentais belong in this category. The North American cantaloupe (*Cucumis melo* reticulatus), popular in the United States and in some parts of Canada, is named *reticulatus* due to its net-like (or reticulated) skin covering. *Honeydew melons* have sweet, characteristic pale green succulent flesh. In general, melons feature round or oblong shape, measure 4.5- 6.5 inches in diameter and weigh 450 – 850 gm. Internally, flesh colour ranges from orange-yellow to salmon, has soft consistency and juicy texture with a sweet, musky aroma that emanates best in the completely ripe fruits. At the center, there is a hollow cavity filled with small off-white colour seeds encased in a web of mucilaginous netting.

Health Benefits of Cantaloupe

- ☆ Wonderfully delicious with rich flavour, muskmelons are very low in calories (100 g fruit has just 34 calories) and fats. Nonetheless, the fruit is rich in numerous health promoting poly-phenolic plant derived compounds, vitamins, and minerals that are absolute for optimum health.

Cantaloupe (*Cucumis melo*), Fresh, Nutritive Value per 100 g

Principle	*Nutrient Value*	*Percentage of RDA*
Energy	34 Kcal	1.5 per cent
Carbohydrates	8.6 g	6.5 per cent
Protein	0.84 g	1.5 per cent
Total Fat	0.19 g	<1 per cent
Cholesterol	0 mg	0 per cent
Dietary Fiber	0.9 g	2.25 per cent
Vitamins		
Folates	21 µg	5 per cent
Niacin	0.734 mg	4.5 per cent
Pantothenic acid	0.105 mg	2 per cent
Pyridoxine	0.072 mg	5.5 per cent
Riboflavin	0.026 mg	2 per cent
Thiamin	0.017 mg	1 per cent
Vitamin A	3382 IU	112 per cent
Vitamin C	36.7 mg	61 per cent
Vitamin E	0.05 mg	0.5 per cent
Vitamin K	2.5 mcg	2 per cent
Electrolytes		
Sodium	1 mg	0 per cent
Potassium	267 mg	6 per cent
Minerals		
Calcium	9 mg	1 per cent
Copper	41 µg	4.5 per cent
Iron	0.21 mg	2.5 per cent
Magnesium	12 mg	3 per cent
Manganese	0.041 mg	2 per cent
Zinc	0.18 mg	1.5 per cent
Phyto-nutrients		
Carotene-α	2020 µg	—
Crypto-xanthin-β	1 µg	—
Lutein-zeaxanthin	26 µg	—

☆ The fruit is an excellent source of Vitamin A, (100 g provides 3382 IU or about 112 per cent of recommended daily levels) *one of the highest among fruits.* Vitamin A is a powerful antioxidant and is essential for vision. It is also required for maintaining healthy mucus membranes and skin. Consumption of natural fruits rich in vitamin A is known to help to protect from lung and oral cavity cancers.

- ✰ It is also rich in antioxidant flavonoids such as *beta-carotene, lutein, zeaxanthin and cryptoxanthin.* These antioxidants have the ability to help protect cells and other structures in the body from oxygen-free radicals and hence; offer protection against colon, prostate, breast, endometrial, lung, and pancreatic cancers.
- ✰ Total antioxidant strength measured in terms of oxygen radical absorbance capacity (ORAC) of cantaloupe melons is 315 µmol TE/100 g. The value for honeydew melon is 241 µmol TE/100 g.
- ✰ Zeaxanthin, an important dietary carotenoid, selectively absorbed into the retinal macula lutea in the eye where it is thought to provide antioxidant and protective UV light-filtering functions. It thus, offers protection of eyes from "Age-related macular degeneration" (ARMD) disease in the elderly.
- ✰ It is a moderate source of electrolyte, potassium. 100 g fruit provides 267 mg of this electrolyte. Potassium is an important component of cell and body fluids and helps control heart rate and blood pressure. It thus offers protection against stroke, and coronary heart diseases.
- ✰ The fruit also contains moderate levels of B-complex vitamins, such as niacin, pantothenic acid and vitamin C, and minerals like manganese. Consumption of foods rich in vitamin-C helps the human body develop resistance against infectious agents and scavenge harmful oxygen-free radicals. Manganese is used by the body as a co-factor for the antioxidant enzyme, *superoxide dismutase*. Commercially, muskmelons are being used to extract an enzyme superoxide dismutase (SOD), which plays a vital role as strong first-line antioxidant defenses inside the human body.

10

Cherimoya

Sweet, pulpy, and fragrant rich cherimoya is among the most delicious tropical fruits of Andean valley origin. The greenish-yellow, conical fruits are from the evergreen trees belonging to the family *Annonaceae*, of the genus of *Annona*. It is thought to be native to Loja region of Ecuador, bordering Peru of the low rising tropical forests of Central Andean Mountains. Scientific name: *Annona cherimola*.

Annona is a small-sized tree, grows to about 15 to 30 feet tall with dense foliage, and bears fruits after 4-5 years of plantation. At their natural habitat, cherimoya flowers are pollinated by insects (*Coleoptera, Hemiptera*). However, in the cultivated farms, pollination is largely done artificially to obtain good yields. The fruit is irregularly oval or conical in shape, features leathery dark-green skin with polygonal indentations. *Annona reticulata* (custard apple) features small round or conical protuberances. The fruit measures about 10-20 cm in length and 10 cm in diameter and weighs about 350 g to 500 g and in some varieties weighing several pounds. Ripe fruits turn pale green to light brown colour, and emanate fragrant sweet aroma that can be appreciated from a distance. Inside it features, cream colour pulp with black colour smooth seeds embedded all over. Seeds and skin are inedible.

- ☆ **Sugar apple** (*Annona squamosa*) is small sized tropical fruit with the polygonal mosaic patterns on the surface. It has rich taste and flavour similar to that of cherimoya.
- ☆ **Custard apple** (*Annona reticulata*) is another small size fruit commonly grown in India, Pakistan and other South Asian regions. It features smooth conical protuberances (carpels) on the surface. It has similar taste and flavour as cherimoya; however, has less pulp but more seeds for its size.
- ☆ **Atemoya** is hybrid of cherimoya (*Annona cherimola*) and sugar apple (*A. squamosa*). It has better temperature and humidity tolerance suitable for cultivation in tropical climates.

☆ **Guanabana** (*Annona muricata*) also known as *soursop* (graviola), is much larger in size than the cherimoya. It is easily distinguished by its size, spiky surface and fibrous, tart flesh.

Health Benefits of Cherimoya

☆ Very sweet and pleasant flavour annonas contain an impressive list of essential nutrients, vitamins, anti-oxidants and minerals.

Cherimoya Fruit (*Annona cherimola*), Fresh, Nutritive Value per 100 g

Principle	*Nutrient Value*	*Percentage of RDA*
Energy	75 Kcal	4 per cent
Carbohydrates	17.71 g	13.5 per cent
Protein	1.57 g	3 per cent
Total Fat	0.68 g	3 per cent
Cholesterol	0 mg	0 per cent
Dietary Fiber	3 g	8 per cent
Vitamins		
Folates	23 µg	6 per cent
Niacin	0.644 mg	4 per cent
Pantothenic acid	0.345 mg	7 per cent
Pyridoxine	0.257 mg	20 per cent
Riboflavin	0.131 mg	10 per cent
Thiamin	0.101 mg	8 per cent
Vitamin A	5 IU	<1 per cent
Vitamin C	12.6 mg	21 per cent
Vitamin E	0.27 mg	2 per cent
Electrolytes		
Sodium	7 mg	0.5 per cent
Potassium	287 mg	6 per cent
Minerals		
Calcium	10 mg	1 per cent
Copper	0.069 mg	8 per cent
Iron	0.27 mg	3 per cent
Magnesium	17 mg	4 per cent
Manganese	0.093 mg	4 per cent
Phosphorus	26 mg	4 per cent
Zinc	0.16 mg	1 per cent
Phyto-nutrients		
Carotene-β	2 µg	—
Crypto-xanthin-β	1 µg	—
Lutein-zeaxanthin	6 µg	—

- ☆ The fruit has calories equivalent to that of mangoes. 100 g of fresh fruit pulp provide about 75 calories. It is, however, contain no saturated fats or cholesterol. It characteristically contains a good amount of dietary fiber (3 g per100 g) that helps prevent absorption of cholesterol in the gut. The fiber also helps protect the mucous membrane of the colon from exposure to toxic substances by binding cancer-causing chemicals in the colon.
- ☆ Cherimoya contains several poly-phenolic antioxidants. Among them, the most prominent in annona family fruits are *Annonaceous acetogenins.* Acetogenin compounds such as *asimicin, bullatacinare* etc are powerful cytotoxins and have been found to have anti-cancer, anti-malarial, and anti-helminthes properties.
- ☆ It is very good in *vitamin-C.* Vitamin C is a powerful natural antioxidant. Consumption of fruits rich in vitamin C helps the human body develop resistance against infectious agents and scavenge harmful, pro-inflammatory free radicals from the body.
- ☆ In addition, cherimoya fruit is a good source of B-complex vitamins, especially vitamin B-6 (pyridoxine). 100 g fresh fruit provides 0.257 mg or 20 per cent of daily-recommended levels. Pyridoxine helps keep-up GABA neuro chemical in the brain. High GABA levels calm down nervous irritability, tension, and headache ailments.
- ☆ Further, it has a well balanced sodium-potassium ratio. A good potassium level in the body helps control heart rate and blood pressure; thus, counters the bad influences of sodium. It also contains more minerals weight per weight than many common fruits like apples, rich in copper, magnesium, iron and manganese.

11

Cherry

Wonderfully delicious, cherry fruit is packed with full of health-benefiting nutrients and unique antioxidants. Cherries are native to Eastern Europe and Asia Minor regions. Botanically, the fruit is a "drupe" (stone fruit), belonging to the broad *Rosaceae* family of small tree fruits in the genus, *Prunus*. Some of the common "drupe" family fruits are plums, peaches, apricots etc. Although several species of cherries exist, two popular cultivars are wild or sweet-cherry, and sour or tart-cherry. While sweet cherries belong to the species of *Prunus avium*, tart variety belongs to that of *Prunus cerasus*. Cherries are drupe fruits with a central "stony-hard" seed surrounded by edible fleshy fruit measuring 2 cm in diameter. Externally the fruits have bright "shiny" red or purple colour with very thin peel. The West Indian cherry fruit known as acerola (*Malpighia emarginata*) is native to West Indian islands and grown in Mexico, Texas regions in North America. Acerola belongs to tropical fruit-bearing shrub or small tree in the family Malpighiaceae and contains 2-3 tiny seeds. Acerola contains exceptionally high levels of vitamin-C and vitamin-A than North American and European cherries.

Health Benefits of Cherry Fruit

- ☆ Cherries are one of the very low calorie fruits; however, are rich source of nutrients, vitamins, and minerals. Both sweet as well as tart cherries are packed with numerous health benefiting compounds that are essential for wellness.
- ☆ Cherries are pigment rich fruits. These pigments, in fact, are polyphenolic flavonoid compounds known as anthocyanin glycosides. Anthocyanins are red, purple or blue pigments found in many fruits and vegetables, especially concentrated in their skin, known to have powerful anti-oxidant properties.

Sweet and Tart Cherry Fruits, Fresh, Raw, Nutritive Value per 100 g

Principle	*Nutrient Value per 100g*		*Percentage of RDA*	
Cherry Type	*Sweet*	*Tart*	*Sweet*	*Tart*
Energy	63 cal	50 Kcal	3 per cent	2.5 per cent
Carbohydrates	16.1 g	12.18 g	12 per cent	9 per cent
Protein	1.06 g	1.00 g	2 per cent	2 per cent
Total Fat	0.2 g	0.3 g	1 per cent	1.5 per cent
Cholesterol	0 g	0 g	0 per cent	0 per cent
Dietary Fiber	2.1 g	1.6 g	5.5 per cent	4 per cent
Vitamins				
Folates	4 µg	8 µg	1 per cent	2 per cent
Niacin	0.154 mg	0.400 mg	1 per cent	2.5 per cent
Pantothenic acid	0.199 mg	0.143 mg	4 per cent	3 per cent
Pyridoxine	0.049 mg	0.044 mg	4 per cent	3.5 per cent
Riboflavin	0.033 mg	0.040 mg	2.5 per cent	3 per cent
Thiamin	0.027 mg	0.030 mg	2 per cent	2.5 per cent
Vitamin C	7 mg	10 mg	11 per cent	17 per cent
Vitamin A	640IU	1283 IU	21 per cent	43 per cent
Vitamin E	0.07 mg	0.07 mg	0.5 per cent	0.5 per cent
Vitamin K	2.1 µg	2.1 µg	2 per cent	2 per cent
Electrolytes				
Sodium	0 mg	3mg	0 per cent	0 per cent
Potassium	222 mg	179mg	5 per cent	4 per cent
Minerals				
Calcium	13 mg	16 mg	1.3 per cent	1.6 per cent
Copper	0.060 mg	0.104 mg	7 per cent	11.5 per cent
Iron	0.36 mg	0.32 mg	4.5 per cent	4 per cent
Magnesium	11 mg	9mg	3 per cent	2 per cent
Manganese	0.070 mg	0.112mg	3 per cent	5 per cent
Phosphorus	21 mg	15 mg	3 per cent	2 per cent
Zinc	0.07 mg	0.10 mg	0.5 per cent	0.1 per cent
Phyto-nutrients				
Carotene-α	0 µg	0 µg	—	—
Carotene-β	38 µg	770 µg	—	—
Crypto-xanthin-β	0 µg	0 µg	—	—
Lutein-zeaxanthin	85 µg	85 µg	—	—

☆ Scientific studies have shown that anthocyanins in the cherries are found to act like anti-inflammatory agents by blocking the actions of cyclooxygenase-1, and 2 enzymes. Thus, consumption of cherries has potential health effects against chronic painful episodes such as gout arthritis, fibromyalgia (painful muscle condition) and sports injuries.

- Research studies also suggest that anti-oxidant compounds in tart cherries help the human body to fight against cancers, ageing and neurological diseases, and pre-diabetes condition.
- Cherry fruits are very rich in stable anti-oxidant melatonin. Melatonin can cross the blood-brain barrier easily and produces soothing effects on the brain neurons, calming down nervous system irritability, which helps relieve neurosis, insomnia and headache conditions.
- Further, they are also mild source of zinc, moderate sources of iron, potassium, and manganese and good source of copper. Potassium is a heart-healthy mineral; an important component of cell and body fluids that regulate heart rate and blood pressure.
- The fruits, especially tart cherries are exceptionally rich in health promoting flavonoid poly phenolic anti-oxidants such as *lutein, zeaxanthin* and *beta carotene*. These compounds act as protective scavengers against harmful free radicals and reactive oxygen species (ROS) that play a role in ageing, cancers and various disease processes.
- Anti-inflammatory property of cherries has been found effective in reducing heart-disease risk factors by scavenging action against free radicals.
- Acerola or West Indian cherry has exceptionally very high levels of vitamin-C (1677.6 mg per 100 g or 2796 per cent of RDA) and vitamin-A (767 IU per 100 g).

12

Chokeberry

Chokeberry is one of its own kinds of berry packed with essential phyto-nutrients, vitamins and antioxidants. These humble tiny berries from the wild natural shrub have recently grabbed the attention of fitness lovers and food scientists alike for their exceptional nutritive value. Botanically, it belongs to the *Rosaceae* family within the genus: Aronia. Scientific name: *Aronia melanocarpa*. At least *two* species of chokeberries are cultivated widely, black and red. The plant bears numerous small, about 1 cm size fruits with relatively thick, pigmented skin. Red berries are sweeter in taste than black varieties; the latter are slightly bitter in taste; however, black and blue colour berries are rather rich sources *anthocyanin* class of anti-oxidants.

Health Benefits of Chokeberry

- ✰ Chokeberries are low in calories and fats. 100 g of fresh berries contain 47 calories. Nonetheless, they are one of the nature's richest sources of flavonoid anthocyanin antioxidants. In addition, the berries contain handsome levels of minerals, and vitamins, and dietary fiber obtained through their peel.
- ✰ The oxygen radical absorbency capacity or ORAC (measurement of antioxidant strength of food items) demonstrates chokeberry with one of the highest values yet recorded among berries-16,062 micro-moles of Trolox Equivalents (TE) per 100 g.
- ✰ Black colour berries consist of significantly high amounts of phenolic flavonoid phyto-chemicals called anthocyanins. Total anthocyanin content is 1480 mg per 100 g of fresh berries, and proanthocyanidin concentration is 664 mg per 100 g (Wu et al. 2004, 2006). Scientific studies have shown that consumption of berries on a regular basis offers potential health benefits against cancer, ageing and neurological diseases, inflammation, diabetes, and bacterial infections. (- By Dr. Paul Gross, 2007-07-09).

Chokeberries (*Aronia melanocarpa*), Nutrition Value per 100 g, ORAC Value- 16,062

Principle	*Nutrient Value*	*Percentage of RDA*
Energy	47 Kcal	2.5 per cent
Carbohydrates	9.6 g	7 per cent
Protein	1.4 g	2.5 per cent
Total Fat	0.5 g	2 per cent
Cholesterol	0 mg	0 per cent
Dietary Fiber	5.3 g	14 per cent
Vitamins		
Folates	25 μg	6 per cent
Vitamin A	214 IU	7 per cent
Vitamin C	21 mg	35 per cent
Vitamin E	1.17 mg	8 per cent
Vitamin K	19.8 μg	17 per cent
Electrolytes		
Sodium	1 mg	0 per cent
Potassium	162 mg	3 per cent
Minerals		
Calcium	30 mg	3 per cent
Iron	0.62 mg	8 per cent
Magnesium	20 mg	5 per cent
Manganese	0.646 mg	32 per cent
Zinc	0.53 mg	5 per cent
Phyto-nutrients		
Carotene-α	0 μg	—
Carotene-β	128 μg	—
Lutein-zeaxanthin	118 μg	—

- ☆ Laboratory analyses of anthocyanins in chokeberries have identified the following individual chemicals: *cyanidin-3-galactoside, quercetin, peonidin, delphinidin, petunidin, epicatechin, caffeic acid, pelargonidin* and *malvidin.* These flavonoid poly-phenolic antioxidants have proven health benefits through scavenging dangerous oxygen-free radicals from the body.
- ☆ Cancer research on anthocyanins where in black chokeberry preparations were first used to inhibit chemically induced cancer in the rat esophagus was found to reduce the disease severity by 30-60 per cent and that of the colon cancer by up to 80 per cent. Effective at both the initiation and promotion/progression stages of tumor development, these berries are a practical research tool and hold a promising therapeutic source, since they contain the highest amount of anthocyanins among native North American berries [J. Agric. Food Chem. 50 (12): 3495–500].

- They are also rich in flavonoid anti-oxidants such as *carotenes, luteins* and *zeaxanthins*. Zeaxanthins as photo-filtering effects on UV rays and thus protects eyes from age-related macular disease in the elderly (ARMD).
- Further, they are an also good source of many antioxidant vitamins like vitamin-C, vitamin A, vitamin E, beta-carotene and folate and minerals like potassium, iron and manganese. 100 g of fresh berries provide about 35 per cent of daily-recommended levels of vitamin C.

13

Cranberry

Unique, wild and natural by habitat, cranberries are rich in phyto-nutrients (naturally derived plant compounds), particularly proanthocyanidin antioxidants, which are essential for all-round wellness. The berries are indeed containing numerous chemical substances that may offer protection from tooth cavities, urinary tract infection, and inflammatory diseases. The berry-plant is described as an evergreen dwarf, creeping shrub, or a low-lying trailing vine belonging within the family of *Ericaceae,* in the genus: Vaccinium, and subgenus: *Oxycoccos.* Scientific name: *Vaccinium macrocarpon.* In their natural habitat, the plant grows vigorously in acidic sandy bogs throughout the cooler parts of the Europe, Northern states in the United States and Canada. The plant is actually a dwarf, creeping shrub, or vine, which runs up to 2 m long and 10 to 20 cm in height. It features slender, wiry, not so thick, woody stems bearing small, evergreen leaves. Cranberry season generally lasts from October until December. The fruit is small, round, red colour berry. Each berry features four centrally situated tiny seeds enclosed inside the capsules. The berry is very acidic in taste, having ph in the range of 2.3 to 2.5.

Health Benefits of Cranberries

- ✰ Delicious, tart cranberries have significantly high amounts of phenolic flavonoid phytochemicals called *pro-anthocyanidins* (PAC's). Scientific studies have shown that consumption of berries have potential health benefits against cancer, ageing and neurological diseases, inflammation, diabetes, and bacterial infections.
- ✰ Antioxidant compounds in cranberries such as oligomeric proanthocyanidins (OPC's), anthocyanidin flavonoids, cyanidin, peonidin and quercetin may prevent cardiovascular disease by counteracting against cholesterol plaque formation in the heart and blood vessels. Further, these compounds help the human body lower LDL cholesterol levels and increase HDL-good cholesterol levels in the blood.

Cranberries (*Vaccinium macrocarpon*), Fresh, ORAC score 9584, Nutritive Value per 100 g

Principle	*Nutrient Value*	*Percentage of RDA*
Energy	46 Kcal	2.3 per cent
Carbohydrates	12.2 g	9 per cent
Protein	0.4 g	1 per cent
Total Fat	0.13 g	<1 per cent
Cholesterol	0 mg	0 per cent
Dietary Fiber	4.6 g	12 per cent
Vitamins		
Folates	1 μg	<1 per cent
Niacin	0.101 mg	1 per cent
Pantothenic acid	0.295 mg	6 per cent
Pyridoxine	0.057 mg	4 per cent
Riboflavin	0.020 mg	2 per cent
Thiamin	0.012 mg	1 per cent
Vitamin A	60 IU	2 per cent
Vitamin C	13.3 mg	22 per cent
Vitamin E	1.20 mg	8 per cent
Vitamin K	5.1 μg	4 per cent
Electrolytes		
Sodium	2 mg	0 per cent
Potassium	85 mg	2 per cent
Minerals		
Calcium	8 mg	1 per cent
Copper	0.061 mg	7 per cent
Iron	0.25 mg	3 per cent
Magnesium	6 mg	1.5 per cent
Manganese	0.360 mg	16 per cent
Phosphorus	13 mg	2 per cent
Selenium	0.1 μg	0 per cent
Zinc	0.10 mg	1 per cent
Phyto-nutrients		
Carotene-β	36 μg	—
Crypto-xanthin-β	0 μg	—
Lutein-zeaxanthin	91 μg	—

☆ Research studies show that cranberry juice consumption offers protection against gram-negative bacterial infections such as *E. coli* in the urinary system by inhibiting bacterial-attachment to the bladder and urethra.

- ✰ Consumption of cranberries turns urine acidic. This, together with the bacterial anti-adhesion property of cranberry juice helps prevent the formation of alkaline (calcium ammonium phosphate) stones in the urinary tract by working against proteus bacterial-infections.
- ✰ Further, the berries prevent plaque formation on the tooth surface by interfering with the ability of another gram-negative bacterium, *Streptococcus mutans*, to stick to the surface. It thus helps prevent the development of cavities in a way similar to the action in preventing urinary tract infections.
- ✰ In addition, the berries are an also good source of many vitamins like vitamin C, vitamin A, β-carotene, lutein, zeaxanthin, and folate and minerals like potassium, and manganese.
- ✰ Oxygen Radical Absorbance Capacity or ORAC (measurement of antioxidant strength of food items) demonstrates cranberry at an ORAC score of 9584 μmol TE units per 100 g, one of the highest in the category of edible berries.

14

Dates

What Nutrients are there in Dates?

A lot indeed! Here are sweet, delicious fruits from the tropical oasis, brimming with much-needed minerals and energy to help you stay fit and healthy. Botanically; they are the fruits grow on the palm tree belonging to the family of *Arecaceae* in the genus: *Phoenix* and scientifically named as Phoenix dactylifera. The tree is believed to originate in the lands on the banks of Nile and Euphrates Rivers of ancient Egypt and Mesopotamia. Date palm is now grown extensively for its edible fruits under warmer climates across all the continents.

The date fruit is a "drupe" in which outer fleshy part (exocarp and mesocarp) surrounds a shell (the pit or stone) of hardened endocarp with a seed inside. The fruit is oval to cylindrical in shape, 3–7 cms long, and 2–3 cms diameter, and when ripe, range from bright red to depending on the variety. There are many varieties of date palm cultivated. 'Amir Hajj', 'Saidy', 'Khadrawy' and 'Medjool' is some of the important varieties that are popular for their superior quality.

Health Benefits of Dates

- ☆ Wonderfully delicious, dates are one of the most popular fruits packed with an impressive list of essential nutrients, vitamins, and minerals that are required for normal growth, development and overall well-being.
- ☆ Fresh dates compose of soft, easily digestible flesh and simple sugars like fructose and dextrose. When eaten, they replenish energy and revitalize the body instantly. For these qualities, they are being used to break the fast during Ramadan month since ancient times.
- ☆ The fruit is rich in dietary fiber, which prevents LDL cholesterol absorption in the gut. Additionally, the fiber works as a bulk laxative. It, thus, helps to protect the colon mucous membrane by decreasing exposure time and as well as binding to cancer-causing chemicals in the colon.

Dates (*Phoenix dactylifera*), Medjool, Nutritive Value per 100 g

Principle	*Nutrient Value*	*Percentage of RDA*
Energy	277 Kcal	14 per cent
Carbohydrates	74.97 g	58 per cent
Protein	1.81g	3 per cent
Total Fat	0.15 g	<1 per cent
Cholesterol	0 mg	0 per cent
Dietary Fiber	6.7 g	18 per cent
Vitamins		
Folates	15 µg	4 per cent
Niacin	1.610 mg	10 per cent
Pantothenic acid	0.805 mg	16 per cent
Pyridoxine	0.249 mg	19 per cent
Riboflavin	0.060 mg	4.5 per cent
Thiamin	0.050 mg	4 per cent
Vitamin A	149 IU	5 per cent
Vitamin C	0 mg	0 per cent
Vitamin K	2.7 µg	2 per cent
Electrolytes		
Sodium	1 mg	0 per cent
Potassium	696 mg	16 per cent
Minerals		
Calcium	64 mg	6.5 per cent
Copper	0.362 mg	40 per cent
Iron	0.90 mg	11 per cent
Magnesium	54 mg	13 per cent
Manganese	0.296 mg	13 per cent
Phosphorus	62 mg	9 per cent
Zinc	0.44 mg	4 per cent
Phyto-nutrients		
Carotene-β	89 µg	—
Crypto-xanthin-β	0 µg	—
Lutein-zeaxanthin	23 µg	—

- ✰ They contain health benefiting flavonoid polyphenolic antioxidants known as tannins. Tannins are known to possess anti-infective, anti-inflammatory, and anti-hemorrhagic (prevent easy bleeding tendencies) properties.
- ✰ They are moderate sources of vitamin-A (contains 149 IU per 100 g), which is known to have antioxidant properties and essential for vision. Additionally, it is also required maintaining healthy mucus membranes

and skin. Consumption of natural fruits rich in vitamin A is known to help to protect from lung and oral cavity cancers.

- ☆ They compose antioxidant flavonoids such as β-carotene, lutein, and zeaxanthin. These antioxidants found to have the ability to protect cells and other structures in the body from harmful effects of oxygen-free radicals. Thus, eating dates found to offer some protection from colon, prostate, breast, endometrial, lung, and pancreatic cancers.
- ☆ Zeaxanthin is an important dietary carotenoid that selectively absorbed into the retinal macula lutea, where it thought to provide antioxidant and protective light-filtering functions. It thus offers protection against age-related macular degeneration, especially in elderly populations.
- ☆ Dates are an excellent source of iron, contains 0.90 mg/100 g of fruits (about 11 per cent of RDI). Iron, being a component of hemoglobin inside the red blood cells, determines the oxygen-carrying capacity of the blood.
- ☆ Further, they are very good in potassium. 100 g contains 696 mg or 16 per cent of daily-recommended levels of this electrolyte. Potassium is an important component of cell and body fluids that help controlling heart rate and blood pressure. They, thus, offers protection against stroke and coronary heart diseases.
- ☆ They are also rich in minerals like calcium, manganese, copper, and magnesium. Calcium is an important mineral that is an essential constituent of bone and teeth, and required by the body for muscle contraction, blood clotting, and nerve impulse conduction. Manganese is used by the body as a co-factor for the antioxidant enzyme, superoxide dismutase. Copper is required for the production of red blood cells. Magnesium is essential for bone growth.
- ☆ Further, the fruit has adequate levels of B-complex group of vitamins as well as vitamin K. It contains very good amounts of pyridoxine (vitamin B-6), niacin, pantothenic acid, and riboflavin. These vitamins are acting as cofactors help body metabolize carbohydrates, protein, and fats. Vitamin K is essential for many coagulant factors in the blood as well as in bone metabolism.

15

Durian Fruit

One of its own kinds, Durian fruit is delicious, soft, succulent and very popular for its unique characteristics. Durian is widely revered as the "King of Fruits" in the South-East Asian countries. The tree is tropical in origin and belongs to the genus *Durio,* family of *Malvaceae;* a large family of plant species which also include some of the interesting relatives such as hibiscus, okra, etc. It is botanically known as *Durio zibethinus.*

The exotic durian is native to Brunei, Indonesia and Malaysia. While there are 30 known species of *Durio* exist only nine of them have been identified as producing edible fruits. Durian tree starts bearing fruits after four or five years after plantation. The tree grows up to 50 meters in height depending on the species. Durian is a seasonal fruit; its season lasts typically from June until August, which coincides with that of the other tropical specific fruits like mangosteen, jackfruit, and mango. The fruit is distinctive for its large size, unique odor, and formidable thorn-covered husk. It can reach up to 30 cm (12 in) long and 15 cm (6 in) in diameter, and typically weighs one to four kilograms (two to seven lb). In shape, it varies from round to oblong; colour of its husk (rind) is green with brown; and its flesh features creamy-yellow to saffron colour depending upon the species.

The flesh or pulp can be consumed at various stages of ripeness, and is used as flavour base in a wide variety of culinary and sweet preparations in Southeast Asian cuisines. Durian seeds are small, round to oval shape and appear like jackfruit seeds. Many discard them, though they can be eaten after boiled thoroughly. The seeds feature a bland taste akin to jackfruit seeds. Edible flesh emits a distinctive odor which can be described as strong and penetrating even when the husk is intact; this unusual stinky flavour and intense odor of the fruit have prompted many people to express diverse and peculiar opinions ranging from deep appreciation to disgust about durian.

Health Benefits of Durian Fruit

☆ Durian, like other tropical fruits such as Banana, avocado, and jackfruit, is rich in energy, minerals and vitamins. 100 g fresh fruit provides 147 calories.

Durian Fruit (*Durio zibethinus*), Nutrition Value per 100 g

Principle	*Nutrient Value*	*Percentage of RDA*
Energy	147 Kcal	7 per cent
Carbohydrates	27.09 g	21 per cent
Protein	1.47 g	2.5 per cent
Total Fat	5.33 g	20 per cent
Cholesterol	0 mg	0 per cent
Dietary Fiber	3.8 g	10 per cent
Vitamins		
Folates	36 mcg	9 per cent
Niacin	1.074 mg	7 per cent
Pantothenic acid	0.230 mg	4.5 per cent
Pyridoxine	0.316 mg	24 per cent
Riboflavin	0.200 mg	15 per cent
Thiamin	0.374 mg	31 per cent
Vitamin A	44 IU	1.5
Vitamin C	19.7 mg	33 per cent
Electrolytes		
Sodium	2 mg	0 per cent
Potassium	436 mg	9.5 per cent
Minerals		
Calcium	6 mg	0.6 per cent
Copper	0.207 mg	23 per cent
Iron	0.43 mg	5 per cent
Magnesium	30 mg	7.5 per cent
Manganese	0.325 mg	14 per cent
Phosphorus	39 mg	6 per cent
Zinc	0.28 mg	2.5 per cent
Phyto-nutrients		
Carotene-α	6 mcg	—
Carotene-β	23 mcg	—
Lutein-zeaxanthin		

☆ The fruit is made of soft, easily digestible flesh made of simple sugars like fructose and sucrose that when eaten replenish energy and revitalize the body instantly. Though it contains a relatively high amount of fats among fruits, it is free from saturated fats and cholesterol.

- Durian is rich in dietary fiber, which makes it a good bulk laxative. The fiber content helps to protect the colon mucous membrane by decreasing exposure time and as well as binding to cancer-causing chemicals in the colon.
- The durian fruit is a good source of antioxidant vitamin-C (about 33 per cent of RDA). Consumption of foods rich in vitamin-C helps the body develop resistance against infectious agents and scavenge harmful free radicals.
- The fruit is an excellent source of health benefiting B-complex groups of vitamins; a rare feature among fruits, such as niacin, riboflavin, pantothenic acid (vitamin B-5), pyridoxine (vitamin B-6) and thiamin (vitamin B-1). These vitamins are essential for the body as it requires them from external sources to replenish.
- Further, it also contains a good amount of minerals like manganese, copper, iron and magnesium. Manganese is used by the body as a co-factor for the antioxidant enzyme, superoxide dismutase. Copper is required in the production of red blood cells. Iron is required for red blood cell formation.
- Fresh durian fruit is a very rich source of potassium. Potassium is an important electrolyte of cell and body fluids that helps controlling heart rate and blood pressure.
- Additionally, it also contains high levels of essential amino acid, tryptophan (also known as "nature's sleeping pill") which in the body metabolizes into serotonin and melatonin; these neuro-chemicals have important functions like sleep induction and in the treatment of epilepsy.

16

Fig

Delicious, sweet fig fruit is one of the prime fruits enjoyed since antiquity in the human history. Fig is naturally rich in much health benefiting phyto-nutrients, anti-oxidants and vitamins. Dried figs, in fact, are concentrated source of minerals and vitamins. The fully ripe fig has bell or pear shape with succulent flesh. Botanically figs belong to the mulberry family (Moraceae), of the genus: *Ficus*. Scientific name: *Ficus carica.*

Fig fruit tree is native to temperate climate of Asia Minor or present day Turkey and today is grown as important fruit of commerce in the eastern Mediterranean region, USA, Spain. However, it is also cultivated as a fruit tree in home gardens. During each season, the tree bears several hundred pear-shaped fruits twice a year, which vary in size and colour depending on the variety. Interiorly, the fig fruit features numerous club shaped ovaries protruding towards central hollow cavity. In their natural habitat, "caprifigs" are pollinated by a tiny gall wasp (*Blastophaga grossorum*) that enters the flower cluster through a small opening in the apex. Several cultivars exist; some of the popular varieties commonly grown in the USA are Brown Turkey, Conadria, Kadota, and Black mission. However, since the wasp is not present in North America, most of these fruits do not have true seeds, and they develop by parthenogenesis (without pollination).

Health Benefits of Figs

- ☆ Fig fruit is low in calories. 100 g fresh fruits provide only 74 calories. However, they contain health benefiting soluble dietary fiber, minerals, vitamins, and pigment anti-oxidants that contribute immensely for optimum health and wellness.
- ☆ Dried figs are an excellent source of minerals, vitamins and anti-oxidants. In fact, dried fruits are concentrated sources of energy. 100 g dried figs provide 249 calories.

Fig Fruit (*Ficus carica*), Nutrition Value per 100 g

Principle	*Nutrient Value*	*Percentage of RDA*
Energy	74 Kcal	4 per cent
Carbohydrates	19.18 g	15 per cent
Protein	0.75 g	1.5 per cent
Total Fat	0.30 g	1 per cent
Cholesterol	0 mg	0 per cent
Dietary Fiber	2.9 g	7 per cent
Vitamins		
Folates	6 µg	1.5 per cent
Niacin	0.400 mg	2.5 per cent
Pantothenic acid	0.300 mg	6 per cent
Pyridoxine	0.113 mg	9 per cent
Riboflavin	0.050 mg	4 per cent
Thiamin	0.060	5 per cent
Vitamin A	142 IU	5 per cent
Vitamin C	2 mg	3 per cent
Vitamin E	0.11 mg	1 per cent
Vitamin K	4.7 µg	4 per cent
Electrolytes		
Sodium	1 mg	0 per cent
Potassium	232 mg	5 per cent
Minerals		
Calcium	35 mg	3.5 per cent
Copper	0.070 mg	8 per cent
Iron	0.37 mg	5 per cent
Magnesium	17 mg	4 per cent
Manganese	0.128 mg	5.5 per cent
Selenium	0.2 µg	<1 per cent
Zinc	0.15 mg	1 per cent
Phyto-nutrients		
Carotene-β	85 µg	—
Lutein-zeaxanthin	9 µg	—

- ✰ Fresh figs, especially black mission, are good in poly-phenolic flavonoid anti-oxidants such as *carotenes, lutein, tannins, chlorogenic acid* etc. Their anti-oxidant value is comparable to that of apples at 3200 umol/100 g.
- ✰ In addition, fresh fruits contain adequate levels of some of the anti-oxidant vitamins such as vitamins A, E, and K. Altogether these phyto-chemical compounds in fig fruit help scavenge harmful oxygen derived free radicals

from the body and thereby protect us from cancers, diabetes, degenerative diseases and infections.

- Furthermore, research studies suggest that chlorogenic acid in these berries help lower blood sugar levels and control blood-glucose levels in type-II diabetes mellitus (Adult onset) condition.
- Fresh, as well as dried figs contain good levels of B-complex group of vitamins such as niacin, pyridoxine, folates, and pantothenic acid. These vitamins function as co-factors for metabolism of carbohydrates, proteins, and fats.
- Dried figs are excellent sources of minerals like calcium, copper, potassium, manganese, iron, selenium and zinc. 100 g of dried figs contain 640 mg of potassium, 162 mg of calcium, 2.03 mg of iron and 232 mg of potassium. Potassium is an important component of cell and body fluids that helps controlling heart rate and blood pressure. Copper is required in the production of red blood cells. Iron is required for red blood cell formation as well for cellular oxidation.

17

Gooseberry/Aonla

Gooseberries are small, round to oval berries of European origin. They grow wild all over temperate climates of Europe, North America, and Siberia. Botanically, the berries are close relative of currants and belong to the same family of *Grossulariaceae*, of the genus: *Ribes*. Rich in antioxidant polyphenolics and vitamins, the berries comes in different colour, flavour, and shapes. Scientific name: *Ribes uva-crispa* L.

Gooseberries (*R. uva-crispa* L.) are one of the four wild Ribes species (*R. alpinum* L., *R. rubrum* L. and *R. petraeum* Wulf.) growing in the Northern Hemisphere. As in currants, gooseberry grows best in regions where summer is humid but winter severe and chilling. The plant is a fast growing, small deciduous shrub growing about 4-6 ft tall featuring sharp thorns all along its woody branches. The bush begins fruiting 2-3 years after plantation. Berries come in many shapes, colours, and taste. They can be round, oval, pear-shaped or elongated, green, white, yellow, purple, red-brown or black colour, sweet and tart, surface smooth or hairy with conspicuous veins and 15-30 tiny edible seeds. In general, the berries measure 1-2 cm in width and weigh 4 g to 10 g. *Indian gooseberries* known as amla in the subcontinent belongs to different family of Euphorbiaceae. Their scientific name is *Phyllanthus emblica*. Indian gooseberry features round to transversely spherical shape with light green colour and excessively acidic and bitter (astringent) in taste. *Cape gooseberry* (*Physalis peruviana*) also known as Peruvian cherry in the US, is native to South American Andes region. The berries are small, round, orange-yellow colour, encased inside a Chinese lantern like papery thin husk.

Health Benefits of Gooseberries

- ☆ Gooseberries are low in calories; 100 g of fresh berries provide only 44 calories. As in blackcurrants, they too have significantly high amounts of phenolic phytochemicals especially flavones and anthocyanins. These compounds have been found to have numerous health-benefiting effects against cancer, ageing, inflammation, and neurological diseases.

Gooseberries (*Ribes uva-crispa* L.), Nutrition Value per 100 g.
ORAC Value (Anti-oxidant level) 3277 umol TE/100 g.

Principle	*Nutrient Value*	*Percentage of RDA*
Energy	44 Kcal	2 per cent
Carbohydrates	10.18 g	8 per cent
Protein	0.88 g	1.5 per cent
Total Fat	0.58 g	3 per cent
Cholesterol	0 mg	0 per cent
Dietary Fiber	4.3 g	11 per cent
Vitamins		
Folates	6 mcg	1.5 per cent
Niacin	0.300 mg	2 per cent
Pantothenic acid	0.286 mg	6 per cent
Pyridoxine	0.080 mg	6 per cent
Riboflavin	0.030 mg	2 per cent
Thiamin	0.040 mg	3 per cent
Vitamin A	290 IU	10 per cent
Vitamin C	27.7 mg	46 per cent
Electrolytes		
Sodium	1 mg	0 per cent
Potassium	198 mg	4 per cent
Minerals		
Calcium	25 mg	2.5 per cent
Copper	0.070 mg	8 per cent
Iron	0.31 mg	4 per cent
Magnesium	10 mg	2.5 per cent
Manganese	0.144 mg	6 per cent
Phosphorus	27 mg	4 per cent
Zinc	0.12 mg	1 per cent

- ✩ They have moderately well in anti-oxidant values. At 3277 umol TE/100g, gooseberries have oxygen radical absorbance capacity (ORAC) value comparable to that of red currants (3387umol TE/100g).
- ✩ The berries are moderate source of vitamin-C. 100 g of fresh berries provide 27.7 mcg or 46 per cent of daily-recommended intake values of vitamin C. Research studies have shown that consumption of fruits rich in vitamin C helps body develop immunity against infectious agents and help scavenge harmful oxygen free radicals from the body.
- ✩ They have small amounts of vitamin A. 100 g berries has 290 IU or 10 per cent of RDA of this vitamin. Vitamin A is required for maintaining integrity

of mucus membranes and skin, and essential component of visual cycle. In addition, consumption of natural fruits rich in vitamins and flavonoid anti-oxidants has been found to protect from lung and oral cavity cancers.

- Fresh berries contain small amounts of essential vitamins such as pyridoxine (vitamin B-6), pantothenic acid (vitamin B-5), folates, and thiamin (vitamin B-1). Some of these vitamins are essential in the sense that the body requires them for metabolism from external sources to replenish.
- Furthermore, gooseberries contain adequate levels of minerals such as copper, calcium, phosphorus, manganese, magnesium, and potassium.
- Indian gooseberries (amla) are exceptionally rich in vitamin-C. However, their anti-oxidant properties come from other anti-oxidant compounds tannins (emblicanin, punigluconin, pedunculagin etc).

18

Grape

Widely popular, grapes are regarded in many cultures as "the queen of fruits," since centuries. These tiny berries are the storehouse of numerous health promoting phyto-nutrients such as poly-phenolic antioxidants, vitamins, and minerals. No wonder why many of us include them as an integral part of our diet, be it in the form of fresh table fruits, juice, or in salads. Botanically, they are small round berries; grow in clusters on a perennial and deciduous woody vine of the genus: *Vitis*. Grapes are native to Europe and Mediterranean regions but now widely cultivated all around the world. In structure, each berry features semi-translucent flesh encased in a smooth, thin skin. Some varieties contain edible seeds, while others are seedless. The colour to the fruit is because of the presence of poly-phenolic pigments in them. Red or purple berries are rich in anthocyanins while white-green berries contain more of tannins, especially, catechin. Interestingly, these antioxidant compounds are densely concentrated on the skin and seeds. The three main species of grapes grown around the world are; European (*Vitis vinifera*), North American (*Vitis labrusca* and *Vitis rotundifolia*), and French hybrids.

- ☆ Some of popular green cultivars are Thompson seedless, sugarone, and calmeria.
- ☆ Red varieties include emperor, red globe, cardinal, and flame seedless.
- ☆ Concord and zinfandel are some of flavourful blue-black cultivars.

Commercially, many cultivars of grapes are grown for different purposes either eaten as table fruit, fresh or dried (raisin, currant, sultana) *or* in wine production.

Health Benefits of Grapes

- ☆ Grapes are rich in polyphenolic phytochemical compound resveratrol. Resveratrol is one of the powerful anti-oxidant, which has been found to play a protective role against cancers of colon and prostate, coronary heart disease (CHD), degenerative nerve disease, Alzheimer's disease and viral/ fungal infections.

Grapes, Red or Green (European type, Thompson seedless), Nutritive Value per 100 g, ORAC Value 3,277

Principle	*Nutrient Value*	*Percentage of RDA*
Energy	69 Kcal	3.5 per cent
Carbohydrates	18 g	14 per cent
Protein	0.72 g	1 per cent
Total Fat	0.16 g	0.5 per cent
Cholesterol	0 mg	0 per cent
Dietary Fiber	0.9 g	2 per cent
Vitamins		
Folates	2 µg	0.5 per cent
Niacin	0.188 mg	1 per cent
Pantothenic acid	0.050 mg	1 per cent
Pyridoxine	0.086 mg	7.5 per cent
Riboflavin	0.070 mg	5 per cent
Thiamin	0.069 mg	6 per cent
Vitamin A	66 IU	3 per cent
Vitamin C	10.8 mg	18 per cent
Vitamin E	0.19 mg	1 per cent
Vitamin K	14.6 µg	12 per cent
Electrolytes		
Sodium	0 per cent	1 mg
Potassium	191 mg	4 per cent
Minerals		
Calcium	10 mg	1 per cent
Copper	0.127 mg	14 per cent
Iron	0.36 mg	4.5 per cent
Magnesium	7 mg	2 per cent
Manganese	0.071 mg	3 per cent
Zinc	0.07 mg	0.5 per cent
Phyto-nutrients		
Carotene-α	1 µg	—
Carotene-β	39 µg	—
Crypto-xanthin-β	0 µg	—
Lutein-zeaxanthin	72 µg	—

- Resveratrol reduces stroke risk by altering the molecular mechanisms in the blood vessels. It does so firstly by reducing susceptibility of blood vessel damage through decreased activity of angiotensin (a systemic hormone causing blood vessel constriction that would otherwise elevate blood pressure) and secondly, through increased production of the vasodilator substance, nitric oxide (a beneficial compound that causes relaxation of blood vessels).
- Anthocyanins are another class of polyphenolic anti-oxidants present abundantly in the red grapes. These phyto-chemicals have been found to have an anti-allergic, anti-inflammatory, anti-microbial, as well as anti-cancer activity.
- Catechins, a type of flavonoid tannin group of anti-oxidants, found in the white/green varieties have also shown to possess these health-protective functions.
- In addition, the berries are very low in calories. 100 g fresh grapes just provide 69 calories but zero cholesterol levels.
- Grapes are rich source of micronutrient minerals like copper, iron and manganese. Copper and manganese are an essential co-factor of antioxidant enzyme, superoxide dismutase. Iron is especially concentrated more in raisins. In addition, 100 g of fresh grapes contain about 191 mg of health benefiting electrolyte, potassium.
- They are an also good source of vitamin-C, vitamin-A, vitamin-K, carotenes, B-complex vitamins such as pyridoxine, riboflavin, and thiamin.

19

Grapefruit

Refreshing and delicious grapefruit is rich in phytonutrients such as vitamin A, beta-carotene, and *lycopene*. It is revered as fruit from *the "paradise"* for its unique health-promoting as well as disease-healing properties, especially among health-conscious fitness freaks. The plant is a subtropical citrus tree and botanically belongs to the large *Rutaceae* family of citrus fruits of the genus: *Citrus*. Scientific name: *Citrus paradisi*.

Grapefruit is an evergreen tree, grown for its fruit, was first discovered in the forests of Caribbean island, *Barbados*. It is now one of the widely cultivated fruits in the United States, particularly in Florida, California, and the other semi-tropical Southen states. The fruit, in fact, is a natural hybridization of pomelo and orange. The tree grows up to 15 m in length and bears numerous yellow-orange skinned fruits in clusters during each season.

Each fruit is largely oblate, ranges in diameter from 3-4 in and weigh up to 150 g. characteristically the fruit has slightly thick and tough skin than that in oranges. Inside, its flesh is segmented as in oranges. The fruit is very juicy, acidic, and varying in colour depending on the cultivars, which could be white, pink and red pulps of varying sweetness. While some varieties are seedless, there may be up to 50 white, elliptical, pointed seeds about 1/2 in, in length.

Health Benefits of Grapefruit

☆ Delicious, grapefruit is very low in calories, consists of just 42 calories per 100 g. Nonetheless; it is rich in dietary insoluble fiber pectin, which by acting as a bulk laxative helps to protect the colon mucous membrane by decreasing exposure time to toxic substances in the colon as well as binding to cancer-causing chemicals in the colon.

Grapefruit (*Citrus paradisi*), Fresh, Nutritive Value per 100 g

Principle	*Nutrient Value*	*Percentage of RDA*
Energy	42 Kcal	2.3 per cent
Carbohydrates	10.7 g	8 per cent
Protein	0.77 g	1 per cent
Total Fat	0.14 g	<1 per cent
Cholesterol	0 mg	0 per cent
Dietary Fiber	1.70 g	4 per cent
Vitamins		
Folates	13 µg	3 per cent
Niacin	0.204 mg	1.5 per cent
Pantothenic acid	0.262 mg	5 per cent
Pyridoxine	0.053 mg	4 per cent
Riboflavin	0.031 mg	2.5 per cent
Thiamin	0.043 mg	4 per cent
Vitamin A	1150 IU	38 per cent
Vitamin C	31.2 mg	52 per cent
Vitamin E	0.13 mg	1 per cent
Vitamin K	0 µg	0 per cent
Electrolytes		
Sodium	0 mg	0 per cent
Potassium	135 mg	3 per cent
Minerals		
Calcium	22 mg	2 per cent
Copper	0.032 mg	4 per cent
Iron	0.08 mg	1 per cent
Magnesium	9 mg	2 per cent
Manganese	0.022 mg	1 per cent
Phosphorus	18 mg	2.5 per cent
Selenium	0.1 µg	0 per cent
Zinc	0.07 mg	1 per cent
Phyto-nutrients		
Carotene-β	686 µg	—
Crypto-xanthin-β	73 µg	—
Lycopene	1419 µg	—

☆ Pectin has also been shown to reduce blood cholesterol levels by decreasing re-absorption of cholesterol binding bile acids in the colon.

☆ The fruit contains very good levels of vitamin-A (provides about 1150 IU per 100g), and flavonoid antioxidants such as naringenin, and naringin.

Besides, it is a moderate source of lycopene, beta-carotene, xanthin and lutein. Studies suggest that these compounds have antioxidant properties and are essential for vision. The total antioxidant strength measured in terms of oxygen radical absorbance capacity (ORAC) of grapefruit is 1548 µmol TE/100 g.

- Further, vitamin A is also required maintaining healthy mucus membranes and skin. Consumption of natural fruits rich in vitamin-A, and flavonoids helps to protect from lung and oral cavity cancers.
- It is an excellent source of antioxidant *vitamin-C*; providing about 52 per cent of DRI. Vitamin-C is a powerful natural anti-oxidant and helps the body develop resistance against infectious agents and scavenge harmful free radicals. It, furthermore, is required for the maintenance of healthy connective tissue and aids in early wound healing. It also facilitates dietary iron absorption from the intestine.
- 100 g of fresh fruit contains about 135 mg of potassium electrolyte. Potassium is an important component of cell and body fluids, helps controlling heart rate and blood pressure through countering sodium effects.
- Red varieties of grapefruits are especially rich in the most powerful flavonoid antioxidant, lycopene. Studies have shown that lycopene protects skin damage from UV rays, and offers protection against prostate cancer.
- Additionally, it contains moderate levels of B-complex group of vitamins such as folates, riboflavin, pyridoxine, and thiamin in addition to some resourceful minerals such as iron, calcium, copper, and phosphorus.

20

Guava

Guava is another tropical fruit rich in nutrition. With its unique flavour, taste, and health-promoting qualities, the fruit easily fits in the new functional foods category, often called "super-fruits." It is an evergreen, tropical shrub or low-growing small tree probably originated in Middle Americas. Guavas actually thrive in both humid and dry climates and can tolerate brief periods of cold spells, but can survive only a few degrees of frost. Adaptability makes it a favourite commercial crop in some tropical areas. Botanically, this wonderful fruit belongs within the family of *Myrtaceae* of the genus: *Psidium*. Scientific name: *Psidium guajava.*

During each season, the guava tree bears numerous round, ovoid or pear-shaped fruits that are about 5-10 cm long and weigh around 50–200 g. Different cultivars are grown all over the world which, vary widely in flavour, pulp colour, and seediness. The fruit is soft when ripe with sweet musky aroma and creamy in texture. Internally, the flesh varies in colour depending up on the cultivar and may be white, pink, yellow, or red. Ripe fruits have rich flavour with sweet-tart taste. Each fruit contains numerous tiny, semi-hard edible seeds, concentrated especially at its center.

Health Benefits of Guava Fruit

- ☆ Guavas are low in calories and fats but contain several vital vitamins, minerals, and antioxidant poly-phenolic and flavonoid compounds that play a pivotal role in prevention of cancers, anti-aging, immune-booster, etc.
- ☆ The fruit is very rich source of soluble dietary fiber (5.4 g per 100 g of fruit, about 14 per cent of DRA), which makes it a good bulk laxative. The fiber content helps protect the colon mucous membrane by decreasing exposure time to toxins as well as binding to cancer-causing chemicals in the colon.
- ☆ Guava-fruit is an excellent source of antioxidant vitamin-C. 100 g fresh fruit provides 228 mg of this vitamin, more than three times the DRI

(daily-recommended intake). Outer thick rind contains exceptionally higher levels of vitamin C than central pulp.

Guava (*Psidium guajava*), Fresh, Nutritive Value per 100 g

Principle	*Nutrient Value*	*Percentage of RDA*
Energy	68 Kcal	3.5 per cent
Carbohydrates	14.3 g	11.5 per cent
Protein	2.55 g	5 per cent
Total Fat	0.95 g	3 per cent
Cholesterol	0 mg	0 per cent
Dietary Fiber	5.4 g	14 per cent
Vitamins		
Folates	49 µg	12.5 per cent
Niacin	1.084 mg	7 per cent
Pantothenic acid	0.451 mg	9 per cent
Pyridoxine	0.110 mg	8.5 per cent
Riboflavin	0.040 mg	3 per cent
Thiamin	0.067 mg	5.5 per cent
Vitamin A	624 IU	21 per cent
Vitamin C	228 mg	396 per cent
Vitamin E	0.73 mg	5 per cent
Vitamin K	2.6 µg	2 per cent
Electrolytes		
Sodium	2 mg	0 per cent
Potassium	417 mg	9 per cent
Minerals		
Calcium	18 mg	2 per cent
Copper	0.230 mg	2.5 per cent
Iron	0.26 mg	3 per cent
Magnesium	22 mg	5.5 per cent
Manganese	0.150 mg	6.5 per cent
Phosphorus	11 mg	2 per cent
Selenium	0.6 mcg	1 per cent
Zinc	0.23 mg	2 per cent
Phyto-nutrients		
Carotene-β	374 µg	—
Crypto-xanthin-β	0 µg	—
Lycopene	5204 µg	—

- ☆ Scientific studies shown that regular consumption of fruits rich in vitamin C helps the body develop resistance against infectious agents and scavenge cancer causing harmful free radicals from the body. Further, the vitamin is required for collagen synthesis within the body. Collagen is the main structural protein in the human body required for maintaining the integrity of blood vessels, skin, organs, and bones.
- ☆ The fruit is a very good source of vitamin-A, and flavonoids like beta-carotene, lycopene, lutein and cryptoxanthin. The compounds are known to have antioxidant properties and are essential for optimum health. Further, vitamin-A is also required for maintaining healthy mucus membranes and skin. Consumption of natural fruits rich in carotene is known to protect from lung and oral cavity cancers.
- ☆ 100 g of pink guava fruit provides 5204 µg of lycopene, nearly twice the amount that in tomatoes. (100 g tomato contains 2573 µg of lycopene). Studies suggest that *a lycopenen pink guava prevents* skin damage from UV rays and offers protection from prostate cancer.
- ☆ Fresh fruit is a very rich source of potassium. It contains more potassium than other fruits like banana weight per weight. Potassium is an important component of cell and body fluids that helps controlling heart rate and blood pressure.
- ☆ Further, the fruit is also a moderate source of B-complex vitamins such as pantothenic acid, niacin, vitamin-B6 (pyridoxine), vitamin E and K, as well as minerals like magnesium, copper, and manganese. Manganese is used by the body as a co-factor for the antioxidant enzyme, *superoxide dismutase*. Copper is required for the production of red blood cells.

21

Jackfruit

Jackfruit is one of its own kind tropical fruits, recognized for its fruity flavour that can be appreciated from a distance. The fruit is unique for its size and delicious sweet taste. In common with other tropical fruits such as durian, banana, etc., it is also rich in energy, dietary fiber, minerals, and vitamins and free from saturated fats or cholesterol, making it one of the healthy summer seasons treated to relish. Botanically, this popular Asian tropical fruit belongs to the family of *moraceae*, genus: *Artocarpus* and is closely related to figs, mulberry, and breadfruit. Scientific name: *Artocarpus heterophyllus.*

The jackfruit tree is believed to be indigenous to the Southwestern rain forests of India. It is widely cultivated in the tropical regions of Indian subcontinent, Thailand, Malaysia, Indonesia and Brazil for its fruit, seeds, and wood. The tree grows best under tropical humid and rainy climates but rarely survives cold and frosty conditions. Jackfruit is a huge tree that grows to as high as 30 meters, higher than the mango tree. During the season, each tree bears as many as 250 large fruits, supposed to be the largest tree-borne fruits in the world. The fruit varies widely in size, weigh from 3 to 30 kg, and has oblong or round shape measuring 10 cm to 60 cm in length, 25 to 75 cm in diameter. The unripe fruits are green; when ripen, might turn to light brown colour and spread a strong sweet, fruity smell. As in durian fruit, its outer surface is covered with the blunt thorns like projections, which become soft as the fruit ripe. Its interior consists of orange-yellow coloured edible bulbs. Each bulb consists of sweet-flavoured sheaths that enclose a smooth, oval, light-brown colour seed. Jackfruit seed is 2 to 4 cm long and 1 to 3 cm thick and is white and crisp within. There may be as many as 100 to 500 edible bulbs embedded in a single fruit interspersed between thin bands of fibers. Almost all the parts of the tree secrete white sticky latex like milk (juice) when injured.

Health Benefits of Jackfruit

☆ 100 g of edible jackfruit bulbs provide 95 calories. The fruit is made of soft, easily digestible flesh (bulbs) with simple sugars like fructose and sucrose that when eaten replenishes energy and revitalizes the body instantly.

Jack Fruit (*Artocarpus heterophyllus*), Nutritive Value per 100 g

Principle	*Nutrient Value*	*Percentage of RDA*
Energy	95 Kcal	5 per cent
Carbohydrates	23.5 g	18 per cent
Protein	1.72 g	3 per cent
Total Fat	0.64 g	3 per cent
Cholesterol	0 mg	0 per cent
Dietary Fiber	1.5 g	4 per cent
Vitamins		
Folates	24 μg	6 per cent
Niacin	0.920 mg	6 per cent
Pyridoxine	0.329 mg	25 per cent
Riboflavin	0.055 mg	4 per cent
Thiamin	0.105 mg	9 per cent
Vitamin A	110 IU	3.5 per cent
Vitamin C	13.7 mg	23 per cent
Vitamin E	0.34 mg	2 per cent
Electrolytes		
Sodium	3 mg	0 per cent
Potassium	303 mg	6.5 per cent
Minerals		
Calcium	34 mg	3.4 per cent
Iron	**0.60 mg**	**7.5 per cent**
Magnesium	37 mg	9 per cent
Manganese	0.197 mg	8.5 per cent
Phosphorus	36 mg	5 per cent
Phosphorus	21 mg	3 per cent
Selenium	0.6 mg	1 per cent
Zinc	0.42 mg	4 per cent
Phyto-nutrients		
Carotene-β	61 μg	—
Crypto-xanthin-β	5 μg	—
Lutein-zeaxanthin	157 μg	—

- Jackfruit is rich in dietary fiber, which makes it a good bulk laxative. The fiber content helps to protect the colon mucous membrane by decreasing exposure time and as well as binding to cancer-causing chemicals in the colon.
- Fresh fruit has small amounts of vitamin-A, and flavonoid pigments such as carotene-β, xanthin, lutein and cryptoxanthin-β. Together, these compounds play vital roles in antioxidant and vision functions. Vitamin A is also required for maintaining integrity of mucus membranes and skin. Consumption of natural fruits rich in vitamin-A, and carotenes has been found to protect from lung and oral cavity cancers.
- Addition, jackfruit is a good source of antioxidant vitamin-C, provides about 13.7 mg or 23 per cent of RDA. Consumption of foods rich in vitamin C helps the body develop resistance against infectious agents and scavenge harmful free radicals.
- It is one of the rare fruits that is rich in B-complex group of vitamins. It contains very good amounts of vitamin B-6 (pyridoxine), niacin, riboflavin, and folic acid.
- Fresh fruit is a good source of potassium, magnesium, manganese, and iron. Potassium is an important component of cell and body fluids that helps controlling heart rate and blood pressure.

22

Kiwifruit

Kiwifruit, also known as *Chinese gooseberry,* is one of the delicious fruits with full of promising health promoting phyto-chemicals, vitamins and minerals. This widely recognized, wonderfully unique fruit is native to eastern Chinese "Shaanxi" province. This exotic fruit is also recognized as China's national fruit. The kiwifruit plant is a semitropical, deciduous, large woody vine belonging within the family of *Actinidiaceae;* Genus, *Actinidia.* Scieintific name: *Actinidia chinensis.*

The journey of this humble gooseberry from China's mainland to Oceania, and thence to worldwide is quite interesting! During early years of twentieth-century kiwifruit seeds were carried to New Zealand by missionaries where it has become naturalized. From New Zealand, the fruit spread all across the world, and now grown at a commercial scale in many European and as far as in California regions in North America.

During each season which lasts from September until November, The kiwi vine bears numerous ovals shaped, fuzzy, brown coloured fruits.

Ovals shaped, fuzzy, brown coloured fruits. Each fruit measures approximately the size of large hen's egg, and weighs up to 125 g. internally; its flesh is soft, juicy, emerald green with rows of tiny, black, edible seeds. Fruit texture is similar to strawberry or sapodilla and the flavour resembles a blend of strawberry, apple and pineapple fruits. Several other cultivars are grown apart from the common kiwifruit variety. "Hardy kiwi fruit" (*A. arguta*) also known as *baby kiwi,* are much smaller than "Fuzzy kiwi fruit" (*A. deliciosa*). Their size is similar to that of a large grape with smooth, edible skin. Inside, "hardy kiwi fruit" resembles "fuzzy kiwi-fruit" in colour, texture, but usually has intense flavour and sweetness. "Gold Kiwifruit," developed by hybrid technique by agricultural research department in New Zealand, has a smooth, sparse hairs, bronze skin, a pointed cap at one end and distinctive golden-yellow flesh with less tart and more tropical flavour than green-kiwifruit. Commercially, it has a higher market value than common green kiwifruit.

Health Benefits of Kiwi Fruit

☆ Kiwifruit is a very rich source of soluble dietary fiber (3.8 g per 100 g of fruit OR 10 per cent of RDA), which makes it a good bulk laxative. The fiber content helps to protect the colon mucous membrane by decreasing exposure time to toxins as well as binding to cancer-causing chemicals in the colon.

Kiwi Fruit or Chinese Gooseberry (*Actinidia chinensis*), Fresh, Nutrient Value per 100 g

Principle	*Nutrient Value*	*Percentage of RDA*
Energy	61 Kcal	3 per cent
Carbohydrates	14.66 g	11 per cent
Protein	1 g	2 per cent
Total Fat	0.52 g	2 per cent
Cholesterol	0 mg	0 per cent
Dietary Fiber	3 g	8 per cent
Vitamins		
Folates	25 µg	6 per cent
Niacin	0.341 mg	2 per cent
Riboflavin	0.025 mg	2 per cent
Thiamin	0.027 mg	2 per cent
Vitamin A	87 IU	3 per cent
Vitamin C	92.7 mg	154 per cent
Vitamin E	1.46 mg	10 per cent
Vitamin K	40.3 µg	34 per cent
Electrolytes		
Sodium	3 mg	0 per cent
Potassium	312 mg	7 per cent
Minerals		
Calcium	34 mg	3.5 per cent
Copper	0.130 mg	14 per cent
Iron	0.31 mg	4 per cent
Magnesium	17 mg	4 per cent
Manganese	0.098 mg	4 per cent
Zinc	0.14 mg	1 per cent
Phyto-nutrients		
Carotene-β	52 µg	—
Crypto-xanthin-β	0 µg	—
Lutein-zeaxanthin	122 µg	—

- The fruit is an excellent source of antioxidant vitamin-C; providing about 154 per cent of the DRI (daily-recommended intake). Consumption of foods rich in vitamin-C helps the body develop resistance against infectious agents and scavenge harmful free radicals.
- Kiwi fruit contains very good levels of vitamin-A, vitamin-E, vitamin-K and flavonoid anti-oxidants such as beta-carotene, lutein and xanthin. Vitamin K has a potential role in the increase of bone mass by promoting osteotrophic activity in the bone. It also has established role in Alzheimer's disease patients by limiting neuronal damage in the brain. Total antioxidant strength measured in terms of *oxygen radical absorbance capacity* (ORAC) of kiwifruit (gold, raw) is 1210 µmol TE/100 g.
- Research studies have shown that certain substances in kiwi-fruit functions as blood thinner function similar to aspirin; thus, it helps prevent clot formation inside the blood vessels and protect from stroke and heart-attack risk.
- Kiwi-fruit seeds are an excellent source of omega-3 fatty acids. Research studies show that consumption of foods rich in omega-3 fatty acids may reduce the risk of coronary heart disease, stroke, and help prevent the development of ADHD, autism, and other developmental disorders in children.
- Fresh kiwi fruit is a very rich source of heart-healthy electrolyte "potassium." 100 g contains 312 mg or 7 per cent of daily-recommended levels of this electrolyte. Potassium is an important component of cell and body fluids that help regulate heart rate and blood pressure by countering malefic effects of sodium.
- It also contains good amounts of minerals like manganese, iron and magnesium. Manganese is used in the body as a co-factor for the powerful antioxidant enzyme, *superoxide dismutase*. Magnesium is an important bone-strengthening mineral like calcium.

23

Kumquat

Delicious, sweet yet tangy, kumquat fruit or cumquat (as the fruit generally recognized in Europe) is a winter/spring season delicacy. Although kumquats taste just like that of citrus fruits, they are distinguished in a way that they can be eaten completely including the peel.

Botanically they belonged to the Rutaceae family; of the genus, Fortunella, named after the botanist Robert fortune, who brought them from China to Europe in the middle of the 19th century. Kumquats are a small sized evergreen tree native to Southeastern parts of mountainous China. Today, they have grown for their delicious fruits and as ornament tree in many parts of the world, including USA. A mature kumquat tree bears several hundred olive-sized brilliant orange colour fruits in winter. On the Interior, the fruit resembles tiny orange with juicy segments firmly adherent each other and with the rind. The pulp has 1-2 seeds placed centrally. The seeds are bitter in taste as in oranges and generally spit out. There exist several varieties of kumquat; however, some four cultivars are grown widely for their fruits.

- ✰ **Nagami kumquat** (*Fortunella margarita*): The fruit is oval and is the most common varieties grown in USA. The fruit features smooth light yellow rind tart flavour.
- ✰ **Marumi kumquat** (*Fortunella japonica*): The fruit is round and has the distinctive sweet flavour.
- ✰ **Meiwa kumquat** (*Fortunella crassifolia*): It is round in shape and larger than other verities. It is popular in Japan as *ninpo* or *neiha kinkan*.
- ✰ **Hong Kong Wild** (*Fortunella hindsii*).

Health Benefits of Kumquat Fruit

- ✰ Kumquat has calorific value equivalent to that of grapes. 100 g of fresh fruits provide only 71 calories. Nonetheless, they are incredibly rich sources of health-benefiting dietary fiber, minerals, vitamins, and pigment antioxidants that contribute immensely towards wellness.

Kumquat Fruit (*Fortunella species*), Nutrition Value per 100 g

Principle	*Nutrient Value*	*Percentage of RDA*
Energy	71 Kcal	3.5 per cent
Carbohydrates	15.90 g	12 per cent
Protein	1.88 g	3 per cent
Total Fat	0.86 g	4 per cent
Cholesterol	0 mg	0 per cent
Dietary Fiber	6.5 g	17 per cent
Vitamins		
Folates	17 µg	4 per cent
Niacin	0.429 mg	2.5 per cent
Pantothenic acid	0.208 mg	4 per cent
Pyridoxine	0.036 mg	3 per cent
Riboflavin	0.090 mg	7 per cent
Thiamin	0.037 mg	3 per cent
Vitamin A	290 IU	10 per cent
Vitamin C	43.9 mg	73 per cent
Vitamin E	0.15 mg	1 per cent
Vitamin K	0 µg	0 per cent
Electrolytes		
Sodium	10 mg	0.5 per cent
Potassium	186 mg	4 per cent
Minerals		
Calcium	62 mg	6 per cent
Copper	0.095 mg	10 per cent
Iron	0.86 mg	11 per cent
Magnesium	20 mg	5 per cent
Manganese	0.135 mg	6 per cent
Selenium	0.0 mcg	0 per cent
Zinc	0.17 mg	1 per cent
Phyto-nutrients		
Carotene-β	0 µg	—
Carotene-α	155 µg	—
Cryptoxanthin-β	193 µg	—
Lutein-zeaxanthin	129 µg	—

- ✰ Kumquats are eaten along with the peel, a unique feature that differentiates them from other citrus family fruits. The peel is rich in many essential oils, anti-oxidants, and fiber. 100 g whole kumquats provide 6.7 g or 17 per cent of daily-recommended levels of fiber composed of tannins, pectin, hemi-cellulose, and other non-starch polysaccharides (NSP).
- ✰ Fresh kumquats are packed with numerous health benefiting poly-phenolic flavonoid anti-oxidants such as carotenes, lutein, zeaxanthin, tannins etc. The kumquat peel is composed many important essential oils in it, including limonene, pinene, α-bergamotene, caryophyllene, α-humulene, and α-muurolene. Together, these compounds impart special citrus aroma to the fruit.
- ✰ Further, fresh fruits contain adequate levels of some of the anti-oxidant vitamins such as vitamin A, C and E. Altogether these phyto-chemical compounds in kumquat fruit help scavenge harmful oxygen derived free radicals from the body and thereby protect us from cancers, diabetes, degenerative diseases and infections.
- ✰ Like oranges, kumquats also very rich in vitamin C. 100 g fruit provides 47.9 or 73 per cent of RDA (Recommended daily allowances). Vitamin-C is one of the powerful natural anti-oxidant, which has many essential biological roles like collagen synthesis and wound healing; anti-viral and anti-cancer activity; and helps prevent from neuro-degenerative diseases, arthritis, diabetes.etc. by removing oxidant free-radicals from the body. Furthermore, vitamin C felicitates iron absorption in the food by reducing it from ferrous to ferric form in the stomach.
- ✰ Cumquat contains good levels of B-complex group of vitamins such as thiamin, niacin, pyridoxine, folates and pantothenic acid. These vitamins function as co-factors for metabolism of carbohydrates, proteins, and fats.
- ✰ In addition, kumquats are modest sources of minerals like calcium, copper, potassium, manganese, iron, selenium, and zinc. Calicum is the chief element required for bone and teeth formation. Copper is required in the production of red blood cells. Iron is required for red blood cell formation as well for cellular oxidation.

24

Lemon

Juicy, acidic and flavourful lemon is the most widely used citrus fruit worldwide. Lime is a close related species, relatively smaller in size and possesses a thinner skin. It is the smallest among citrus fruits; nevertheless, contains more health benefiting nutrients than other citrus family fruits like oranges, pomelo etc. Botanically, this citrus fruit belongs to the family of *Rutaceae* of the genus, *Citrus* (which also includesorange, pomelo, tangerine (mandarin orange), and grapefruit). Scientific name: *Citrus limon*.

Lemons are thought to be originated in the Himalayan foothills of North-East India, from where they spread all across the Middle East, Europe, Africa, and as far as Americas. Limes are smaller in size and less sour than lemons. Put together, both these fruits perhaps are now the largest consumables among fruits all over the world. Like other citrus plants, lemons are small, spreading, evergreen trees growing up to 10-12 feet in most cultivated plantations. They flourish well in temperate and tropical environments, whereas, cold and frosty conditions would affect their growth adversely. Stems are often armed with sharp, stout thorns. Fully-grown plant bears fragrant, white flowers in short cymes.

Fruits belonging to the citrus group are described as "hesperidium," (A hesperidium is a scientific term to describe the fruit structure belonging to the citrus group. In fact, the fruit is a modified berry with tough, leathery rind. Lemon's peel contains many volatile oil glands in pits. Interior flesh is composed of segments, called carpels, made up of numerous juice-filled vesicles that are actually specialized hair cells). Mature fruits measure about 5-8 cm in diameter and weigh about 50- 80 g in weight.

Health Benefits of Lemon

- ☆ Lemons are packed with numerous health benefiting nutrients. The fruit is low in calories, 29 calories per 100 g, one of the lowest among the citrus groups.

☆ It contains no saturated fats or cholesterol, but is rich in dietary fiber (7.36 per cent of RDA). Lemon is one of the very low glycemic fruits.

Lemon (*Citrus limon*), Fresh, without Peel, Nutrition Value per 100 g

Principle	*Nutrient Value*	*Percentage of RDA*
Energy	29 Kcal	1.5 per cent
Carbohydrates	9.32 g	7 per cent
Protein	1.10 g	2 per cent
Total Fat	0.30 g	1 per cent
Cholesterol	0 mg	0 per cent
Dietary Fiber	2.80 g	7 per cent
Vitamins		
Folates	11 µg	3 per cent
Niacin	0.100 mg	1 per cent
Pantothenic acid	0.190 mg	4 per cent
Pyridoxine	0.080 mg	6 per cent
Riboflavin	0.020 mg	1.5 per cent
Thiamin	0.040 mg	3.5 per cent
Vitamin C	53 mg	88 per cent
Vitamin A	22 IU	1 per cent
Vitamin E	0.15 mg	1 per cent
Vitamin K	0 µg	0 per cent
Electrolytes		
Sodium	2 mg	0 per cent
Potassium	138 mg	3 per cent
Minerals		
Calcium	26 mg	3 per cent
Copper	37 µg	4 per cent
Iron	0.60 mg	7.5 per cent
Magnesium	8 mg	2 per cent
Manganese	0.030 mg	1 per cent
Zinc	0.06 mg	0.5 per cent
Phyto-nutrients		
Carotene-β	3 µg	—
Carotene-α	1 µg	—
Crypto-xanthin-β	20 µg	—
Lutein-zeaxanthin	11 µg	—
Lycopene	0 µg	—

- ☆ Its acidic taste is because of citric acid. Citric acid constitutes up to 8 per cent in its juice. Citric acid is a natural preservative, aids digestion and helps dissolve kidney stones.
- ☆ Lemons, like other fellow citrus fruits, are an excellent source of vitamin C (ascorbic acid); provides about 88 per cent of daily recommended intake. Ascorbic acid is a powerful water soluble natural anti-oxidant. This vitamin is helpful in preventing scurvy. Besides, consumption of foods rich in vitamin-C helps the human body develop resistance against infectious agents and scavenge harmful, pro-inflammatory free radicals from the blood.
- ☆ Lemons, like oranges, contain a variety of phytochemicals. Hesperetin, naringin, and naringenin are flavonoid glycosides commonly found in citrus fruits. Naringenin is found to have a bio-active effect on human health as antioxidant, free radical scavenger, anti-inflammatory, and immune system modulator. This substance has also been shown to reduce oxidant injury to DNA in the cells *in-vitro* studies.
- ☆ Additionally, they also compose a minute levels of vitamin A, and other flavonoid anti-oxidants such as *α, and β-carotenes, β-cryptoxanthin, zeaxanthin and lutein.* These compounds are known to have antioxidant properties. Vitamin A also required for maintaining healthy mucus membranes and skin and is also essential for vision. Consumption of natural fruits rich in flavonoids helps the body to protect from lung and oral cavity cancers.
- ☆ Total ORAC value, which measures the anti-oxidant strength of 100 g of fresh lemon juice, is 1225 μmol TE (Trolex equivalents).
- ☆ The fruit is also a good source of B-complex vitamins such as pantothenic acid, pyridoxine, and folates. These vitamins are essential in the sense that body requires them from external sources to replenish.
- ☆ Further, they contain a healthy amount of minerals like iron, copper, potassium, and calcium. Potassium in an important component of cell and body fluids helps control heart rate and blood pressure.

Citrus fruits, as such, have long been valued for their wholesome nutritious and antioxidant properties. It is scientifically established that citrus fruits, especially lemons and oranges, by virtue of their richness in vitamins and minerals, have many proven health benefits. Moreover, it is now beginning to be appreciated that the other biologically active, non-nutrient compounds found in citrus fruits such as phyto-chemical antioxidants, and soluble as well as insoluble dietary fiber is helpful in reduction in the risk for cancers, many chronic diseases like arthritis, and from obesity and coronary heart diseases.

25

Loquat Fruits

Succulent, tangy and sweet, wonderfully delicious loquat fruits are rich in vitamins, minerals, and anti-oxidants. This unique fruit is originated in the mountainous, evergreen rain forest of Southeastern China, from where it spread all across the world, including Japan. Some of the common names of loquat include *Japanese plum, Japanese medlar or Maltese plum,* etc. The plant is an evergreen, large shrub or small tree belonging to the family of *Rosaceae* of the genus; Eriobotrya and named botanically as *Eriobotrya japonica*. It is closely related to the "apples."

Loquat fruits begin appearing in the trees by the end of winter season. Mature fruits are ready to harvest by June in Japan; however, their harvesting season may vary from region to region. The fruits are oval to pear shape, appear in bunches of 5-20 and measure about 3 cm in width and 3-5 cm in length. In general, the fruits are allowed to ripen on the tree before harvesting. Ripe fruits have soft texture. Externally, the skin surface is fuzzy but smooth with yellow to orange colour. Internally, the flesh is either white or golden-yellow depending on the cultivar type. Each fruit contains 3-5, centrally placed large, brown seeds. The seeds are inedible and may contain toxic cyanogen-glycosides. Being a member of *Rosaceae* family of fruits, loquats have similar taste and flavour as that of apples; tart, and sweet with pleasant aroma. Loquat's leaves are also used in traditional medicines and as herbal tea in many parts of the world.

Health Benefits of Loquat Fruit

- ☆ Delicious, loquats are very low in calories; provide just 47 calories per 100 g, however, rich in insoluble dietary fiber, pectin. Pectin retains moisture in the colon and thus functions as bulk laxative and by this way, it helps to protect the colon mucous membrane by decreasing exposure time to toxic substances as well as binding to cancer-causing chemicals in the colon.

Loquat Fruit (*Eriobotrya japonica*), Fresh, Nutrition Value per 100 g

Principle	*Nutrient Value*	*Percentage of RDA*
Energy	47 Kcal	2.4 per cent
Carbohydrates	12.14 g	9 per cent
Protein	0.43 g	2 per cent
Total Fat	0.20 g	1 per cent
Cholesterol	0 mg	0 per cent
Dietary Fiber	1.70 g	4 per cent
Vitamins		
Folates	14 µg	3.5 per cent
Niacin	0.180 mg	1 per cent
Pyridoxine	0.100 mg	8 per cent
Riboflavin	0.024 mg	2 per cent
Thiamin	0.019 mg	2 per cent
Vitamin A	1528 IU	51 per cent
Vitamin C	1 mg	2 per cent
Electrolytes		
Sodium	1 mg	0 per cent
Potassium	266 mg	6 per cent
Minerals		
Calcium	16 mg	1.6 per cent
Copper	0.040 mg	4.5 per cent
Iron	0.28 mg	3.5 per cent
Magnesium	13 mg	3 per cent
Manganese	0.148 mg	6.5 per cent
Phosphorus	27 mg	4 per cent
Selenium	0.6 µg	1 per cent
Zinc	0.05 mg	0.5 per cent

- ☆ Pectin has also been shown to reduce blood cholesterol levels by decreasing its re-absorption in the colon by binding bile acids resulting in its excretion from the body.
- ☆ Loquat fruit is an excellent source of vitamin-A (provides about 1528 IU or 51 per cent of daily recommended levels of this vitamin per 100g), and phenolic flavonoid antioxidants such as chlorogenic acid, neo-chlorogenic acid, hydroxybenzoic acid, feruloylquinic acid, protocatechuic acid, epicatechin, coumaric acids and ferulic acid. Ripen fruits have more chlorogenic-acid concentrations.
- ☆ Vitamin A maintains integrity of mucus membranes, and skin. Lab studies have shown that consumption of natural fruits rich in vitamin-A, and flavonoids helps to protect from lung and oral cavity cancers.

- Fresh fruit is very rich in potassium and some B-complex vitamins such as folates, vitamin B-6 and niacin and contain small amounts of vitamin-C. Potassium is an important component of cell and body fluids, helps controlling heart rate and blood pressure.
- Furthermore, the fruit is also a good source of iron, copper, calcium, manganese, and other minerals. Manganese is used by the body as a co-factor for the antioxidant enzyme, *superoxide dismutase*. Copper is required in the production of red blood cells. Iron is required for as a cofactor in cellular oxidation as well for red blood cell formation.

26

Lychee

Delicious and juicy lychee or *"Litchi"* heralds you the arrival of summer. Besides sweet and nutritious, the berries have cooling effect on the human body. Botanically, this exotic fruit belongs to the family of *Sapindaceae* and named scientifically as *Litchi chinensis*. The *L. chinensis* is a tropical fruit tree native to the low elevations of the *Kwangtung* and *Fukien proyinces* of South China. It is a slow growing, medium-sized evergreen tree with dense, round-topped foliage, and smooth, gray, brittle trunk and branches. It may reach 40-50 feet height. Litchis, not only eye-catching in spring when the huge sprays of flowers adorn the tree but also is a stunning sight for nature lovers when the tree is full of berries.

In structure, the fruit is a drupe; oval, heart-shaped or nearly round, measures about 3–5 cm long and 3 cm in diameter and weigh about 10 g. In appearance, the fruit has close resemblances with *longan* and *rambutan*fruits. Its outer surface is covered with rough leathery rind or peel featuring pink colour. The peel can be easily removable in the ripe fruits. Inside, the pulp consists of edible portion or aril that is white, translucent, sweet, and juicy. The fruit has sweet, fragrant flavour and delicious to savor. The pulp has single, glossy brown nut-like seed, 2 cm long, and 1–1.5 cm in diameter. The seeds, like in sapodilla, are not poisonous but should not be eaten. Fresh lychees are readily available in the markets from July to October, about 120-140 days after flowering.

Health Benefits of Lychee

- ✰ Lychee fruit contains 66 calories per 100 g, comparable to that in the table-grapes. It has no saturated fats or cholesterol, but composes of good amounts of dietary fiber, vitamins, and antioxidants.
- ✰ Research studies suggest that oligonol, a low molecular weight polyphenol, is found abundantly in lychee fruit. Oligonol is thought to have anti-oxidant and anti-influenza virus actions. In addition, it helps improve blood flow

in organs, reduce weight, and protect skin from harmful UV rays. (Takuya Sakurai (Kyorin University, Japan), Biosci. Biotechnol. Biochem, 72(2), 463-476, 2008).

Lychee Fruit (*Litchi chinensis*), Fresh, Nutritive Value per 100 g

Principle	*Nutrient Value*	*Percentage of RDA*
Energy	66 kcal	3.3 per cent
Carbohydrates	16.53 g	12.7 per cent
Protein	0.83 g	1.5 per cent
Total Fat	0.44 g	2 per cent
Cholesterol	0 mg	0 per cent
Dietary Fiber	1.3 g	3.5 per cent
Vitamins		
Folates	14 µg	3.5 per cent
Niacin	0.603 mg	3.5 per cent
Choline	7.1 mg	1 per cent
Pyridoxine	0.100 mg	9 per cent
Riboflavin	0.065 mg	5 per cent
Thiamin	0.011 mg	1 per cent
Vitamin A	0 mg	0 per cent
Vitamin C	71.5 mg	119 per cent
Vitamin E	0.07 mg	0.5 per cent
Vitamin K	0.4 µg	0.3 per cent
Electrolytes		
Sodium	1 mg	0 per cent
Potassium	171 mg	3.5 per cent
Minerals		
Calcium	5 mg	0.5 per cent
Copper	0.148 mg	16 per cent
Iron	0.31 mg	4 per cent
Magnesium	10 mg	2.5 per cent
Manganese	0.055 mg	2.5 per cent
Phosphorus	31 mg	4.5 per cent
Selenium	0.6 µg	1 per cent
Zinc	0.07 mg	0.5 per cent
Phyto-nutrients		
Carotene-β	0 µg	—
Crypto-xanthin-β	0 µg	—
Lutein-zeaxanthin	0 µg	—

- Litchi, like citrus fruits, is an excellent source of vitamin C; 100 g fresh fruits provide 71.5 mg or 119 per cent of daily-recommended value. Studies suggest that consumption of fruits rich in vitamin C helps the human body develop resistance against infectious agents and scavenge harmful, pro-inflammatory free radicals.
- Further, it is a very good source of B-complex vitamins such as thiamin, niacin, and folates. These vitamins are essential since they function by acting as co-factors to help the body metabolize carbohydrates, protein, and fats.
- Litchi also contains a very good amount of minerals like potassium and copper. Potassium is an important component of cell and body fluids help control heart rate and blood pressure; thus, it offers protection against stroke and coronary heart diseases. Copper is required in the production of red blood cells.

27

Mango Fruit

The king of the fruits mango fruit is one of the most popular, nutritionally rich fruits with unique flavour, fragrance, taste, and heath promoting qualities making it a common ingredient in new functional foods often labeled "super fruits."

Mango is one of the delicious seasonal fruits grow in the tropics. The tree is believed to be originating in the sub-Himalayan plains of Indian subcontinent. Botanically, this exotic fruit belongs within the family of*Anacardiaceae,* a family that also includes numerous species of tropical-fruiting trees in the flowering plants such as cashew, pistachio etc. Scientific name: *Mangifera indica*.

Mango is a tropical tree cultivated in many regions of India and now distributed wide across the world in many continents. Usually, fruits grow at the end of a long, string like stem, with sometimes more than one fruit to a stem. Each fruit measures 5 to 15 cms in length and about 4 to 10 cms in width, and has typical "mango" shape, or sometimes oval or round. Its weight ranges from 150 gm to around 750 gm. Outer skin is smooth and is green in un-ripe mangoes but turns into golden yellow, bright, yellow or orange-red when ripen depending upon the cultivar. Fresh mango season lasts from April until August. Mango comes in different shapes and sizes depending upon cultivar types. Internally, the flesh is juicy, orange-yellow in colour with numerous soft fibrils radiating from its centrally placed flat, oval-shaped stone (enveloping a single large kidney-shaped seed). Its flavour is pleasant and rich, and tastes sweet with mild tartness. A high-quality mango fruit should feature no or very less fiber content and minimal sour taste. Mango seed (stone) may either has a single embryo, or sometimes polyembryonic.

Health Benefits of Mangoes

- ☆ Mango fruit is rich in pre-biotic dietary fiber, vitamins, minerals, and *poly-phenolic flavonoid* antioxidant compounds.

Mango Fruit (*Mangifera indica*), Fresh, Nutrition Value per 100 g

Principle	*Nutrient Value*	*Percentage of RDA*
Energy	70 Kcal	3.5 per cent
Carbohydrates	17 g	13 per cent
Protein	0.5 g	1 per cent
Total Fat	0.27 g	1 per cent
Cholesterol	0 mg	0 per cent
Dietary Fiber	1.80 g	4.5 per cent
Vitamins		
Folates	14 µg	3.5 per cent
Niacin	0.584 mg	3.5 per cent
Pantothenic acid	0.160 mg	1 per cent
Pyridoxine (vit. B-6)	0.134 mg	10 per cent
Riboflavin	0.057 mg	4 per cent
Thiamin	0.058 mg	5 per cent
Vitamin C	27.7 mg	46 per cent
Vitamin A	765 IU	25.5 per cent
Vitamin E	1.12 mg	7.5 per cent
Vitamin K	4.2 µg	3.5 per cent
Electrolytes		
Sodium	2 mg	0 per cent
Potassium	156 mg	3 per cent
Minerals		
Calcium	10 mg	1 per cent
Copper	0.110 mg	12 per cent
Iron	0.13 mg	1.5 per cent
Magnesium	9 mg	2 per cent
Manganese	0.027 mg	1 per cent
Zinc	0.04 mg	0 per cent
Phyto-nutrients		
Carotene-β	445 µg	—
Carotene-α	17 µg	—
Crypto-xanthin-β	11 µg	—
Lutein-zeaxanthin	0 µg	—
Lycopene	0 µg	—

☆ According to new research study, mango fruit has been found to protect against colon, breast, leukemia and prostate cancers. Several trial studies suggest that *polyphenolic anti-oxidant* compounds in mango are known to offer protection against breast and colon cancers.

- ☆ Mango fruit is an excellent source of **Vitamin-A** and flavonoids like ***beta-carotene, alpha-carotene,*** and ***beta-cryptoxanthin.*** 100 g of fresh fruit provides 765 mg or 25 per cent of recommended daily levels of vitamin A. Together; these compounds are known to have antioxidant properties and are essential for vision. Vitamin A is also required for maintaining healthy mucus membranes and skin. Consumption of natural fruits rich in carotenes is known to protect the body from lung and oral cavity cancers.
- ☆ Fresh mango is a good source of potassium. 100 g fruit provides 156 mg of potassium while just 2 mg of sodium. Potassium is an important component of cell and body fluids that helps controlling heart rate and blood pressure.
- ☆ It is also a very good source of vitamin-B6 (pyridoxine), vitamin-C and vitamin-E. Consumption of foods rich in vitamin C helps the body develop resistance against infectious agents and scavenge harmful oxygen-free radicals. Vitamin B-6 or pyridoxine is required for GABA hormone production within the brain. It also controls homocystiene levels within the blood, which may otherwise be harmful to blood vessels resulting in CAD and stroke.
- ☆ Further, it composes moderate amounts of copper. Copper is a co-factor for many vital enzymes, including *cytochrome c-oxidase* and *superoxide dismutase* (other minerals function as co-factors for this enzyme are manganese and zinc). Copper is also required for the production of red blood cells.
- ☆ Additionally, mango peel is also rich in phytonutrients, such as the pigment antioxidants like carotenoids and polyphenols.

28

Mangosteen

Unique for its appearance and flavour, mangosteen is often revered as queen of the tropical fruits, especially in the oriental regions. This exotic, purple colour tropical-fruit is quite popular for its snow-white, juicy, delicious arils all over the Asian countries and in recent years grabbing the attention of European and American fruit lovers as well. Botanically the fruit belongs to the family of *Clusiaceae*, of the genus: *Garcinia*. Scientific name: *Garcinia mangostana.*

Mangosteen plant is an evergreen, erect tree reaching about 20- 60 ft in height. It commonly found in tropical rainforests of Indonesia, Malaysia, and Philippines as well as in some cultivated lands in Sri Lanka, and India, where annual precipitation and relative humidity are favourable for its growth. Fresh purple fruits are available in the markets from June until October. Each tree bears several deep purple coloured round shaped fruits, capped with light green calyx at the stem end. Fully matured fruit measures about 3-7 cm in diameter. Its outer tough rind is about 7-12 mm thick, contains bitter yellow latex that stains clothes black.

Internally; the fruit features 4 to 10 juicy, snow-white colour, soft, fleshy triangular segments as in oranges. Each segment may contain 1-4 off white coloured seeds. Seeds are inedible and bitter in taste. The flavour of fruit is sweet, fragrant, and delicious. Different species of *Garcinia* genus are grown all around the Southeast Asia, Africa, and South America all along the tropical belt. Cherapu or Button mangosteen (*Garcinia prainiana*) is native to Malaysia. It features somewhat flat, mandarin orange-like shape with thin rind that can be peeled very easily by hand again as in oranges.

- ☆ **Lemon drop mangosteens** (*Garcinia madruno*) are small size (2-4 cm diameter) fruits, native to Malayan peninsula.
- ☆ **African mangosteen or Imbe** (*Garcinia livingstonei*) is native to West Africa. They are slightly larger than lemon drop type.

Health Benefits of Mangosteen

☆ Delicious and juicy, mangosteen is one of the popular tropical fruits. It comprises an impressive list of essential nutrients, which are required for normal growth and development and overall nutritional well-being.

Mangosteen (*Garcinia mangostana*), Fresh, Nutrition value per 100 g

Principle	*Nutrient Value*	*Percentage of RDA*
Energy	63 Kcal	3 per cent
Carbohydrates	15.6 g	12 per cent
Protein	0.50 g	1 per cent
Total Fat	0.4 g	2 per cent
Cholesterol	0 mg	0 per cent
Dietary Fiber	5.10 g	13 per cent
Vitamins		
Folates	31 µg	8 per cent
Niacin	0.286 mg	2 per cent
Pantothenic acid	0.032 mg	<1 per cent
Pyridoxine	0.041 mg	3 per cent
Riboflavin	0.054 mg	4 per cent
Thiamin	0.054 mg	4.5 per cent
Vitamin A	35 IU	1 per cent
Vitamin C	7.2 mg	12 per cent
Electrolytes		
Sodium	7 mg	0.5 per cent
Potassium	48 mg	1 per cent
Minerals		
Calcium	5.49 mg	0.5 per cent
Copper	0.069 mg	7 per cent
Iron	0.17 mg	2 per cent
Magnesium	13.9 mg	3.5 per cent
Manganese	0.10 mg	4 per cent
Phosphorus	9.21 mg	1 per cent
Zinc	0.12 mg	1 per cent
Phyto-nutrients		
Carotene-α	1 µg	—
Carotene-β	16 µg	—
Cryptoxanthin-β	9 µg	—

☆ This fruit is very low in calories (63 calories per 100 g) and contains no saturated fats or cholesterol. Nonetheless, it is rich in dietary fiber (100 g provides about 13 per cent of RDA), which can be very important for individuals who are concerned about putting on excess body weight.

- Mangosteen is good in vitamin C and provides about 12 per cent of RDA per 100 g. Vitamin-C is a powerful water soluble anti-oxidant. Consumption of fruits rich in vitamin C helps the body develop resistance against flu-like infectious agents and scavenge harmful, pro-inflammatory free radicals.
- Fresh fruit is a moderate source of B-complex vitamins such as thiamin, niacin and folates. These vitamins are acting as cofactors the help body metabolize carbohydrates, protein, and fats.
- Further, it also contains a very good amount of minerals like copper, manganese and magnesium. Potassium is an important component of cell and body fluids and helps control heart rate, and blood pressure; thus, it offers protection against stroke and coronary heart diseases.

29

Mulberry

Refreshingly succulent, tart and sweet mulberries are indeed rich in numerous health benefiting flavonoid phyto-nutrients. Botanically, the berries are obtained from the silkworm tree belonging to the *Moraceae* family within the genus: *Morus.* Scientific name: *Morus nigra.* L. in Spanish, the berries are known as *moras.*

More than hundred species of morus exist. In taxonomy, species generally are identified not by the colour of the fruits (berries) but by the colour of flower buds and leaves. So, a morus plant can exhibit different colour berries (black, purple, red, white, etc.) in the same plant.

Three species have been recognized for their economic importance.

- ✰ **The white mulberry** (*Morus alba*) is native to eastern and central China.
- ✰ **The red or American mulberry** (*Morus rubra*) is native to eastern United States.
- ✰ **Black mulberry** (*Morus nigra*) is native to western Asia.

Mulberries are large, deciduous trees native to warm, temperate, and subtropical regions of Asia, Africa, and the Americas. Technically, the mulberry fruit is an aggregation of small fruits arranged longitudinally around the central axis as in blackberry or loganberries. Each fruit measures 2-5 cm long. In most species, these berries are purple-red when ripen; however, they can be white, red, purple or multiple colours in the same fruit.

Health Benefits of Mulberries

- ✰ Delicious, fleshy, succulent mulberries are low in calories (just 43 calories per 100 g). They contain health promoting phyto-nutrient compounds like polyphenol pigment antioxidants, minerals and vitamins that are essential for optimum health.

Mulberry (*Morus nigra* L.), Fresh, Taw, Nutrition Value per 100 g

Principle	*Nutrient Value*	*Percentage of RDA*
Energy	43 Kcal	2 per cent
Carbohydrates	9.80 g	7.5 per cent
Protein	1.44 g	2.5 per cent
Total Fat	0.39 g	2 per cent
Cholesterol	0 mg	0 per cent
Dietary Fiber	1.7 g	4.5 per cent
Vitamins		
Folates	6 μg	1.5 per cent
Niacin	0.620 mg	4 per cent
Pyridoxine	0.050 mg	4 per cent
Riboflavin	0.101 mg	8 per cent
Vitamin A	25 IU	1 per cent
Vitamin C	36.4 mg	61 per cent
Vitamin E	0.87 mg	6 per cent
Vitamin K	7.8 μg	6.5 per cent
Electrolytes		
Sodium	10 mg	0.75 per cent
Potassium	194 mg	4 per cent
Minerals		
Calcium	39 mg	4 per cent
Copper	60 μg	6.5 per cent
Iron	1.85 mg	23 per cent
Magnesium	18 mg	4.5 per cent
Selenium	0.6 μg	1 per cent
Zinc	0.12 mg	1 per cent
Phyto-nutrients		
Carotene-β	9 μg	—
Carotene-α	12 μg	—
Lutein-zeaxanthin	136 μg	—

☆ Mulberries have significantly high amounts of phenolic flavonoid phytochemicals called anthocyanins. Scientific studies have shown that consumption of berries have potential health effects against cancer, ageing and neurological diseases, inflammation, diabetes, and bacterial infections.

☆ The berries contain resveratrol, another polyphenol flavonoid antioxidant. Resveratrol protects against stroke risk by altering molecular mechanisms in the blood vessels; reducing their susceptibility to damage through decreased activity of angiotensin (a systemic hormone causing blood vessel constriction that would elevate blood pressure) and increased production of the vasodilator hormone, *nitric oxide*.

- In addition, these berries are an excellent source of vitamin-C (36.4 mg per 100, about 61 per cent of RDI), which is also a powerful natural antioxidant. Consumption of foods rich in vitamin-C helps the body develop resistance against infectious agents, counter inflammation and scavenge harmful free radicals.
- Further, the berries also contain small amounts of vitamin A, vitamin E and in addition to the above-mentioned antioxidants. Consumption of mulberry provides another group of health promoting flavonoid polyphenolic antioxidants such as *lutein, zeaxanthin,* β*-carotene and* α*-carotene* in small but notably significant amounts. Altogether, these compounds help act as protect from harmful effects of oxygen-derived free radicals and reactive oxygen species (ROS) that play a role in ageing and various disease processes.
- Zeaxanthin, an important dietary carotenoid selectively concentrates into the retinal macula lutea, where it thought to provide antioxidant functions and protects the retina from the harmful ultraviolet rays through light-filtering actions.
- Mulberries are an excellent source of iron, which is a rare feature among berries, contains 1.85 mg/100 g of fruits (about 23 per cent of RDI). Iron, being a component of hemoglobin inside the red blood cells, determines the oxygen-carrying capacity of the blood.
- They also good source of minerals like potassium, manganese, and magnesium. Potassium is an important component of cell and body fluids that helps controlling heart rate and blood pressure. Manganese is used by the body as a co-factor for the antioxidant enzyme, superoxide dismutase.
- They are rich in B-complex group of vitamins and vitamin K. Contain very good amounts of vitamin B-6, niacin, riboflavin and folic acid. These vitamins are function as co-factors and help body in the metabolism of carbohydrates, proteins and fats.

30

Olives

Olives are oval-shaped fruits obtained from the *Olea europaea* tree; a naturalized, medium-size tree of Mediterranean origin. The fruits, and oil extracted from them, have been part of important food sources for the natives around Mediterranean Sea since centuries. Ancient Greeks believed that growing olives would bring peace and prosperity since they rely upon it to derive most of the essentialities. Botanically, the fruits belong to the family of *Oleaceae*, in the genus: *Olea*.

Olive tree is a slow growing, evergreen tree with gnarled trunk with grey, fissured branches and foliage of gray-green, fleshy, narrow leaves. It may reach up to 50 feet in height in their natural habitat and live for more than 500 years. It grows well in frost free, dry, and well-drained soils. It starts bearing fruits when the plant reaches about 3-4 years of age, however, some cultivars may take much longer time before their first harvest. The tree blooms in the spring which subsequently develops into small round to oval green colour berries by summer. Olive is a drupe or stone-fruit with a central single seed, surrounded by edible pulp. Their size and shape greatly varies depending on the cultivar type. On an average, a fruit weighs about 3-5 g. Raw fruits are green, which turn yellow to dark as it ripens further. Its fruits are generally picked at stages, whether they destined to be used as table fruits or pressed for oil.

Health Benefits of Olives

- ✰ Traditionally, olives have been viewed as very healthy food. The fruit provides calories; contain significant amounts of plant-derived anti-oxidants, minerals, phyto-sterols, and vitamins.
- ✰ Olives are a moderate source of calories; 100 g of fruits provide just 115 calories. Their calorie content basically comes from fats. Nonetheless, the fruit composes healthy fat in the form of mono-unsaturated fatty acids (MUFA) like *oleic acid* (18:1) and *palmitoleic acid* (16:1) that help lower LDL or "bad cholesterol" and increase HDL or "good cholesterol" in the blood.

Olives (*Olea europaea*), Ripe, Canned, (Small to extra large)
Nutritional value per 100 g

Principle	*Nutrient Value*	*Percentage of RDA*
Energy	115 Kcal	5.75 per cent
Carbohydrates	6.26 g	5 per cent
Protein	0.84 g	1.5 per cent
Total Fat	10.68 g	50 per cent
Cholesterol	0 mg	0 per cent
Dietary Fiber	3.2 g	8 per cent
Vitamins		
Folates	0 µg	0 per cent
Niacin	0.037 mg	<1 per cent
Pantothenic acid	0.015 mg	<1 per cent
Pyridoxine	0.009 mg	0 per cent
Riboflavin	0 mg	0 per cent
Thiamin	0.003 mg	0 per cent
Vitamin A	403 IU	13.5 per cent
Vitamin C	0.9	1 per cent
Vitamin E	1.65 mg	11 per cent
Vitamin K	1.4 µg	1 per cent
Electrolytes		
Sodium	735 mg	49 per cent
Potassium	8 mg	17 per cent
Minerals		
Calcium	88 mg	9 per cent
Copper	0.251 mg	28 per cent
Iron	3.30 mg	41 per cent
Magnesium	4 mg	1 per cent
Manganese	0.020 mg	1 per cent
Phosphorus	3 mg	<1 per cent
Selenium	0.9 µg	1.5 per cent
Zinc	0.22 mg	2 per cent
Phyto-nutrients		
Carotene-β	237 µg	—
Crypto-xanthin-β	9 µg	—
Lutein-zeaxanthin	510 µg	—
Phytosterols	221 mg	—
Lipids (fatty acids)		
Total saturated	1.415 g	—
Total Mono-unsaturated	7.888 g	—
Total Poly-unsaturated	0.911 g	—

Research studies suggest that Mediterranean diet, which is rich in monounsaturated fatty acids help to prevent coronary artery disease and strokes by favouring healthy blood lipid profile.

- ☆ Olive fruit contains tyrosol phenolic compounds such as oleuropein and oleocanthal. These compounds are responsible for its bitter and pungent taste. Oleocanthal, oleurpein, and its derivative hydroxytyrosol are nature's most powerful anti-oxidants. Together with vitamin E and carotenoids, they play a vital role fighting against cancer, inflammation, coronary artery disease, degenerative nerve diseases, diabetes...etc.
- ☆ Studies suggest that *oleocanthal* has ibuprofen (NSAID) like ant-inflammatory activities. Mediterranean diet that uses olive and its oil may be responsible in part for the lower incidences of coronary artery disease.
- ☆ Olive contains a good amount of vitamin E. 100 g cured, and canned fruits provide 1.65 mg (11 per cent of RDA) of α-tocopherol. Vitamin E is a powerful lipid soluble antioxidant, required for maintaining the integrity of cell membrane of mucus membranes and skin by protecting it from harmful oxygen-free radicals.
- ☆ In addition, the fruits contain good amounts of minerals like calcium, copper, iron, manganese, and zinc. Further, they are small sources of B-complex vitamins such as niacin, choline, and pantothenic acid.
- ☆ Oil expressed from these fruits is recognized as one of the healthiest edible oils since it contains less saturated fat, and composes linoleic (omega-6) and linolenic acid (omega-3) essential fatty acids at the recommended 8:1 ratio.

31

Orange

Delicious and juicy orange fruit contains an impressive list of essential nutrients, vitamins, minerals for normal growth and development and overall well-being. Botanically; orange is the citrus fruit belonging to the family of Rutaceae of the genus; Citrus (which also includes pomelo, tangerine (mandarin orange) and grapefruit). Scientific name: *Citrus sinensis.*

Orange is a tropical to semitropical, evergreen, small flowering tree growing to about 5 to 8 m tall and bears seasonal fruits that measure about 3 inches in diameter and weigh about 100-150 g. Oranges are classified into two general categories, *sweet* and *bitter*, with the former being the type most commonly consumed. Popular sweet-varieties include *Valencia, Navel, Persian variety,* and *blood orange*. Tangerines are related varieties of oranges distinguished by loose, easily peeled shin (pericarp) and sweet juicy flesh (arils). They are also known as mandarin oranges in Europe and *Satsumas* in Japan. Just as oranges, these too belong to the *Rutaceae* (citrus Family) and known scientifically as *Citrus reticulata*. Fruits belonging to the citrus group are described as "hesperidium," (A hesperidium is a scientific term to describe the fruit structure belonging to the citrus group. In fact, the fruit is a *modified berry*with tough, leathery rind. Orange peel contains many volatile oil glands in pits. Interior flesh is composed of segments, called *carpels*, made up of numerous fluid-filled vesicles that are actually specialized hair cells).

Health Benefits of Oranges

☆ Nutrients in oranges are plentiful and diverse. The fruit is low in calories, contains no saturated fats or cholesterol, but is rich in dietary fiber, *pectin*, which is very effective in persons with excess body weight. Pectin, by its action as a bulk laxative, helps to protect the mucous membrane of the colon by decreasing its exposure time to toxic substances as well as by binding to cancer-causing chemicals in the colon. Pectin has also been shown to reduce blood cholesterol levels by decreasing its re-absorption in the colon by binding to *bile acids* in the colon.

Oranges (*Citrus sinensis*), Fresh, Nutritive Value per 100 g

Principle	*Nutrient Value*	*Percentage of RDA*
Energy	47 Kcal	2.5 per cent
Carbohydrates	11.75 g	9 per cent
Protein	0.94 g	1.5 per cent
Total Fat	0.12 g	0.5 per cent
Cholesterol	0 mg	0 per cent
Dietary Fiber	2.40 g	6 per cent
Vitamins		
Folates	30 μg	7.5 per cent
Niacin	0.282 mg	2 per cent
Pantothenic acid	0.250 mg	5 per cent
Pyridoxine	0.060 mg	4.5 per cent
Riboflavin	0.040 mg	3 per cent
Thiamin	0.100 mg	8 per cent
Vitamin C	53.2 mg	90 per cent
Vitamin A	225 IU	7.5 per cent
Vitamin E	0.18 mg	1 per cent
Vitamin K	0 μg	0 per cent
Electrolytes		
Sodium	0 mg	0 per cent
Potassium	169 mg	3.5 per cent
Minerals		
Calcium	40 mg	4 per cent
Copper	39 μg	4 per cent
Iron	0.10 mg	1 per cent
Magnesium	10 mg	2.5 per cent
Manganese	0.024 mg	1 per cent
Zinc	0.08 mg	1 per cent
Phyto-nutrients		
Carotene-β	71 μg	—
Carotene-α	11 μg	—
Crypto-xanthin-β	116 μg	—
Lutein-zeaxanthin	129 μg	—
Lycopene	0 μg	—

- ✰ Oranges, like other citrus fruits, is an excellent source of vitamin C (provides 53.2 mg per 100 g, about 90 per cent of DRI); Vitamin C is a powerful natural antioxidant. Consumption of foods rich in vitamin C helps the body develop resistance against infectious agents and scavenge harmful, pro-inflammatory free radicals from the blood.
- ✰ Orange fruit contains a variety of phytochemicals. Hesperetin, naringin, and naringenin are flavonoids found in citrus fruits. Naringenin is found to have a bio-active effect on human health as antioxidant, free radical scavenger, anti-inflammatory, and immune system modulator. This substance has also been shown to reduce oxidant injury to DNA in vitro studies. Total antioxidant strength (ORAC) of oranges (navel variety) is 1819 µmol TE/100 g.
- ✰ Oranges also contain very good levels of *vitamin A*, and other flavonoid antioxidants such as *alpha* and *beta-carotenes*, *beta-cryptoxanthin*, *zeaxanthin* and *lutein*. These compounds are known to have antioxidant properties. Vitamin A is also required for maintaining healthy mucus membranes and skin and is essential for vision. Consumption of natural fruits rich in flavonoids helps the body to protect from lung and oral cavity cancers.
- ✰ It is also a very good source of B-complex vitamins such as thiamin, pyridoxine, and folates. These vitamins are essential in the sense that body requires them from external sources to replenish.
- ✰ Orange fruit also contains a very good amount of minerals like potassium and calcium. Potassium is an important component of cell and body fluids that helps control heart rate and blood pressure through countering sodium actions.

Citrus fruits, as such, have long been valued for their wholesome nutritious and antioxidant properties. It is scientifically established that citrus fruits, especially oranges, by virtue of their richness in vitamins and minerals, have many proven health benefits. Moreover, it is now beginning to be appreciated that the other biologically active, non-nutrient compounds in the citrus fruits such as *phyto-chemical antioxidants*, *soluble* and *insoluble dietary fiber* helps in cutting risk for cancers, chronic diseases like arthritis, obesity, and coronary heart diseases.

32

Papaya

Exotic papaya fruit or *pawpaw*is packed with numerous health benefiting nutrients. The fruit is one of the favourites of fruit lovers for its nutritional, digestive, and medicinal properties. Papaya tree is grown extensively all over the tropical regions and cultivated for its fruits and latex *papain*, an enzyme that is used in food industry. Botanically, the plant belongs in the family of *Caricaceae* of genus; *Carica* and scientifically named as *Carica papaya.*

Papaya tree bears many spherical or pear-shaped fruits that can have about 6-20 inches in length and 4-12 inches in diameter. The average-sized papaya weighs about a pound. The fruit is said to ripen when it feels soft to thumb pressure, and its skin has attained amber to orange hue. Inside, the fruit features numerous black seeds, encased in a mucin coat, in the hollow central cavity. The flesh is orange in colour with either yellow or pink hues, soft in consistency and has a deliciously sweet, musky taste with rich flavour. Babaco fruit is closely related to papaya, has similar appearance and flavour as papaya.

Health Benefits of Papaya Fruit

- ☆ The papaya fruit is very low in calories (just 39 calories/100 g) and contains no cholesterol; however, is a rich source of phyto-nutrients, minerals, and vitamins.
- ☆ Papayas contain soft, easily digestible flesh with a good amount of soluble dietary fiber that helps to have normal bowel movements; thereby reducing constipation.
- ☆ Fresh, ripe fruit is one of the fruits with the highest vitamin-C content (provides 61.8 mg or about 103 per cent of DRI, more than that of in oranges, or lemons). Research studies have shown that vitamin C has many important functions like free radicals scavenging, immune booster, and anti-inflammatory actions.

Papaya Fruit (*Carica papaya*), Fresh, Nutritive Value per 100 g

Principle	*Nutrient Value*	*Percentage of RDA*
Energy	39 Kcal	2 per cent
Carbohydrates	9.81 g	7.5 per cent
Protein	0.61 g	1 per cent
Total Fat	0.14 g	0.5 per cent
Cholesterol	0 mg	0 per cent
Dietary Fiber	1.80 g	4.5 per cent
Vitamins		
Folates	38 µg	9.5 per cent
Niacin	0.338 mg	2 per cent
Pantothenic acid	0.218 mg	4 per cent
Pyridoxine	0.019 mg	1.5 per cent
Riboflavin	0.032 mg	2.5 per cent
Thiamin	0.027 mg	2 per cent
Vitamin A	1094 IU	36 per cent
Vitamin C	61.8 mg	103 per cent
Vitamin E	0.73 mg	5 per cent
Vitamin K	2.6 µg	2 per cent
Electrolytes		
Sodium	3 mg	0 per cent
Potassium	257 mg	5 per cent
Minerals		
Calcium	24 mg	2.5 per cent
Iron	0.10 mg	1 per cent
Magnesium	10 mg	2.5 per cent
Phosphorus	5 mg	1 per cent
Zinc	0.07 mg	0.5 per cent
Phyto-nutrients		
Carotene-β	276 µg	—
Crypto-xanthin-β	761 µg	—
Lutein-zeaxanthin	75 µg	—

☆ It is also an excellent source of Vitamin-A (provides 1094 IU/100 g) and flavonoids like β-*carotene, lutein, zeaxanthin* and *cryptoxanthin*. Vitamin A is required for maintaining healthy mucus membranes and skin and is essential for visual sight. These compounds are known to have antioxidant properties; help act as protective scavengers against oxygen-derived free radicals and reactive oxygen species (ROS) that play a role in ageing and various disease processes. Consumption of natural fruits rich in carotenes has known to protect the body from lung and oral cavity cancers.

- ☆ Papaya fruit is also rich in many essential B-complex vitamins such as Folic acid, pyridoxine (vitamin B-6), riboflavin, and thiamin (vitamin B-1). These vitamins are essential in the sense that body requires them from external sources to replenish and play a vital role in metabolism.
- ☆ Fresh papaya also contains a good amount of potassium (257 mg per 100 g) and calcium. Potassium is an important component of cell and body fluids and helps controlling heart rate and blood pressure countering effects of sodium.
- ☆ Papaya seeds have been the proven natural remedy for many ailments in the traditional medicines. The seeds are found application as anti-inflammatory, anti-parasitic, and analgesic, and used to treat stomachache and ringworm infections.

33

Passion Fruit

Pleasantly sweet and tart, passion fruit or granadilla is rich in health benefiting plant nutrients. Botanically this exotic fruit belongs to the family of *Passifloraceae*, of the genus *Passiflora*. Scientific name: *Passiflora edulis*. Passions are native to subtropical wild regions of South America probably originated in Paraguay. The plant is an avid climber (vine) which grows on anything it can grab through tendrils.

The passiflora plant requires well-drained fertile soil and good moisture to flourish. It grows quickly and reaches about 15-20 feet per annum once established. Its average life span is about 5-7 years. Over five hundred cultivate types exist; however, two main type purple and yellow passion fruits are widely cultivated. During each season, the vine bears greenish-white fragrant flowers. The fruit features round to oval shape, 4 to 8 centimeters in diameter, have a tough shell mangosteen-like rind. Average weight is about 35-50 g. Inside, the fruit consists of membranous sacs containing light orange-coloured, pulpy juice with numerous small, hard, dark-brown or black, pitted seeds. Yellow passions are generally larger than the purple varieties, but the pulp of the purple fruit is less acid, richer in aroma and flavour, and has a higher proportion of juicy pulp.

Health Benefits of Passion Fruit

- ✰ Delicious, passion fruit is rich source of antioxidants, minerals, vitamins and fiber. 100 g fruit contains about 97 calories.
- ✰ The fruit is a very good source of dietary fiber. 100 g fruit pulp contains 10.4 g or 27 per cent of fiber. Good fiber in the diet helps remove cholesterol from the body. In addition dietary insoluble fiber by acting as a bulk laxative helps protect the colon mucous membrane by decreasing exposure time to toxic substances in the colon as well as binding to cancer-causing chemicals in the colon.

Passion Fruit (*Passiflora edulis*), Fresh, Nutritive Value per 100 g

Principle	*Nutrient Value*	*Percentage of RDA*
Energy	97 Kcal	5 per cent
Carbohydrates	23.38 g	18 per cent
Protein	2.20 g	4 per cent
Total Fat	0.70 g	3 per cent
Cholesterol	0 mg	0 per cent
Dietary Fiber	10.40 g	27 per cent
Vitamins		
Folates	14 µg	3 per cent
Niacin	1.500 mg	9 per cent
Pyridoxine	0.100 mg	8 per cent
Riboflavin	0.130 mg	10 per cent
Thiamin	0.00 mg	0 per cent
Vitamin A	1274 IU	43 per cent
Vitamin C	30 mg	50 per cent
Vitamin E	0.02 µg	<1 per cent
Vitamin K	0.7 mg	0.5 per cent
Electrolytes		
Sodium	0 mg	0 per cent
Potassium	348 mg	7 per cent
Minerals		
Calcium	12 mg	1.2 per cent
Copper	0.086 mg	9.5 per cent
Iron	1.60 mg	20 per cent
Magnesium	29 mg	7 per cent
Phosphorus	68 mg	10 per cent
Selenium	0.6 µg	1 per cent
Zinc	0.10 µg	1 per cent
Phyto-nutrients		
Carotene-β	743 µg	—
Crypto-xanthin-β	41 µg	—
Lycopene	0 µg	—

- ✰ Passion fruit is good in vitamin C, providing about 30 mg per 100 g. Vitamin-C (ascorbic acid) is a powerful water soluble anti-oxidant. Consumption of fruits rich in vitamin C helps the body develop resistance against flu-like infectious agents and scavenge harmful, pro-inflammatory free radicals.
- ✰ The fruit contains very good levels of vitamin-A (provides about 1274 IU per 100 g), and flavonoid antioxidants such as β-carotene and

cryptoxanthin-β. Current research studies suggest that these compounds have antioxidant properties, and along with vitamin A are essential for good eye-sight.

- ☆ Vitamin A is also required maintaining healthy mucus membranes and skin. Consumption of natural fruits rich in vitamin-A, and flavonoids helps to protect from lung and oral cavity cancers.
- ☆ Fresh granadilla is very rich in potassium. 100 g fruit pulp has about 348 mg of potassium. Potassium is an important component of cells and body fluids, and helps regulate heart rate and blood pressure.
- ☆ Furthermore, granadilla is a very good source of minerals. Iron, copper, magnesium and phosphorus are present in adequate amounts in the fruit.

34

Peaches

Delicious and widely popular peaches are native to China from where they spread to rest of the world via ancient silk route. They belong to the genus: Prunus, family of *Rosaceae*. Scientific name: *Prunus persica*. Technically, the fruit is a "drupe" having similar features of the other members of the genus: Prunus including plums, nectarine, almonds, and damson.

The *Prunus persica* plant is described as a small, deciduous tree that grows up to 25 to 30 feet tall and widely grown in the United States, Europe, and China for its juicy fruits. There are many cultivars of peaches are grown all over the world, which differ in their colour, size and growing characteristics based upon their country of origin. Generally, the peach tree bears numerous, almost uniform sized fruits between May and September months. Its fruit is about the size of small apple, measuring about 7-10 cm in diameter and weigh about 130 g. It features fuzzy outer surface with longitudinal depressions extending from the stem end to the tip. Depending up on the seed that is free or firmly attached to the pulp, they are classified into free-stone variety or clinging seed variety. Flat or donut peaches are flattened, disc like freestone fruits drawn in at the center. They have similar peach flavour and taste. Internally, pulp is juicy sweet and tart in taste with pleasant aroma. Depending upon the variety, it features white to creamy-yellow colour with a centrally placed single seed enclosed inside the hard shell. Seeds are inedible.

Health Benefits of Peaches

- ☆ Wonderfully delicious peaches are low in calories (100 g just provide 39 calories) and contain no saturated fats. Nonetheless, they are packed with numerous health promoting compounds, minerals and vitamins.
- ☆ The total measured anti-oxidant strength **(ORAC value)** of 100 g peach fruit is 1814 TE (Trolex equivalents).

Peaches (*Prunus persica*), Fresh, Nutritive Value per 100 g

Principle	*Nutrient Value*	*Percentage of RDA*
Energy	39 Kcal	2 per cent
Carbohydrates	9.54 g	7 per cent
Protein	0.91 g	1.5 per cent
Total Fat	0.25 g	1 per cent
Cholesterol	0 mg	0 per cent
Dietary Fiber	1.5 g	4 per cent
Vitamins		
Folates	4 µg	1 per cent
Niacin	0.806 mg	5 per cent
Pantothenic acid	0.153 mg	3 per cent
Pyridoxine	0.025 mg	2 per cent
Riboflavin	0.031 mg	2.5 per cent
Thiamin	0.024 mg	2 per cent
Vitamin A	326 IU	11 per cent
Vitamin C	6.6 mg	11 per cent
Vitamin E	0.73 mg	5 per cent
Vitamin K	2.6 µg	2 per cent
Electrolytes		
Sodium	0 mg	0 per cent
Potassium	190 mg	4 per cent
Minerals		
Calcium	6 mg	0.6 per cent
Copper	0.068 mg	7.5 per cent
Iron	0.25 mg	3 per cent
Magnesium	9 mg	2 per cent
Manganese	0.61 mg	3 per cent
Phosphorus	11 mg	2 per cent
Zinc	0.17 mg	1.5 per cent
Phyto-nutrients		
Carotene-β	162 µg	—
Crypto-xanthin-β	67 µg	—
Lutein-zeaxanthin	91 µg	—

☆ Fresh peaches are a moderate source of antioxidant, vitamin C. Vitamin-C has anti-oxidant effects and is required for connective tissue synthesis in the body. Consumption of foods rich in vitamin C helps the body develop resistance against infectious agents, and help scavenges harmful free radicals.

- ☆ Fresh fruits are an also moderate source of vitamin A and β-carotene. β-carotene is a pro-vitamin, which converts into vitamin A in the body. Vitamin A is essential for vision. It is also required for maintaining healthy mucus membranes and skin. Consumption of natural fruits rich in vitamin A is known to offer protection from lung and oral cavity cancers.
- ☆ They are rich in many vital minerals such as potassium, fluoride and iron. Iron is required for red blood cell formation. Fluoride is a component of bones and teeth and is essential for prevention of dental caries. Potassium is an important component of cell and body fluids that help regulate heart rate and blood pressure.
- ☆ Peaches contain health promoting flavonoid poly phenolic antioxidants such as *lutein, zeaxanthin* and β-*cryptoxanthin*. These compounds help act as protective scavengers against oxygen-derived free radicals and reactive oxygen species (ROS) that play a role in ageing and various disease processes.

35

Pears

Sweet, delicious and rich flavoured pears offer crunchiness of apples yet juicy as peach and nectarine. They are widely popular, especially in the whole of the northern hemisphere, for their unique nutrient qualities. Botanically, it is a "pome fruit" produced by the tree belonging to the family of *Rosaceae,* in the genus: *Pyrus*. Pome fruit family also includes apple, loquat, medlar etc. Scientific name: *Pyrus communis*.

Pears are broadly classified based upon their place of origin as *Asian-pears* and *European-pears.* Asian varieties feature crispy texture and firm consistency that do not change after harvesting or storage makes them ready-to-eat. Whereas, European types generally become soft and juicy when allowed to ripen. In structure, pear fruit has bell or "pyriform" shape; around 5-6 inches long and weigh about 200 gm. Fresh fruit is firm in consistency with mild 'apple' flavour. Externally, its skin is very thin and depending upon the cultivar type; the fruit may have green, red-orange or yellow-orange colours. Internally, it's off white colour pulp is crunchy and juicy. However, in fully matured ones, the pulp turns to grainy texture with gritty sensation while cutting with a knife. Center of the fruit is more or less similar to apple with centrally located tiny inedible seeds.

1. Some of the popular **Asian varieties** are Ichiban Nashi, Shinsui, Shinsieki and Nijisseki.
2. Popular **European types** are Bartlett and Comice.

Health Benefits of Pears

☆ Pears fruit is packed with health benefiting nutrients such as dietary fiber, anti-oxidants, minerals and vitamins, which are necessary for optimum health. Total measured antioxidant strength (ORAC value) in the fruit is 2941 μmol TE/100 g.

Pears (*Pyrus communis*), Fresh, Nutritive Value per 100 g

Principle	*Nutrient Value*	*Percentage of RDA*
Energy	58 Kcal	3 per cent
Carbohydrates	13.81 g	11 per cent
Protein	0.38 g	<1 per cent
Total Fat	0.12 g	0.5 per cent
Cholesterol	0 mg	0 per cent
Dietary Fiber	3.10 g	8 per cent
Vitamins		
Folates	7 μg	2 per cent
Niacin	0.157 mg	1 per cent
Pantothenic acid	0.048 mg	1 per cent
Pyridoxine	0.028 mg	2 per cent
Riboflavin	0.025 mg	2 per cent
Thiamin	0.012 mg	1 per cent
Vitamin A	23 IU	1 per cent
Vitamin C	4.2 mg	7 per cent
Vitamin E	0.12 mg	1 per cent
Vitamin K	4.5 μg	4 per cent
Electrolytes		
Sodium	1 mg	0 per cent
Potassium	119 mg	2.5 per cent
Minerals		
Calcium	9 mg	1 per cent
Copper	0.082 mg	9 per cent
Iron	0.17 mg	2 per cent
Magnesium	7 mg	2 per cent
Manganese		2 per cent
Phosphorus	11 mg	2 per cent
Zinc	0.10 mg	1 per cent
Phyto-nutrients		
Carotene-β	12 μg	—
Crypto-xanthin-β	2 μg	—
Lutein-zeaxanthin	45 μg	—

☆ Pears are a good source of dietary fiber. 100 g fruit provides 3.1 g or 8 per cent of fiber per 100g. Regular eating of this fruit may offer protection against colon cancer. Most of the fiber in them is non soluble polysaccharide (NSP), which functions as a good bulk laxative in the gut. Additionally, the gritty fiber content binds to cancer-causing toxins and chemicals in the colon, protecting its mucous membrane from contact with these compounds.

- In addition, pear fruit is one of the very low calorie fruits, provides 58 calories per 100g. Just a few sections a day in the diet can bring significant reduction in weight and blood LDL cholesterol levels.
- They contain good quantities of *vitamin* C. Fresh fruits provide about 7 per cent of RDA per 100 g.
- They are moderate sources of antioxidant flavonoids phyto-nutrients such as *beta-carotene, lutein and zeaxanthin*. These compounds, along with vitamin C and A, help the body protected from harmful free radicals.
- The fruit is a good source of minerals such as copper, iron, potassium, manganese and magnesium as well as B-complex vitamins such as folates, riboflavin and pyridoxine (vitamin B-6).
- Although not well documented, pears are among the least allergenic of all fruits and are therefore recommended by health practitioners as a safe alternative in the preparation of food products for allergy sufferers.
- Pears have suggested in various traditional medicines being useful in treating colitis, chronic gallbladder disorders, arthritis and gout.

36

Persimmon

Sweet, delicious persimmon fruits are rich in health promoting nutrients such as vitamins, minerals, and anti-oxidants that are vital for optimum health. Botanically, the fruit belongs to the family of *Ebenaceae* of the genus: *Diospyros*. Scientific name: *Diospyros virginiana*.

This delicate oriental fruit is native to China. It spread to Japan very long ago and later was introduced to California during the middle of nineteenth century. Persimmon is a multi-trunked or single-stemmed deciduous tree, which grows up to 25 ft. in height. They grow best in areas that have moderate winters and relatively mild summers. Persimmon trees classified broadly into two general categories: those that bear "astringent fruit" (whilst unripe) and those that bear "non-astringent" fruits. An astringent cultivar, which is commonly cultivated in Japan known as *"Hachiya,"* is high in *tannins* and must be allowed to ripen fully until it attains jelly-soft consistency before fit to eat. A non-astringent persimmon, on the other hand, contains less tannin and can be eaten while it is crispy as in apples. Astringency can be removed by treating the fruit with carbon dioxide or alcohol. During each season, the tree bears numerous fruits that vary by cultivar from spherical to heart to flatten or squash in shape. They also greatly vary in size from as little as a few ounces to more than a pound. The colour of the fruit varies from light yellow-orange to dark orange-red. The entire fruit is edible except for the seed and calyx.

- ☆ **Astringent varieties:** Eureka, Hachiya, Saijo, Tamopan, Tanenashi, Triumph, etc.
- ☆ **Non-astringent varieties:** Fuyu (Fuyugaki), Gosho/Giant Fuyu/O'Gosho, Imoto, Izu, Jiro, Maekawajiro, Okugosho, Suruga, etc.

Health Benefits of Persimmon Fruit

- ☆ The fruit is low in calories (provides 70 calories/100g) and fats but is rich source of dietary fiber.

Persimmon Fruit (*Diospyros virginiana*), Japanese, Fresh, Nutrition Value per 100 g

Principle	*Nutrient Value*	*Percentage of RDA*
Energy	70 Kcal	3.5 per cent
Carbohydrates	18.59 g	14 per cent
Protein	0.58 g	1 per cent
Total Fat	0.19g	1 per cent
Cholesterol	0 g	0 per cent
Dietary Fiber	3.6 g	9.5 per cent
Vitamins		
Choline	7.6 mg	1.5 per cent
Folates	8 µg	2 per cent
Niacin	0.100 mg	1 per cent
Pyridoxine	0.100 mg	7.5 per cent
Riboflavin	0.020 mg	1.5 per cent
Thiamin	0.030 mg	2.5 per cent
Vitamin C	7.5 mg	12.5 per cent
Vitamin A	81 IU	3 per cent
Vitamin E	0.73 mg	5 per cent
Vitamin K	2.6 µg	2 per cent
Electrolytes		
Sodium	1 mg	0 per cent
Potassium	161 mg	2.5 per cent
Minerals		
Calcium	8 mg	0.8 per cent
Copper	0.113 mg	12.5 per cent
Iron	0.15 mg	2 per cent
Magnesium	9 mg	2 per cent
Manganese	0.355 mg	15 per cent
Phosphorus	17 mg	4.5 per cent
Zinc	0.11 mg	1 per cent
Phyto-nutrients		
Carotene-α	0 µg	—
Carotene-β	253 µg	—
Crypto-xanthin-β	1447 µg	—
Lutein-zeaxanthin	834 µg	—
Lycopene	159 µg	—

☆ Persimmons contain health benefiting phyto-nutrients flavonoid poly-phenolic anti-oxidants like catechins and gallocatechins as well as important anti-tumor compound betulinic acid. Catechins are known to have anti-infective, anti-inflammatory and anti-hemorrhagic (prevents bleeding from small blood vessels) properties.

☆ Fresh persimmons contain anti-oxidant compounds like vitamin-A, *beta-carotene, lycopene, lutein, zeaxanthin and cryptoxanthin.* Together, these compounds functions as protective scavengers against oxygen-derived free radicals and reactive oxygen species (ROS) that plays a role in ageing and various disease processes.

☆ Zeaxanthin, an important dietary carotenoid, selectively absorbed into the retinal macula lutea in the eyes where it is thought to provide antioxidant and protective light-filtering functions; thus, helps prevent "Age-related macular related macular disease"(ARMD) in the elderly.

☆ They are also a very good source of vitamin-C, another powerful antioxidant (especially native Chinese and American persimmons; provide 80 per cent of DRI). Regular consumption of foods rich in vitamin C helps the body develop resistance against infectious agents and scavenge harmful, pro-inflammatory free radicals.

☆ The fruit is good in many valuable B-complex vitamins such as folic acid, pyridoxine (vitamin B-6), thiamin.etc. These vitamins act as co-factors for numerous metabolic enzymatic functions in the body.

☆ Fresh Persimmon fruits also contain healthy amounts of minerals like potassium, manganese (15 per cent of DRI), copper (12 per cent of DRI), and phosphorus. Manganese is a co-factor for the enzyme, *superoxide dismutase,* which is a very powerful free radical scavenger. Copper is a co-factor for many vital enzymes, including *cytochrome c-oxidase and superoxide dismutase* (other minerals function as cofactors for this enzyme are manganese, and zinc). Copper is also required for the production of red blood cells.

37

Pineapple

One of the most prized and popular fruits, pineapple or "ananas" has an interesting history to narrate. The ananas plant is actually native to Paraguay in South America. It spread by the local Indians up through South and Central America and to the West Indies. Later, it was brought to Spain when Columbus discovered Americas' in 1493, from where; it spread to rest of the world by the sailors (just like tomatoes) who carried it for protection from *scurvy* wherever they went. Scientifically, it is known as *"Ananas comosus"* and belongs to the family of *Bromeliaceae,* of the genus; Ananas.

Pineapple is a tropical, perennial, drought-tolerant plant that grows up to 5-8 ft in height and spreads around about three to four ft. It is essentially a short, stout stem with a rosette of waxy long, needle-tipped leaves. The plant bears several long, oval to cylindrical fruits during each season from March until June.

The fruit is described as compound (multiple) fruit that develops from many small fruits fused together around the central core. Its pulp is juicy and fleshy with the stem serving as a supporting fibrous core. The outer rough, tough, scaly rind may be dark green, yellow, orange-yellow or reddish when the fruit is ripe. Internal juicy flesh may range from creamy white to yellow and has a mix of sweet and tart taste with rich flavour. Each fruit measures in size up to 12 inches long and weighs 1 to 8 pounds or more.

Health Benefits of Pineapple Fruit

- ☆ Fresh pineapple is low in calories. Nonetheless, ii is a storehouse of several unique health promoting compounds, minerals and vitamins that are essential for optimum health.
- ☆ 100 g fruit provides just about 50 calories equivalent to that of apples. Its flesh contains no saturated fats or cholesterol; however, it is rich source of soluble and insoluble dietary fiber like *pectin*.

Pineapple (*Ananas comosus*), Fresh, Nutritive Value per 100 g

Principle	*Nutrient Value*	*Percentage of RDA*
Energy	50 Kcal	2.5 per cent
Carbohydrates	13.52 g	10 per cent
Protein	0.54 g	1 per cent
Total Fat	0.12 g	<1 per cent
Cholesterol	0 mg	0 per cent
Dietary Fiber	1.40 g	4 per cent
Vitamins		
Folates	18 μg	4.5 per cent
Niacin	0.500 mg	4 per cent
Pyridoxine	0.112 mg	9 per cent
Riboflavin	0.018 mg	1.5 per cent
Thiamin	0.079 mg	6.5 per cent
Vitamin A	58 IU	2 per cent
Vitamin C	47.8 mg	80 per cent
Vitamin E	0.02 mg	<1 per cent
Vitamin K	0.07 μg	0.5 per cent
Electrolytes		
Sodium	1 mg	0 per cent
Potassium	109 mg	2.5 per cent
Minerals		
Calcium	13 mg	1.3 per cent
Copper	0.110 mg	12 per cent
Iron	0.29 mg	3.5 per cent
Magnesium	12 mg	3 per cent
Manganese	0.927 mg	40 per cent
Phosphorus	8 mg	1 per cent
Selenium	0.1 μg	<1 per cent
Zinc	0.12 mg	1 per cent
Phyto-nutrients		
Carotene-β	35 μg	—
Crypto-xanthin-β	0 mcg	—

☆ Pineapple fruit contains a proteolytic enzyme bromelain that digests food by breaking down protein. Bromelain also has anti-inflammatory, anti-clotting and anti-cancer properties. Studies have shown that consumption of pineapple regularly helps fight against arthritis, indigestion and worm infestation.

- ☆ Fresh pineapple is an excellent source of antioxidant vitamin; vitamin C. 100 g fruit contains 47.8 or 80 per cent of this vitamin. Vitamin C is required for the collagen synthesis in the body. Collagen is the main structural protein in the body required for maintaining the integrity of blood vessels, skin, organs, and bones. Regular consumption of foods rich in vitamin C helps the body protect from scurvy; develop resistance against infectious agents (boosts immunity) and scavenge harmful, pro-inflammatory free radicals from the body.
- ☆ It also contains small amount Vitamin A (provides 58 IU per 100 g) and beta-carotene levels. These compounds are known to have antioxidant properties. Vitamin A is also required maintaining healthy mucus membranes, skin and essential for vision. Studies have suggested that consumption of natural fruits rich in flavonoids helps the human body to protect from lung and oral cavity cancers.
- ☆ In addition, this fruit is rich in B-complex group of vitamins like folates, thiamin, pyridoxine, riboflavin and minerals like copper, manganese and potassium. Potassium is an important component of cell and body fluids, helps controlling heart rate and blood pressure. Copper is a helpful cofactor for red blood cell synthesis. Manganese is a co-factor for the enzyme superoxide dismutase, which is a very powerful free radical scavenger.

38

Plums

Wonderfully delicious and juicy plums are botanically belonging within the family of Rosaceae. The fruit is a drupe belonging to the genus: *Prunus* which also includes peaches, nectarine, almonds and damson. Scientific name: *Prunus domestica.* The plant is best described as small tree or large shrub and widely cultivated in the United States, Europe, Japan and China. Many cultivars of plums are grown all over the world, which differ in their colour, size and growing characteristics based upon their country of origin. Generally, each variety of the plum trees bears numerous, almost uniform sized fruits between May and September months. Each berry is about the size of medium-sized tomato, measuring about 5-6 cm in diameter and weigh about 50-70 g. It has central depression at the stem side. Internally, the pulp is juicy and varies widely from creamy yellow, crimson, light-blue or light-green in colour depending upon the cultivar type. There is centrally placed single, smooth but hard stone-like seed. Seeds are inedible. It has sweet and tart taste pulp with pleasant aroma. Some common cultivars of plums are: cherry plum, damson, blackthorn plum.

Health Benefits of Plums

- ☆ Delicious, fleshy, succulent plums are low in calories (46 calories per 100 g) and contain no saturated fats; however, contain numerous health promoting compounds, minerals and vitamins.
- ☆ Certain health benefiting compounds present in the plum fruits, such as dietary fiber, sorbitol, and isatin are known to help regulate the functioning of the digestive system and thereby relieve constipation conditions.
- ☆ Total antioxidant strength of plums measured in terms of ORAC (Oxygen radical absorbance capacity) is 6259 µmol TE/100 g. Fresh berries are a moderate source of vitamin C, which is also a powerful natural antioxidant. Consumption of foods rich in vitamin C helps the body develop resistance against infectious agents, counter inflammation and scavenge harmful free radicals.

Plums (*Prunus domestica*), fresh, Nutrition Value per 100 g

Principle	*Nutrient Value*	*Percentage of RDA*
Energy	46 Kcal	2.3 per cent
Carbohydrates	11.42 g	8 per cent
Protein	0.70 g	1 per cent
Total Fat	0.28 g	1 per cent
Cholesterol	0 mg	0 per cent
Dietary Fiber	1.40 g	3.5 per cent
Vitamins		
Folates	5 µg	1 per cent
Niacin	0.417 mg	3 per cent
Pantothenic acid	0.135 mg	3 per cent
Pyridoxine	0.029 mg	2 per cent
Riboflavin	0.026 mg	2 per cent
Thiamin	0.028 mg	2 per cent
Vitamin A	345 IU	11.5 per cent
Vitamin C	9.5 mg	16 per cent
Vitamin E	0.26 mg	2 per cent
Vitamin K	6.4 µg	5 per cent
Electrolytes		
Sodium	1 mg	0 per cent
Potassium	157 mg	3 per cent
Minerals		
Calcium	6 mg	0.6 per cent
Copper	0.057 mg	6 per cent
Iron	0.17 mg	2 per cent
Magnesium	7 mg	2 per cent
Manganese	0.052 mg	2 per cent
Phosphorus	16 mg	2 per cent
Selenium	1.0 µg	2 per cent
Zinc	0.10 mg	1 per cent
Phyto-nutrients		
Carotene-β	190 µg	—
Crypto-xanthin-β	35 µg	—
Lutein-zeaxanthin	73 µg	—

☆ Fresh plums, especially yellow Mirabelle type, are a moderate source of vitamin A and beta carotene. Vitamin A is essential for good eye sight. It is also required for maintaining healthy mucus membranes and skin. Consumption of natural fruits rich in vitamin A has found to protect from lung and oral cavity cancers.

- The fruit is also good in health promoting flavonoid poly phenolic antioxidants such as lutein, cryptoxanthin and zeaxanthin in significant amounts. These compounds help act as scavengers against oxygen-derived free radicals and reactive oxygen species (ROS) that play a role in ageing and various disease processes. Zeaxanthin, an important dietary carotenoid selectively absorbed into the retinal macula lutea where it is thought to provide antioxidant and protective UV light-filtering functions.
- Plums are plentiful in minerals like potassium, fluoride and iron. Iron is required for red blood cell formation. Potassium is an important component of cell and body fluids that helps controlling heart rate and blood pressure.
- In addition, the berries are moderate sources in B-complex groups of vitamins such as niacin, vitamin B-6 and pantothenic acid. These vitamins are acting as cofactors help the body metabolize carbohydrates, proteins and fats. They also provide about 5 per cent RDA levels of vitamin K. Vitamin K is essential for many clotting factors function in the blood as well as in bone metabolism and help reduce Alzheimer's disease in the elderly.

39

Pomegranate

Pomegranate is among the most popular, nutritionally rich fruit with unique flavour, taste, and heath promoting characteristics. Along with berries, and some tropical exotics such as mango, it too has novel qualities of functional foods often called as "super fruits." Botanically, it is a small size fruit-bearing deciduous tree belonging within the *Lythraceae* family, of genus: *Punica*. The fruit is thought to originate in the Sub-Himalayan range of North India. Scientific name: *Punica granatum*.

The tree grows to about five and eight meters tall. It is cultivated at a commercial scale in vast regions across Indian sub-continent, Iran, Caucuses and Mediterranean regions for its fruits. Completely grown-up tree bears numerous spherical, bright red, purple, or orange-yellow coloured fruits depending on the cultivar types. Each fruit measures about 6-10 cm in diameter and weighs about 200 gm. Its outer skin or rind is tough and features leathery texture. Interior of the fruit is separated by white, thin, spongy, membranous, bitter tissue into compartments. Such sections, packed as sacs, filled with tiny edible sweet, juicy, pink pulp encasing around a single, angular, soft or hard (in over-mature fruits) seed. Pomegranate is one of the most extensively cultivated fruits for food, juice, flavour, and colour, making it a common ingredient in new functional foods often called "super fruits."

Health Benefits of Pomegranate

- ☆ The fruit is moderate in calories; 100 g provides 83 calories, slightly more than that in the apples. It contains no cholesterol or saturated fats.
- ☆ It is rich source of soluble and insoluble dietary fibers, providing about 4 g per 100 g (about 12 per cent of RDA), which aid in smooth digestive and bowel movements. The fruit is suggested by nutritionists in the diet for weight reduction and cholesterol controlling programmes. Regular inclusion of fruits in the diets boosts immunity, improves circulation, and offers protection from cancers.

Pomegranate (*Punica granatum*), Fresh, Nutrition Value per 100 g

Principle	*Nutrient Value*	*Percentage of RDA*
Energy	83 Kcal	4 per cent
Carbohydrates	18.70 g	14 per cent
Protein	1.67 g	3 per cent
Total Fat	1.17 g	6 per cent
Cholesterol	0 mg	0 per cent
Dietary Fiber	4 g	11 per cent
Vitamins		
Folates	38 μg	9.5 per cent
Niacin	0.293 mg	2 per cent
Pantothenic acid	0.135 mg	3 per cent
Pyridoxine	0.075 mg	6 per cent
Riboflavin	0.053 mg	4 per cent
Thiamin	0.067 mg	5.5 per cent
Vitamin A	0 IU	0 per cent
Vitamin C	10.2 mg	17 per cent
Vitamin E	0.60 mg	4 per cent
Vitamin K	16.4 μg	14 per cent
Electrolytes		
Sodium	3 mg	0 per cent
Potassium	236 mg	5 per cent
Minerals		
Calcium	10 mg	1 per cent
Copper	18 per cent	0.158 mg
Iron	0.30 mg	4 per cent
Magnesium	12 mg	3 per cent
Manganese	0.119 mg	5 per cent
Phosphorus	36 mg	5 per cent
Selenium	0.5 μg	1 per cent
Zinc	0.35 mg	3 per cent
Phyto-nutrients		
Carotene-α	0 μg	—
Crypto-xanthin-β	0 μg	—

☆ Certain ellagitannin compounds such as Granatin B and Punicalagin are found abundantly in the pomegranate juice. Studies suggest that punicalagin and tannins are effective in reducing heart-disease risk factors by scavenging harmful free radicals from the human body.

- ☆ Total antioxidant strength of pomegranate fruit measured in terms of its oxygen radical absorbance capacity (ORAC) is 2341 μmol TE/100 g.
- ☆ The fruit is an also good source of antioxidant vitamin-C, provides about 17 per cent per 100 g of daily requirement. Consumption of fruits rich in vitamin C helps the body develop resistance against infectious agents by boosting immunity.
- ☆ Regular consumption of pomegranate has also been found to be effective against prostate cancer, benign prostatic hyperplasia (BPH), diabetes, and lymphoma.
- ☆ Further, it is an also good source of many vital B-complex groups of vitamins such as pantothenic acid (vitamin B-5), folates, pyridoxine and vitamin K, and minerals like calcium, copper, potassium, and manganese.

40

Quince Fruit

Fragrant rich quince fruit is a member of *Rosaceae* family of pome-fruits. Native to Asia Minor, this once popular delicacy has taken backseat in the modern times of the molecular biotechnology world. Quince is rarely eaten raw but used in cooking where just a small wedge would impart the whole recipe with a pleasant fruity aroma. Binomially, quince is the only fruiting tree in the genus: Cydonia. Scientific name: *Cydonia oblonga.*

Quinces are medium sized semi-tropical deciduous trees, growing to about 10 to 15 feet in height. Pink-white flowers appear in the spring and early summer, which develop into golden colour pear-shaped fruits. The fruit is larger than average apple and bumpy; appear somewhat like large guava, avocado, or as short-necked pear fruit. Its fuzzy surface is smooth as in peaches. The quince fruit weighs about 250-750 g or more in some varieties. Inside, its flesh is light yellow, gritty and has multiple seeds concentrated at the center as in apples. Raw quince has intense fruity smell and together with its bright yellow colour instantly attracts the fruit lover's attention. However, raw fruits, even after ripen, generally astringent and tart.

Health Benefits of Quince

- ☆ Quince is low calorie fruit. 100 g fresh raw fruit provides 57 calories. In addition, it composes several vital poly-phenolic antioxidants than apples and pears. The fruit is the storehouse for phyto-nutrients such as dietary fiber, minerals, and vitamins.
- ☆ Quince pulp along with its peel contains good amounts of fiber. Further, its gritty granules in the pulp are composed of astringent compounds known as tannins namely, catechin and epicatechin. They bind to cancer-causing toxins and chemicals in the colon, protecting its mucous membrane from inflammatory bowel disease (IBD), cancers, and diverticulitis. In addition, it helps reduce body weight and blood LDL cholesterol levels.

Quince Fruit (*Cydonia oblonga*), Fresh, Nutritive Value per 100 g

Principle	*Nutrient Value*	*Percentage of RDA*
Energy	57 Kcal	3 per cent
Carbohydrates	13.81 g	11 per cent
Protein	0.40 g	<1 per cent
Total Fat	0.10 g	0.5 per cent
Cholesterol	0 mg	0 per cent
Dietary Fiber	1.9 g	4 per cent
Vitamins		
Folates	3 µg	1 per cent
Niacin	0.200 mg	1 per cent
Pantothenic acid	0.081 mg	1.5 per cent
Pyridoxine	0.040 mg	3 per cent
Riboflavin	0.030 mg	2 per cent
Thiamin	0.020 mg	2 per cent
Vitamin A	40 IU	1 per cent
Vitamin C	15 mg	25 per cent
Vitamin E	0.12 mg	1 per cent
Vitamin K	4.5 µg	4 per cent
Electrolytes		
Sodium	1 mg	0 per cent
Potassium	119 mg	2.5 per cent
Minerals		
Calcium	11 mg	1 per cent
Copper	0.130 mg	14 per cent
Iron	0.70 mg	9 per cent
Magnesium	8 mg	2 per cent
Phosphorus	11 mg	2 per cent
Selenium	0.6 µg	1 per cent
Zinc	0.04 mg	<1 per cent

- ✫ It has several phenolic compounds such as *caffeoylquinic acid, procyanidin-B2, oligomeric procyanidin, polymeric procyanidin* etc., and essential oils like *furfural, limonene, linalol, vomifoliol, toluene, β-ionone, α-terpineol,* etc. Together; these compounds give quince its unique fragrance.
- ✫ Ripe quince fruit has good concentration of vitamin C. 100 g fruit provides 15 mg or 25 per cent of RDA of vitamin-C. Fruits rich in this vitamin help remove harmful oxygen-free radicals from the body. Vitamin C helps boost immunity, reduce viral episodes and inflammatory conditions.

- The fruit is a good source of minerals such as copper (130 µg or 14 per cent of RDA), iron, potassium, and magnesium as well as B-complex vitamins such as thiamin, riboflavin and pyridoxine (vitamin B-6).
- Although not well documented, quince fruit, like pears, has anti-allergenic and anti-inflammatory properties. The fruit as well its seed's extraction is suggested in the treatment of cystitis, atopic dermatitis, recommended by health practitioners as a safe alternative in the preparation of food products for allergy sufferers.

41

Raspberry

Wonderfully delicious, bright red-coloured raspberry is among the most popular berries grown all over the world. They are rich source of health promoting plant-derived nutrients, minerals, and vitamins that are essential for optimum health. Botanically, the plant is a small shrub belonging to the family *Rosaceae*, of the genus: *Rubus*. It grows very well in temperate regions. The exotic berry is native to Europe but now widely cultivated in many temperate regions all over the world. Chief producers of raspberries are Poland, United States, Germany, and Chile. Botanical name: *Rubus idaeus*.

Several subspecies of raspberries are grown; however, the most important modern commercial red-raspberry cultivars derive from hybrids between *R. idaeus* (*European raspberry*) and *R. strigosus* (*American raspberry*). Technically, the whole berry is an aggregate of small drupe-lets, which are arranged in circular fashion around a hollow central cavity. Each tiny drupelet features small juicy pulp with a single whitish-yellow seed. Raspberries have a taste that varies by cultivar, and ranges from sweet to acidic, a feature quite similar to strawberries. Raspberry has a conical shape, weighs about 3-4 g and contains 80-100 drupelets arranged in circular layers. While the most common type of raspberry (*Rubus idaeus*) is red-pink in colour, hybrids actually come in a range of colours, including black, purple, orange, yellow and white.

Health Benefits of Raspberries

- ✰ Delicious raspberries are low in calories and saturated fats but are rich source of dietary fiber and antioxidants. 100 g berries contain just 52 calories but provide 6.5 g of fiber (16 per cent of daily recommended intake).
- ✰ Raspberries have significantly high levels of phenolic flavonoid phytochemicals such as *anthocyanins, ellagic acid* (*tannin*), *quercetin, gallic acid, cyanidins, pelargonidins, catechins, kaempferol* and *salicylic acid*. Scientific studies show that these antioxidant compounds in these berries have

potential health benefits against cancer, ageing, inflammation, and neuro-degenerative diseases.

Raspberry (*Rubus idaeus*), Nutritive Value per 100 g

Principle	*Nutrient Value*	*Percentage of RDA*
Energy	52 Kcal	2.5 per cent
Carbohydrates	11.94 g	9 per cent
Protein	1.20 g	2 per cent
Total Fat	0.65 g	3 per cent
Cholesterol	0 mg	0 per cent
Dietary Fiber	6.5 g	16 per cent
Vitamins		
Folates	21 mcg	5 per cent
Niacin	0.598 mg	4 per cent
Pyridoxine	0.055 mg	4 per cent
Riboflavin	0.038 mg	3 per cent
Vitamin A	33 IU	1 per cent
Vitamin C	26.2 mg	47 per cent
Vitamin E	1.42 mg	9 per cent
Vitamin K	7.8 mg	6.5 per cent
Electrolytes		
Sodium	1 mg	0 per cent
Potassium	151 mg	3 per cent
Minerals		
Calcium	25 mg	2.5 per cent
Copper	90 mg	10 per cent
Iron	0.69 mg	8.5 per cent
Magnesium	22 mg	5.5 per cent
Manganese	0.670 mg	29 per cent
Zinc	0.42 mg	4 per cent
Phyto-nutrients		
Carotene-β	12 mg	—
Carotene-α	16 mg	—
Lutein-zeaxanthin	136 mg	—

☆ Xylitol is a low-calorie sugar substitute extracted from raspberries. A teaspoonful of xylitol contains just 9.6 calories as compared to that of sugar, which has 15 calories. Xylitol absorbs more slowly in the intestines than sugar and does not contribute to high glycemic index, and thus, can be helpful in diabetics.

- ☆ Fresh raspberries are an excellent source of vitamin-C, which is also a powerful natural antioxidant. 100 g berries provide 26.2 mg or about 47 per cent of DRI of vitamin C. Consumption of fruits rich in vitamin C helps the body develop resistance against infectious agents, counter inflammation, and scavenge harmful free radicals.
- ☆ Raspberry contains anti-oxidant vitamins like vitamin A, and vitamin E. In addition to the above-mentioned antioxidants, is also rich in several other health promoting flavonoid poly phenolic antioxidants such as *lutein, zeaxanthin, and* β*-carotene* in small amounts. Altogether, these compounds help act as protective scavengers against oxygen-derived free radicals and reactive oxygen species (ROS) that play a role in ageing and various disease processes.
- ☆ Raspberry has an ORAC value (oxygen radical absorbance capacity) of about 4900 per 100 grams, crediting it among the top-ranked ORAC fruits.
- ☆ Raspberries contain a good amount of minerals like potassium, manganese, copper, iron and magnesium. Potassium is an important component of cell and body fluids that helps controlling heart rate and blood pressure. Manganese is used by the body as a co-factor for the antioxidant enzyme, *superoxide dismutase*. Copper is required in the production of red blood cells.
- ☆ They are rich in B-complex group of vitamins and vitamin K. The berries contain very good amounts of vitamin B-6, niacin, riboflavin, and folic acid. These vitamins are function as co-factors and help body in the metabolism of carbohydrates, protein and fats.

42

Sapota

Sapodilla or sapota (*chikoo*) is another popular delicious tropical fruit in line with mango, banana, and jackfruit. Its calorie-rich soft, easily digestible pulp composes of simple sugars like fructose and sucrose that when eaten replenishes energy and revitalizes the body instantly. Indeed, it is a vital source of vitamins, minerals and health benefiting anti-oxidant, *tannins.* Sapota is a tropical evergreen fruit-bearing tree belongs to the family of *Sapotaceae* in the genus: *Manilkara*. Scientific name: *Manilkara zapota*.

Originated in the central American rain forests, probably in Mexico and Belize, today it has spread all over the tropical belt and is being grown as a major commercial crop in India, Sri Lanka, Indonesia and Malaysia. The tree grows very fast and is wind and drought resistant suitable for dry arid regions with scanty rains. However, irrigation during summer season results in good fruit yield. Each fruit is a berry, round or oval, measures about 10 cm in diameter, and weigh about 150 g. A tree bears as many as 2000 fruits/year. Sapota fruit is brown in colour with sandy "kiwifruit like" outer surface but without the fuzziness. Internally, the pulp in unripe fruits is white with sticky latex called as saponin. Latex content gradually disappears as the fruit ripens, and its flesh turns brown. The fruit is sweet with smooth or grainy texture and light musky flavour. It contains 3-10 black, smooth, shiny "biconvex/ bean" shaped, inedible seeds located at its center.

Health Benefits of Sapodilla

- ☆ Sapodilla is rich calories, 100 g provides 83 calories. In addition, it is a very good source of dietary fiber (5.6 g/100g), which makes it an excellent bulk laxative. The fiber content helps relieve constipation episodes and help protect the mucous membrane of the colon from cancer-causing toxins by firmly binding to them.

Sapodilla (*Manilkara zapota*), Fresh, Nutritive Value per 100 g

Principle	*Nutrient Value*	*Percentage of RDA*
Energy	83 Kcal	4 per cent
Carbohydrates	19.9 g	15 per cent
Protein	0.44 g	<1 per cent
Total Fat	1.10 g	3.5 per cent
Cholesterol	0 mg	0 per cent
Dietary Fiber	5.3 g	14 per cent
Vitamins		
Folates	14 µg	3.5 per cent
Niacin	0.200 mg	1 per cent
Pantothenic acid	0.252 mg	5 per cent
Pyridoxine	0.037 mg	3 per cent
Riboflavin	0.020 mg	1.5 per cent
Thiamin	0.058 mg	5 per cent
Vitamin A	60 IU	2 per cent
Vitamin C	14.7 mg	24.5 per cent
Electrolytes		
Sodium	12 mg	1 per cent
Potassium	193 mg	4 per cent
Minerals		
Calcium	21 mg	2 per cent
Copper	0.086 mg	9 per cent
Iron	0.80 mg	10 per cent
Magnesium	12 mg	3 per cent
Phosphorous	12 mg	2 per cent
Selenium	0.6 µg	1 per cent
Zinc	0.10 mg	1 per cent

- ☆ The fruit is rich in antioxidant poly-phenolic compound tannin. Tannins are a complex family of naturally-occurring polyphenols that neutralize acids by precipitating proteins. Research studies found that tannins have shown to have potential anti-inflammatory, antiviral, anti-bacterial, and anti-parasitic effects. Hence, these compounds have many useful medicinal applications such as anti-diarrheal, hemostatic (stops bleeding) and as a remedy for hemorrhoids.
- ☆ Furthermore, the anti-inflammatory effect of tannins helps limit conditions like erosive gastritis, reflux-esophagitis, enteritis, and irritating bowel disorders. Some other fruits that are rich in tannins include pomegranate, persimmon, grapes etc.

- ☆ Sapote contains a good amount of antioxidant vitamins like vitamin C (24.5 per cent of recommended daily intake per 100 g of fruit) and vitamin A. Vitamin A is essential for vision. It is also required for maintaining healthy mucus membranes and skin. Consumption of natural fruits rich in vitamin A has been known to protect from lung and oral cavity cancers. So also, consumption of foods containing in vitamin C helps the body develop resistance against infectious agents and scavenge harmful free radicals.
- ☆ Fresh ripen sapodilla is a good source of minerals like potassium, copper, iron and vitamins like folate, niacin and pantothenic acid. These compounds are essential for optimal health as they involve in various metabolic processes in the body as cofactors for the enzymes.

There are many cultivars of sapodilla are grown worldwide like:

- ☆ Brown Sugar variety: Fruit is medium to small, 2 to 2-1/2 inches long, nearly round. Skin is light, scruffy brown. Flesh pale brown, fragrant, juicy, very sweet, and rich, texture slightly granular. Quality is very good.
- ☆ Prolific variety: the fruit is round-conical, 2-1/2 to 3-1/2 inches long and broad. Skin is scruffy, brown, becoming nearly smooth at maturity. Flesh is light-pinkish, mildly fragrant, texture smooth, flavour sweet, quality good. Tree bears early, consistently and heavily.
- ☆ Russel type: The fruit is large, round, 3 to 5 inches in diameter and length. Skin is scruffy brown with gray patches. Flesh is pinkish, mildly fragrant, texture somewhat granular. Flavour is rich and sweet.
- ☆ Tikal: A new seedling selection with excellent flavour. Elliptic in shape, light brown in colour, smaller than Prolific. Ripe very early.

43

Star Fruit

Star fruit, also known as carambola, is a star shaped tropical fruit with sweet and sour flavour. Carambola is native to Malayan peninsula and cultivated in many parts of Southeast Asia, Pacific islands and China. The genus *averrhoa* includes two well-known sorrel (*Oxalidaceae*) families of fruiting trees, carambola and bilimbi (tree cucumber). Although abundant and plentiful, carambola is yet to gain popularity, especially in the western world. Scientific name: *Averrhoa carambola.* The fruit is known as *belimbing manis* in many South East Asian regions and *kamrakh* in India.

Star fruit is a small, bushy evergreen tree that grows very well under hot, humid, tropical conditions. Both sweet and sour varieties begin to yield under cultivable farmlands and ready for harvest when the plants reach 3-4 years old. The plant bears small lilac colour, bell-shaped flowers in clusters, which subsequently develop into oblong shaped fruits with characteristic five angled edges (sides or ribs) that appear like a starfish in cross sections. Carambola fruit features light-green to yellow with attractive smooth waxy surface and weighs about 70-130g. Inside, its crispy, juicy pulp can be sweet or sour type depending upon the amount of oxalic acid concentration. In some seed types, 2-5 tiny edible seeds found at the center of each angled cell.

Health Benefits of Star Fruit

☆ Star fruit is one of the very low calorie exotic fruits. 100 g fruit just provides 31 calories, which is much lower than for any other popular tropical fruits. Nonetheless, it has an impressive list of essential nutrients, antioxidants, and vitamins required for well-being.

☆ The fruit along with its waxy peel provides a good amount of dietary fiber. Fiber helps prevent absorption of dietary LDL cholesterol in the gut. The dietary fibers also help protect the mucous membrane of the colon from exposure to toxic substances by binding to cancer-causing chemicals in the colon.

Star Fruit (*Averrhoa carambola*), Raw, Fresh,
Nutritive value per 100 g, ORAC value-5900

Principle	*Nutrient Value*	*Percentage of RDA*
Energy	31 Kcal	1.5 per cent
Carbohydrates	6.73 g	5 per cent
Protein	1.04 g	2 per cent
Total Fat	0.33 g	1 per cent
Cholesterol	0 mg	0 per cent
Dietary Fiber	2.80 g	7 per cent
Vitamins		
Folates	12 µg	3 per cent
Niacin	0.367 mg	2.25 per cent
Pyridoxine	0.017 mg	1.5 per cent
Riboflavin	0.016 mg	1.25 per cent
Thiamin	0.014 mg	1 per cent
Vitamin A	61 IU	2 per cent
Vitamin C	34.4 mg	57 per cent
Vitamin E	0.15 mg	1 per cent
Vitamin K	0 µg	0 per cent
Electrolytes		
Sodium	2 mg	0 per cent
Potassium	133 mg	3 per cent
Minerals		
Calcium	3 mg	0.3 per cent
Iron	0.08 mg	1 per cent
Magnesium	10 mg	2 per cent
Phosphorus	12 mg	2 per cent
Zinc	0.12 mg	1 per cent

- ✰ Star fruit contains good quantities of vitamin-C. Vitamin C is a powerful natural antioxidant. 100 g of fresh fruit provides 34.7 mg or 57 per cent of daily-required levels of vitamin C. In general, consumption of fruits rich in vitamin C helps the human body develop resistance against infectious agents and scavenge harmful, pro-inflammatory free radicals from the body.
- ✰ Carambola is rich in antioxidant phyto-nutrients polyphenolic *flavonoids*. Some of the important flavonoids are quercetin, epicatechin, and gallic acid. Total polyphenol contents (Folin assay) in star fruit are143 mg/100 g. These compounds help protect from deleterious effects of oxygen derived free radicals by warding them off the body.

- In addition, star fruit is a good source of B-complex vitamins such as folates, riboflavin, and pyridoxine (vitamin B-6). Together, these vitamins help as co-factors for enzymes in metabolism as well as in various synthetic functions inside the body.
- It also contains small amount of minerals and electrolytes like potassium, phosphorus, and zinc and iron. Potassium is an important component of cell and body fluids helps controlling heart rate and blood pressure; thus counters the bad influences of sodium.

44

Strawberry

Delicious and nutrition-rich red coloured strawberries are among the most popular berries. Native to Europe, however, nowadays the berries are widely cultivated as an important commercial crop in many temperate regions all over the world. Botanically, the plant is a runner (creeper), belongs to the family of *Rosaceae*, of the genus: *Fragaria*. Scientific name: *Fragaria ananassa*.

These berries have the taste that varies by cultivar, and ranges from quite sweet to acidic. The berry features red pulp with tiny yellow colour seeds piercing through its surface from inside. Top end has small, green leafy cap and stem that is adorning its crown. Each berry features conical shape, weighs about 25 g and measures about 3 cm in diameter. Although wild varieties are sometimes available in the market, large-scale production uses the modern "plasticulture" system. In this method, raised beds are formed each year, fumigated, and covered with plastic, which prevents weed growth and soiling of berries.

Health Benefits of Strawberries

- ☆ Strawberry is low in calories (32 kcal/100g) and fats but rich source of health promoting phyto-nutrients, minerals and vitamins that are essential for optimum health.
- ☆ Strawberries have significantly high amounts of phenolic flavonoid phyto-chemicals called anthocyanins and ellagic acid. Scientific studies show that consumption of straw berries may have potential health benefits against cancer, ageing, inflammation and neurological diseases.
- ☆ Strawberry has an ORAC value (oxygen radical absorbance capacity, a measure of anti-oxidant strength) of about 3577μmol TE per 100 grams.
- ☆ Fresh berries are an excellent source of vitamin-C (100 g provide 58.8 mg or about 98 per cent of RDI), which is also a powerful natural antioxidant.

Consumption of fruits rich in vitamin C helps the body develop resistance against infectious agents, counter inflammation and scavenge harmful free radicals.

Strawberries (*Fragaria X ananassa*), Nutrition Value per 100 g

Principle	*Nutrient Value*	*Percentage of RDA*
Energy	32 Kcal	1.5 per cent
Carbohydrates	7.7 g	6 per cent
Protein	0.67 g	0.1 per cent
Total Fat	0.30 g	1 per cent
Cholesterol	0 mg	0 per cent
Dietary Fiber	2.0 g	5 per cent
Vitamins		
Folates	24 µg	6 per cent
Niacin	0.386 mg	2.5 per cent
Pantothenic acid	0.125 mg	2.5 per cent
Pyridoxine	0.047 mg	3.5 per cent
Riboflavin	0.022 mg	2 per cent
Vitamin A	12 IU	0.5 per cent
Vitamin C	58.8 mg	98 per cent
Vitamin E	0.29 mg	2 per cent
Vitamin K	2.2 µg	2 per cent
Electrolytes		
Sodium	1 mg	0 per cent
Potassium	153 mg	3 per cent
Minerals		
Calcium	16 mg	1.6 per cent
Iron	0.41 mg	5 per cent
Magnesium	13 mg	3 per cent
Manganese	0.386 mg	17 per cent
Zinc	0.14 mg	1 per cent
Phyto-nutrients		
Carotene-β	7 µg	—
Lutein-zeaxanthin	26 µg	—

- ☆ The fruit is rich in B-complex group of vitamins. It contains very good amounts of vitamin B-6, niacin, riboflavin, pantothenic acid and folic acid. These vitamins are acting as co-factors help the body metabolize carbohydrate, proteins and fats.
- ☆ Strawberries contain vitamin A, vitamin E and health promoting flavonoid poly phenolic antioxidants such as lutein, zeaxanthin, and beta-carotene

in small amounts. These compounds help act as protective scavengers against oxygen-derived free radicals and reactive oxygen species (ROS) that play a role in ageing and various disease processes.

- Furthermore, they contain a good amount of minerals like potassium, manganese, fluorine, copper, iron and iodine. Potassium is an important component of cell and body fluids that helps controlling heart rate and blood pressure. Manganese is used by the body as a co-factor for the antioxidant enzyme, *superoxide dismutase.* Copper is required in the production of red blood cells. Iron is required for red blood cell formation. Fluoride is a component of bones and teeth and is important for prevention of dental caries.

45

Tangerines

Tangerines are related varieties of oranges distinguished by loose, easily peeled shin (pericarp) and sweet juicy flesh (arils). They are also known as mandarin oranges in Europe and *satsumas* in Japan. Just as oranges, they too belong within the *Rutaceae* (citrus Family) and known scientifically as *Citrus reticulata*. Mandarin orange thought to be originating in Southeastern tropical forest of China. Today, it is widely grown in many parts of the world as far as California, as an important commercial crop. The tangerine tree is smaller than oranges with slender branches, and deep-green leaves with pointed ends. The fruit is flat, small compared to "Navel" or "Valencia" oranges. Its loose, deep orange colour skin (pericarp) can be peeled rather easily. Inside, it features extensive fibrous pith, which is loosely attached to the underside of skin and edible flesh. An average-sized fruit has 8-10 juicy segments (arils).

Several Hybrid Varieties of Tangerines Exist

- ☆ **Tangelos**, also known as *honeybell*, are hybrid between tangerine and orange (*Citrus sinensis*) or grapefruit (*Citrus paradisi*). Tangelos, like tangerines, have loose skin and juicy sweet flavoured segments. They are distinguished from oranges by a characteristic knob at the stem end of the fruit.
- ☆ **Tangors** (*Citrus nobilis*) are cross between oranges (*C. sinensis*) and tangerine (*C. reticulata*). They feature large size, and sweet-tart flavour similar to oranges.
- ☆ **Clementines**, another member of citrus family, are smaller, have smooth glossy skin, and very sweet, juicy, almost seedless segments.
- ☆ **Yuzu fruit** or *Japanese citrus fruit* is a hybrid between *C. ichngenesis* and oranges (*C. reticulata*). They are characterized by intense lime-like fragrance and tart flavour. Delicious and juicy orange fruit contains an impressive list of essential nutrients, vitamins, minerals for normal growth and development and overall well-being.

Health Benefits of Tangerines

☆ As in oranges, Tangerines are very low (53 calories/100 g) in calories. Nevertheless, they are valuable sources of flavonoid anti-oxidants like *naringenin, naringin, hesperetin, vitamin A, carotenes, xanthins and luteins;* in fact, several times higher than in the oranges.

Tangerines (*Citrus reticulata*), Fresh, Nutritive Value per 100 g, ORAC value 1620 TE units

Principle	*Nutrient Value*	*Percentage of RDA*
Energy	53 Kcal	2.5 per cent
Carbohydrates	13.34 g	10 per cent
Protein	0.81 g	1.5 per cent
Total Fat	0.31 g	1 per cent
Cholesterol	0 mg	0 per cent
Dietary Fiber	1.8 g	5 per cent
Vitamins		
Folates	16 µg	4 per cent
Niacin	0.376 mg	2.5 per cent
Pantothenic acid	0.216 mg	4 per cent
Pyridoxine	0.078 mg	6 per cent
Riboflavin	0.036 mg	3 per cent
Thiamin	0.058 mg	5 per cent
Vitamin C	26.7 mg	44 per cent
Vitamin A	681 IU	23 per cent
Vitamin E	0.20 mg	1 per cent
Vitamin K	0 µg	0 per cent
Electrolytes		
Sodium	2 mg	<0.5 per cent
Potassium	166 mg	3.5 per cent
Minerals		
Calcium	37 mg	4 per cent
Copper	42 µg	4.5 per cent
Iron	0.15 mg	2 per cent
Magnesium	12 mg	3 per cent
Manganese	0.039 mg	1.5 per cent
Zinc	0.07 mg	<1 per cent
Phyto-nutrients		
Carotene-β	155 µg	—
Carotene-α	101 µg	—
Crypto-xanthin-β	407 µg	—
Lutein-zeaxanthin	138 µg	—
Lycopene	0 µg	—

- ✰ In addition, the citrus fruits are very rich sources of vitamin-C (ascorbic acid), a water-soluble vitamin. Vitamin-C is one of the powerful natural anti-oxidant, which has many essential roles like collagen synthesis, wound healing, anti-viral, anti-cancer activity, and help prevent from neuro-degenerative diseases, arthritis, and cold/fever.etc., by removing oxidant-free radicals from the body. Vitamin C helps absorb iron in the food by reducing ferrous form of the iron elements to ferric form in the gut.
- ✰ Further, they contain natural soluble and insoluble fiber like hemi-cellulose, pectin.etc., which prevents cholesterol absorption in the gut. Adequate fiber in the food aids in smooth bowel movements by acting as a laxative.

Citrus fruits, as such, have long been valued for their wholesome nutritious and antioxidant properties. It is scientifically established that citrus fruits, especially oranges, by virtue of their richness in vitamins and minerals, have many proven health benefits. Moreover, it is now beginning to be appreciated that the other biologically active, non-nutrient compounds found in citrus fruits such as *phyto-chemical antioxidants;* soluble and *insoluble dietary fiber* is helpful in reduction in the risk for cancers, many chronic diseases like arthritis, and from obesity and coronary heart diseases.

46

Watermelon

Want to quench thirst and re-boost your body with anti-oxidant lycopene and vitamin-A? Watermelon has everything you need to beat the scorching summer heat. Wonderfully delicious, thirst-quenching melons are the great source of much-needed water and electrolytes to beat the tropical summer temperatures. Botanically, the fruit belongs to the family of Cucurbitaceae of the genus: citrullus, and is related to the other plants such as cantaloupe, squash, and pumpkin that grow as vines on the ground. It is widely grown across many tropical countries where it is one of the major commercial crops. Botanical name: *Citrullus lanatus*.

Watermelon was originated from southern African countries and from where it spread to all over the tropical and subtropical regions. The plant bears many yellow coloured flowers that may require honeybees for pollination. Externally, the fruit features smooth, deep green or yellow colour thick exterior rind with light-green or gray coloured vertical stripes all over its outer surface. Internally, the flesh is juicy, pink, red, or yellow with numerous small black seeds embedded in the middle third of the flesh. Watermelon has a neutral flavour, and tastes somewhat plain-sweet like light sugar syrup. Varieties of watermelon-fruits are cultivated world over, featuring variation in their size, shape, and colour of the flesh (red, orange, and yellow).

Health Benefits of Watermelon

- ☆ Rich in electrolytes and water content, melons are nature's gift to beat tropical summer thirst.
- ☆ Watermelons are very low in calories (just 30 calories per 100 g) and fats yet very rich source of numerous health promoting phyto-nutrients and anti-oxidants that are essential for optimum health.
- ☆ Watermelon is an excellent source of Vitamin-A, which is a powerful natural anti-oxidant. 100 g fresh fruit provides 569 mg or 19 per cent of daily-required levels of this vitamin. It is essential for vision and immunity.

Vitamin-A is also required for maintaining healthy mucus membranes and skin. Consumption of natural fruits rich in vitamin-A is known to protect from lung and oral cavity cancers.

Watermelon (*Citrullus lanatus*), Fresh, Nutritive Value per 100 g

Principle	*Nutrient Value*	*Percentage of RDA*
Energy	30 Kcal	1.5 per cent
Carbohydrates	7.6 g	6 per cent
Protein	0.6 g	1 per cent
Total Fat	0.15 g	0.5 per cent
Cholesterol	0 mg	0 per cent
Dietary Fiber	0.4 g	1 per cent
Vitamins		
Folates	3 µg	1 per cent
Niacin	0.178 mg	1 per cent
Pantothenic acid	0.221 mg	4.5 per cent
Pyridoxine	0.045 mg	3.5 per cent
Thiamin	0.033 mg	3 per cent
Vitamin A	569 IU	19 per cent
Vitamin C	8.1 mg	13.5 per cent
Vitamin E	0.05 mg	0.5 per cent
Electrolytes		
Sodium	1 mg	0 per cent
Potassium	112 mg	2.5 per cent
Minerals		
Calcium	7 mg	0.7 per cent
Copper	42 µg	4.5 per cent
Iron	0.24 mg	3 per cent
Magnesium	10 mg	2.5 per cent
Manganese	0.038 mg	1.5 per cent
Zinc	0.10 mg	1 per cent
Phyto-nutrients		
Carotene-α	303 µg	—
Crypto-xanthin-β	78 µg	—
Lutein-zeaxanthin	8 µg	—
Lycopene	4532 µg	—

☆ It is also rich in anti-oxidant flavonoids like *lycopene, beta-carotene, lutein, zeaxanthin* and *cryptoxanthin.* These antioxidants are found to offer protection against colon, prostate, breast, endometrial, lung, and pancreatic cancers. Phyto-chemicals present in watermelon like lycopene and carotenoids have the ability to help protect cells and other structures in the body from oxygen-free radicals.

- ☆ Watermelon is an excellent source of carotenoid pigment, Lycopene and indeed, superior to raw red tomato. 100 g of fresh melon provides 4532 μg lycopene, whereas only 2573 μg in tomatoes. Studies suggest that lycopene offer certain protection to skin from harmful UV rays.
- ☆ Watermelon fruit is a good source of potassium; Potassium is an important component of cell and body fluids that helps controlling heart rate and blood pressure; It thus offers protection against stroke and coronary heart diseases.
- ☆ Furthermore, it contains a good amount of vitamin-B6 (pyridoxine), thiamin (vitamin B-1), vitamin-C, and manganese. Consumption of foods rich in vitamin-C helps the body develop resistance against infectious agents and scavenge harmful oxygen-free radicals. Manganese is used by the body as a co-factor for the antioxidant enzyme, *superoxide dismutase*.
- ☆ Total measured antioxidant strength (ORAC value) of watermelon is 142 μmol TE/100 g.

B. Nutritional Value and Health Benefits of Vegetables

Introduction

Why should we get diet rich in vegetable nutrition? What are the health benefits of vegetables?

- ✰ Vegetables, like fruits, are low in fat but contain good amounts of vitamins and minerals. All the Green-Yellow-Orange vegetables are rich sources of calcium, magnesium, potassium, iron, beta-carotene, vitamin B-complex, vitamin-C, vitamin A, and vitamin K.
- ✰ As in fruits, vegetables too are home for many antioxidants that; *firstly*, help protect the human body from oxidant stress, diseases and cancers, and *secondly*; help the body develop the capacity to fight against these by boosting immunity.
- ✰ Additionally, vegetables are packed with soluble as well as insoluble dietary fiber known as non-starch polysaccharides (NSP) such as cellulose, mucilage, hemi-cellulose, gums, pectin.etc. These substances absorb excess water in the colon, retain a good amount of moisture in the fecal matter, and help its smooth passage out of the body. Thus, sufficient fiber offers protection from conditions like hemorrhoids, colon cancer, chronic constipation, and rectal fissures.

Therefore, we have to go for Greens to Help you Stay Fit and Healthy

Vegetable nutrition has widely drawn the attention of fitness conscious as well as food scientists alike for their proven health benefits. Majority of day to day used vegetables are very low in calories. *Celery* is just 16 calories per 100 g. And again here is the long list of vegetables whose calorie is less than 20 per 100 g such as bottle gourd, bitter melon, cabbage, Chinese cabbage, bok-choy, eggplant, endive, spinach, summer squash, Swiss chard; etc. Scientific studies have shown that low-calorie but nutrient-rich foods help the human body stay fit and disease free. Furthermore, human body spends a considerable amount of energy for the metabolism of foods, which is known as BMR or Basal metabolism rate. So just imagine...when you add lots of vegetable nutrition in the diet, in fact, you set to lose more weight than you would gain Right. This is the concept behind the "negative calorie foods."

How much Vegetables Should be in our Daily Diet?

Eat at least 5-7 servings of fresh vegetables every day. Seasonal vegetables should be encouraged. Bring a variety in the choice of vegetable and colour in your diet. Yellow and orange colour vegetables are rich in Vitamin A, α, β carotenes, zeaxanthins and crypto-xanthins, where as dark-green vegetables are a very good source of minerals and phenolic, flavonoid as well as anthocyanin anti-oxidants.

Selection of Vegetables

Whenever possible, go for organic farm vegetables to get maximum health benefits. They are not very expensive if you can find them from the nearby local farm owners. Organic verities tend to be smaller but have rich flavour, possess some good concentration of vitamins, minerals and loaded with numerous health benefiting anti-oxidants.

- In the markets, however, always buy small quantities so that they should last within a day or two. There is no point in eating unfit greens.
- Buy that feature freshness, bright in colour and flavour and feel heavy in your hands.
- Look carefully for blemishes, spots, fungal mold and signs of insecticide spray. Buy whole vegetables instead of section of them (for example, pumpkin).

How to Use Vegetables?

First thing- you need to do immediately after shopping, wash them, especially green leafy vegetables. Rinse in salt water for few minutes and gently swish in cool water until you are satisfied with cleanliness. This way, you ensure they are free from dust, sand and any residual chemical sprays. Use them early while fresh because, firstly, certain vegetables have very short shelf life and secondly, the health benefiting properties of a vegetable declines with time. However, if you need to store them, then place in plastic wrappings or in zip pouches in order to preserve nutrition for short-periods until you use them.

47

Artichoke

Artichoke is one of the popular winter-season edible flower bud of the Mediterranean region known since ancient times for its medicinal and health benefiting qualities. Botanically, it belongs within the thistle family of*Asteraceae*, in the genus; *Cynara*. Scientific name: *Cyanara scolymus.* Globe artichoke grows up to 1.5-2 m tall, with arching, deeply lobed, silvery-green leaves about 0.5 m long. Beautiful light pink flowers develop in a large head from the edible buds. The bud is composed of compactly arranged triangular scales in a whorl fashion around a central "choke." Artichoke globe measures about 6-10 cm in diameter and weighs about 150 g. Fuzzy; immature florets in the centre of the bud constitute its "choke." These are inedible in older, larger flowers. Edible portion of the buds consists primarily of the fleshy lower portions of the involucre bracts (triangular scales) and the base, known as the "heart." Cardoons (*Cynara cardunculus*) are blanched leafy vegetables, closely related to artichokes. Unlike in artichokes where the flower buds are eaten, the leafy-stalks are the edible portions in cardoons.

Several cultivars of artichoke are grown and categorized based upon size, colour, and spine.

- ☆ Green colour, big size,
- ☆ Green colour, medium size,
- ☆ purple colour, big size,
- ☆ Purple colour, medium size,
- ☆ Spine variety.

Health Benefits of Artichoke

- ☆ Artichoke is low in calories and fat; nonetheless, it is a rich source of dietary fiber. It provides 5.4 g per 100 g, about 14 per cent of RDA fiber. Dietary-fiber helps control constipation conditions, decrease bad or "LDL" cholesterol levels by binding to it in the intestines and help prevent colon cancer risks by preventing toxic compounds in the food from absorption.

Artichoke (*Cynara scolymus*), Raw, Nutrition Value per 100 g. ORAC value 6552 TE/100 g

Principle	*Nutrient Value*	*Percentage of RDA*
Energy	47 Kcal	2 per cent
Carbohydrates	10.51 g	8 per cent
Protein	3.27 g	6 per cent
Total Fat	0.15 g	0.5 per cent
Cholesterol	0 mg	0 per cent
Dietary Fiber	5.4 g	14 per cent
Vitamins		
Folates	68 µg	17 per cent
Niacin	1.046 mg	6.5 per cent
Pantothenic acid	0.338 mg	7 per cent
Pyridoxine	0.116 mg	9 per cent
Riboflavin	0.066 mg	5 per cent
Thiamin	0.072 mg	6 per cent
Vitamin C	11.7 mg	20 per cent
Vitamin A	13 IU	0.5 per cent
Vitamin E	0.19 mg	1 per cent
Vitamin K	14.8 µg	12 per cent
Electrolytes		
Sodium	94 mg	6 per cent
Potassium	370 mg	8 per cent
Minerals		
Calcium	44 mg	4 per cent
Copper	0.231 mg	27 per cent
Iron	1.28 mg	16 per cent
Magnesium	60 mg	15 per cent
Manganese	0.256 mg	11 per cent
Phosphorus	90 mg	13 per cent
Selenium	0.2 µg	<0.5 per cent
Zinc	0.49 mg	4.5 per cent
Phyto-nutrients		—
Carotene-α	8 µg	—
Crypto-xanthin	0 µg	—
Lutein-zeaxanthin	464 µg	—

- Artichoke contains bitter principles, *cynarin and sesquiterpene-lactones*. Scientific studies show that these compounds inhibit cholesterol synthesis and increase its excretion in the bile and thus; have overall cholesterol reduction in the blood.
- Fresh artichoke is an excellent source of vitamin folic acid; provide about 68 µg per 100 g (17 per cent of recommended daily allowance). Folic acid acts as a co-factor for enzymes involved in the synthesis of DNA. Scientific studies have proven that adequate levels of folates in the diet during pre-conception period, and during early pregnancy, help prevent neural tube defects in the newborn baby.
- Fresh globes also contain good amounts of anti-oxidant vitamin; vitamin-C (Provides about 20 per cent of recommended levels per 100 g). Regular consumption of foods rich in vitamin C helps the body develop resistance against infectious agents and scavenge harmful, pro-inflammatory free radicals from the body.
- It is one of the vegetable sources for *vitamin K*; provide about 12 per cent of DRI. Vitamin K has potential role bone health by promoting osteotrophic (bone formation) activity. Adequate vitamin-K levels in the diet help limiting neuronal damage in the brain; thus, has established role in the treatment of patients suffering from Alzheimer's disease.
- It is an also good source of antioxidants such as silymarin, caffeic acid, and ferulic acid, which help the body protect from harmful free-radical agents. Total antioxidant strength (ORAC) of artichokes (globe or French) is 6552 µmol TE/100 g.
- It is also rich in B-complex group of vitamins such as niacin, vitamin B-6 (pyridoxine), thiamin, and pantothenic acid that are essential for optimum cellular metabolic functions.
- Further, artichoke is rich source of minerals like copper, calcium, potassium, iron, manganese and phosphorus. Potassium is an important component of cell and body fluids that helps controlling heart rate and blood pressure by countering effects of sodium. Manganese is used by the body as a co-factor for the antioxidant enzyme, *superoxide dismutase*. Copper is required in the production of red blood cells. Iron is required for red blood cell formation.
- Additionally, it contains small amounts of antioxidant flavonoid compounds like *carotene-beta, lutein, and zeaxanthin.*

48

Arugula

Arugula, also known as salad or garden rocket, is a nutritious green-leafy vegetable of Mediterranean origin. It belongs within the *Brassicaceae* family similar as mustard greens, cauliflower, kale etc., and has the scientific name: *Eruca sativa*. Salad rocket is a small, low growing annual herb featuring dandelion like succulent, elongated, lobular leaves with green-veins. In young plant, however, the plain light green leaves appear identical to that of spinach without lobulation. Young, tender leaves feature sweet, nutty, flavour and less peppery taste in contrast to strong, spicy flavour of mature greens. Salad rocket (arugula) is a quick growing, cool season crop. It prefers well-drained, fertile soil and full sun to flourish. In general, arugula grows to about 2-3 feet in height with creamy-white colour edible flowers. Its leaves are ready to harvest within 40 days of sowing the seed.

Health Benefits of Arugula

- As in other greens, arugula is one of very low calorie vegetable. 100 g of fresh leaves provides just 25 calories. Nonetheless, it has many vital phytochemicals, anti-oxidants, vitamins, and minerals that can immensely benefit health.
- Salad rocket has an ORAC value (oxygen radical absorbance capacity, a measure of anti-oxidant strength) of about 1904 µmol TE per 100 grams.
- Rocket salad is rich source of certain phytochemicals such as *indoles, thiocyanates, sulforaphane*, and *iso-thiocyanates*. Together, they have been found to counter carcinogenic effects of estrogen and thus help benefit against prostate, breast, cervical, colon, ovarian cancers by virtue of their cancer-cell growth inhibition, and cytotoxic effects on cancer cells.
- In addition, di-indolyl-methane (DIM), a lipid soluble metabolite of indole has immune modulator, anti-bacterial and anti-viral properties (by potentiating Interferon-Gamma receptors and production). DIM has

currently been found application in the treatment of recurring respiratory papillomatosis caused by the Human Papilloma Virus (HPV) and is in Phase III clinical trials for cervical dysplasia.

Arugula (*Eruca sativa*), Raw, Nutrition Value per 100 g. ORAC Value 1904

Principle	*Nutrient Value*	*Percentage of RDA*
Energy	25 Kcal	1 per cent
Carbohydrates	3.65 g	3 per cent
Protein	2.58 g	5 per cent
Total Fat	0.66 g	3 per cent
Cholesterol	0 mg	0 per cent
Dietary Fiber	1.6 g	4 per cent
Vitamins		
Folates	97 μg	24 per cent
Niacin	0.305 mg	2 per cent
Pantothenic acid	0.437 mg	8 per cent
Pyridoxine	0.073 mg	6 per cent
Riboflavin	0.086 mg	7 per cent
Thiamin	0.044 mg	4 per cent
Vitamin C	15 mg	25 per cent
Vitamin A	2373 IU	79 per cent
Vitamin E	0.43 mg	3 per cent
Vitamin K	108.6 μg	90 per cent
Electrolytes		
Sodium	27 mg	2 per cent
Potassium	369 mg	7.5 per cent
Minerals		
Calcium	160 mg	16 per cent
Copper	0.076 mg	8 per cent
Iron	1.46 mg	18 per cent
Magnesium	47 mg	12 per cent
Manganese	0.321 mg	14 per cent
Phosphorus	52 mg	7.5 per cent
Selenium	0.3 μg	<1 per cent
Zinc	0.47 mg	5 per cent
Phyto-nutrients		
Carotene-β	1424 μg	
Carotene-α	0 μg	
Lutein-zeaxanthin	3555 μg	

- ☆ Fresh rocket is a very good source of folates. 100 g of fresh greens contain 97 µg or 24 per cent of folic acid. When given to the anticipant mothers during their conception time, folate helps prevent neural tube defects in the newborns.
- ☆ Like kale, salad rocket is an excellent source of vitamin A. 100 g fresh leaves contain 1424 µg of beta-carotene, and 2373 IU of vitamin A. Carotenes convert into vitamin A in the body. Studies found that vitamin A and flavonoid compounds in green leafy vegetables help protect from skin, lung and oral cavity cancers.
- ☆ This vegetable also rich in B-complex group of vitamins such as thiamin, riboflavin, niacin, vitamin B-6 (pyridoxine), and pantothenic acid those are essential for optimum cellular enzymatic and metabolic functions.
- ☆ Fresh rocket leaves contain good levels of vitamin C. Vitamin C is a powerful, natural anti-oxidant. Foods rich in this vitamin help the body protect from scurvy disease; develop resistance against infectious agents (boosts immunity) and scavenge harmful, pro-inflammatory free radicals from the body.
- ☆ Salad rocket is one of the excellent vegetable sources for vitamin-K; 100 g provides about 90 per cent of recommended intake. Vitamin K has potential role bone health by promoting osteotrophic (bone formation and strengthening) activity. Adequate vitamin-K levels in the diet help limiting neuronal damage in the brain; thus, has established role in the treatment of patients suffering from *Alzheimer's disease*.
- ☆ Arugula leaves contain adequate levels of minerals, especially copper and iron. In addition, it has small amounts of some other essential minerals and electrolytes such as calcium, iron, potassium, manganese, and phosphorus.

49

Asparagus

Asparagus has been revered by ancient Greek and Romans as a prized delicacy. One of the oldest recorded vegetables; it is thought to have originated along the coastal regions of eastern Mediterranean and Asia Minor areas. Botanically this herbaceous perennial plant belongs to the *Asparagaceae* family. It is closely related to *Liliaceae* family members which includes onion, garlic, tulip, daffodil…etc. Scientific name: *A. officinalis*. This spear vegetable is now grown as major commercial crop in China, Europe, Peru, Australia, and USA.

To cultivate, asparagus crowns are planted by early spring grows into tall branched stems bearing fine fern-like green needles (phylloclades) giving the plant an overall feathery appearance. The plant can also be established by transplanting of 10-12 week-old seedlings or direct seeding. A fully-grown plant reaches about 5 feet in height. Young scaly edible spears emerge from the extensive matted root systems, which are ready for harvesting by early spring.

In general, 7 to 9 inches long young shoots are harvested by either snapping or using a paring knife, cutting close to the ground level. There is no need to cut asparagus shoots far below the soil with a knife. This may injure other erupting buds on the crown. The stump that is left in the soil after snapping dries up and disintegrates. Warm weather makes the spears tip to open prematurely, which may reduce their flavour and quality. Asparagus is normally harvested once a year over a 8 to 10 week period. White or blanched asparagus (spargel) spears are produced by covering erupting shoots with a mound, depriving them of sunlight. This method, like in endive, makes the shoots turn pale through inhibition of photosynthesis. Oftentimes, its blanched spears are preferred in Europe due to their delicate taste and pleasant flavour.

Health Benefits of Asparagus

☆ Asparagus is a very low calorie vegetable. 100 g fresh spears give only 20 calories. More calories will be burnt to digest than gained, the fact, which fits into the category of low calorie or negative-calorie vegetables.

Asparagus (*A. officinalis*), Raw, Nutrition Value per 100 g. ORAC value 2150

Principle	*Nutrient Value*	*Percentage of RDA*
Energy	20 Kcal	1 per cent
Carbohydrates	3.38 g	2.5 per cent
Protein	2.20 g	4 per cent
Total Fat	0.12 g	0.5 per cent
Cholesterol	0 mg	0 per cent
Dietary Fiber	2.1 g	5.5 per cent
Vitamins		
Folates	52 µg	13 per cent
Niacin	0.978 mg	6 per cent
Pantothenic acid	0.274 mg	5 per cent
Pyridoxine	0.091 mg	7 per cent
Riboflavin	0.141 mg	11 per cent
Thiamin	0.143 mg	12 per cent
Vitamin C	5.6 mg	9 per cent
Vitamin A	756 IU	25 per cent
Vitamin E	1.13 mg	7.5 per cent
Vitamin K	41.6 µg	35 per cent
Electrolytes		
Sodium	2 mg	<1 per cent
Potassium	202 mg	4 per cent
Minerals		
Calcium	24 mg	2.5 per cent
Copper	0.189 mg	21 per cent
Iron	1.14 mg	14 per cent
Magnesium	14 mg	1 per cent
Manganese	0.158 mg	7 per cent
Phosphorus	52 mg	7.5 per cent
Selenium	2.3 µg	4 per cent
Zinc	0.54 mg	5 per cent
Phyto-nutrients		
Carotene-β	449 µg	
Carotene-α	9 µg	
Lutein-zeaxanthin	710 µg	

- In addition, the spears contain moderate levels of dietary-fiber. 100 g of fresh spears provide 2.1 g of roughage. Dietary fiber helps control constipation conditions, decrease bad (LDL) cholesterol levels by binding to it in the intestines, and regulate blood sugar levels. Studies suggest that high-fiber diet help cut down colon-rectal cancer risks by preventing toxic compounds in the food from absorption.
- Its shoots have long been used in many traditional medicines to treat conditions like dropsy and irritable bowel syndrome.
- Fresh asparagus spears are the good source of anti-oxidants such as *lutein, zeaxanthin, carotenes,* and *crypto-xanthins*. Together, these flavonoid compounds help remove harmful oxidant free radicals from the body protect it from possible cancer, neuro-degenerative diseases, and viral infections. Their total antioxidant strength, measured in terms of oxygen radical absorbance capacity (ORAC value), is 2150 μmol TE/100 g.
- Fresh asparagus are rich in folates. 100 g of spears provide about 54 μg or 14 per cent of RDA of folic acid. Folates are one of the important co-factors for DNA synthesis inside the cell. Scientific studies have shown that adequate consumption of folates in the diet during pre-conception period and during early pregnancy, help prevents neural tube defects in the newborn baby.
- The shoots are also rich in B-complex group of vitamins such as thiamin, riboflavin, niacin, vitamin B-6 (pyridoxine), and pantothenic acid those are essential for optimum cellular enzymatic and metabolic functions.
- Fresh asparagus also contains fair amounts of anti-oxidant vitamins such as vitamin-C, vitamin-A, and vitamin-E. Regular consumption of foods rich in these vitamins helps the body develop resistance against infectious agents and scavenge harmful, pro-inflammatory free radicals from the body.
- Its shoots are also good source of vitamin K; provide about 35 per cent of DRI. Vitamin K has potential role bone health by promoting osteotrophic (bone formation) activity. Adequate vitamin-K levels in the diet help limiting neuronal damage in the brain; thus, has established role in the treatment of patients suffering from Alzheimer's disease.
- Asparagus is good in minerals, especially copper and iron. In addition, it has small amounts of some other essential minerals and electrolytes such as calcium, potassium, manganese, and phosphorus. Potassium is an important component of cell and body fluids that helps controlling heart rate and blood pressure by countering effects of sodium. Manganese is used by the body as a co-factor for the antioxidant enzyme, *superoxide dismutase*. Copper is required in the production of red blood cells. Iron is required for cellular respiration and red blood cell formation.

50

Basella

Basella or vine spinach is a popular tropical leaf green vegetable commonly grown as pot herb in the home gardens. In the true sense, it is different from English spinach (*Spinacea oleracea*) in that the plant is a creeping vine, and its leaves feature glossy, broad, deep green, thick, and mucilaginous. Commonly found in the home gardens of many south Asian families, it is gaining popularity in some of the tropical and temperate climates of America, Australia and Europe for its succulent, nutritious greens, and tender stems. Vine spinach belongs to the *Basellaceae* family and has two chief cultivars, *Basella alba*, which features green-stems and deep-green leaves, and *Basella rubra* with purplish-stem and deep-green leaves with pink veins.

Vine spinach is noted by regional names in different regions in Asia. Some of the common names for this herb are Ceylon spinach, Malabar spinach, saan choy (Chinese), mong toi (Vietnamese), alugbati (Philippines), pui saag (Bengali), remayong (Malay), etc. It is native to south Asia, probably originated in the monsoon fed tropical regions of Malabar Coast of India and Sri Lanka. The plant is a perennial vine and grown as annual or biennial pot-herb. It prefers hot humid climate and moist, fertile, well-drained soil to flourish. Although its seeds can be sown directly for planting, usually thick cuttings about the length of 20 cm preferred for easy propagation and fast growth. Being a vine, the plant requires trellising for its creeping at a faster rate. It bears white or white-pink colour tiny flowers depending upon the species and deep-purple to black colour berries. Basella alba bears thick, fleshy, broad, oval to heart-shaped leaves all along its vine length. Basella rubra features pink or purplish stems and pink colour veins in the leaves. In either case, leaves and terminal, tender, 8-12 inches stems are ready for harvesting about 35 to 45 days after planting (about 50 days after seeding).

Health Benefits of Basella (Vine Spinach)

- ☆ Vine spinach is a very versatile leaf green vegetable and revered in some East Asian cultures for its wholesome phyto-nutrients profile.

Malabar Spinach (*Basella alba*), Raw, Nutritive Value per 100 g

Principle	*Nutrient Value*	*Percentage of RDA*
Energy	19 Kcal	1 per cent
Carbohydrates	3.40 g	2.5 per cent
Protein	1.80 g	3 per cent
Total Fat	0.30 g	1.5 per cent
Cholesterol	0 mg	0 per cent
Vitamins		
Folates	140 μg	35 per cent
Niacin	0.500 mg	3 per cent
Pantothenic acid	0.053 mg	1 per cent
Pyridoxine	0.240 mg	18 per cent
Riboflavin	0.155 mg	13 per cent
Thiamin	0.050 mg	4 per cent
Vitamin A	8000 IU	267 per cent
Vitamin C	102 mg	170 per cent
Electrolytes		
Sodium	24 mg	1.5 per cent
Potassium	510 mg	11 per cent
Minerals		
Calcium	109 mg	11 per cent
Copper	0.107 mg	12 per cent
Iron	1.20 mg	15 per cent
Magnesium	65 mg	16 per cent
Manganese	0.735 mg	32 per cent
Selenium	0.8 μg	1.5 per cent
Zinc	0.43 mg	4 per cent

☆ Basella is very low in calories and fats (100 g of raw leaves provide just 19 calories). Nonetheless, it contains an incredibly good amount of vitamins, minerals, and antioxidants.

☆ Fresh leaves, particularly of basella rubra, are rich sources of several vital carotenoid pigment anti-oxidants such as β-carotene, lutein, zeaxanthin. Together, these compounds help act as protective scavengers against oxygen-derived free radicals and reactive oxygen species (ROS) that play a healing role in ageing and various disease processes.

☆ Its thick, fleshy leaves are a good source of non-starch polysaccharide, mucilage. In addition to regular fiber (roughage), found in the stem and leaves, mucilage helps in digestion, reduction in cholesterol absorption and prevents bowel movement problems.

- Vine spinach leaves and stem are incredibly rich sources of vitamin A. 100 g fresh leaves provide 8000 IU or 267 per cent of recommended daily allowance (RDA) of this vitamin. Vitamin-A is required for maintaining healthy mucus membranes and skin and is essential for good eye-sight. Consumption of natural vegetables and fruits rich in vitamin-A, and flavonoids has been thought to offer protection from the lung and oral cavity cancers.
- Basella has more vitamin C content than English spinach. 100 g of fresh greens contains 102 mg or 102 per cent of daily recommended levels of vitamin C. Vitamin-C is a powerful antioxidant, which helps the body develop resistance against infectious agents and scavenge harmful oxygen-free radicals.
- Like in spinach, basella too is an excellent source of iron. 100 g fresh leaves contain about 1.20 mg or 15 per cent of daily intake of iron. Iron is an important trace element required by the human body for red blood cell (RBC's) production. Additionally, this element acts as a co-factor for oxidation-reduction enzyme, cytochrome-oxidase, during the cellular metabolism.
- This green leafy vegetable also contains good amounts of many B-complex vitamins such as folate, vitamin-B6 (pyridoxine), and riboflavin. 100 g fresh leaves provide 140 µg or 35 per cent of folates. This vitamin is one of the essential compounds for DNA production and growth. Folate deficiency in during very early stages of pregnancy might results in the neural tube defects in the newborn baby. Anticipating and pregnant women are therefore, advised to include a lot of fresh greens in their diet to help prevent neural tube defects in the offspring.
- Further, basella leaves are very rich sources of minerals like potassium (11 per cent of RDA/100 g), manganese (32 per cent of RDA/100 g), calcium, magnesium, and copper. Potassium is an important component of cell and body fluids that helps controlling heart rate and blood pressure. Manganese and copper are used by the body as a co-factor for the antioxidant enzyme, *superoxide dismutase.*

Similar way to spinach, regular consumption of basella (Malabar spinach) in the diet helps prevent osteoporosis (weakness of bones), iron-deficiency anemia. Besides, it is believed to protect the body from cardiovascular diseases and cancers of colon.

51

Beets

Beets are highly nutritious and "cardiovascular health" friendly root vegetables. Certain unique pigment antioxidants in the root as well as in its top greens have found to offer protection against coronary artery disease and stroke; lower cholesterol levels within the body, and have anti-aging effects. Botanically, the tuberous root vegetable belongs in the *Amarathaceae* family, in the *beta* genus. Its scientific name is *Beta vulgaris.* Swiss chard is another member in the *beta* genus grown for its edible leaves.

Beets are small herbaceous plants with broad dark-green leaves. Its underground taproot matures in 50-60 days of sowing and weighs about 100 to 150 g. If not harvested, it keeps growing in size to more than 500 g and may develop surface cracks, lose taste and become unappetizing because of excess fiber content. Different cultivars exist; red, orange-yellow and white verities. The unique crimson-red colour of red beet is due to betalain pigments, such as *betanin* and *betacyanin*. Yellow varieties are rich in β-*xanthin* pigment. Both root and top leaves of beet are used for consumption. Choggia beet or candy cane variety has alternative red and white concentric whorls.

Health Benefits of Beets

- ☆ Garden beet is very low in calories (provide only 45 kcal/100 g), and contain zero cholesterol and small amount of fat. Its nutrition benefits come particularly from fiber, vitamins, minerals, and unique plant derived anti-oxidants.
- ☆ The root is rich source of phytochemical compound, glycine betaine. Betaine has the property of lowering *homocysteine* levels within the blood. *Homocysteine*, one of highly toxic metabolite, promotes platelet clot as well as atherosclerotic-plaque formation, which, otherwise, can be harmful to blood vessels. High levels of *homocysteine* in the blood result in the development of coronary heart disease (CHD), stroke and peripheral vascular diseases.

Beets (*Beta vulgaris*), Fresh, Raw, Nutrition Value per 100 g

Principle	*Nutrient Value*	*Percentage of RDA*
Energy	43 cal	2 per cent
Carbohydrates	9.56 g	7 per cent
Protein	1.61 g	1 per cent
Total Fat	0.17 g	0.5 per cent
Cholesterol	0 mg	0 per cent
Dietary Fiber	2.80 g	7 per cent
Vitamins		
Folates	109 µg	27 per cent
Niacin	0.334 mg	2 per cent
Pantothenic acid	0.155 mg	3 per cent
Pyridoxine	0.067 mg	5 per cent
Riboflavin	0.057 mg	4 per cent
Thiamin	0.031 mg	2.5 per cent
Vitamin A	33 IU	1 per cent
Vitamin C	4.9 mg	8 per cent
Vitamin E	0.04 mg	0.5 per cent
Vitamin K	0.2 µg	0 per cent
Electrolytes		
Sodium	78 mg	5 per cent
Potassium	325 mg	7 per cent
Minerals		
Calcium	16 mg	1.5 per cent
Copper	0.075 mg	8 per cent
Iron	0.80 mg	10 per cent
Magnesium	23 mg	6 per cent
Manganese	0.329 mg	14 per cent
Zinc	0.35 mg	3 per cent
Phyto-nutrients		
Carotene-β	20 µg	—
Betaine	128.7 mg	—
Lutein-zeaxanthin	0 µg	—

☆ Raw beets are an excellent source of folates. It contains about 109 µg/100 g of this vitamin (Provides 27 per cent of RDA). However, extensive cooking may significantly deplete its level in food. Folates are necessary for DNA synthesis within the cells. When given during peri-conception period folates can prevent neural tube defects in the baby.

- Fresh tubers contain small amounts of vitamin-C; however, its top greens are rather excellent sources of this vitamin. 100 g of beet greens provide 30 mg or 50 per cent of RDA. Vitamin C is one of the powerful natural antioxidants, which helps the human body scavenge deleterious free radicals one of the reasons for cancer development.
- Additionally, the top greens are an excellent source of carotenoids, flavonoid anti-oxidants, and vitamin A; contain these compounds several times more than that of in the roots. Vitamin A is required maintaining healthy mucus membranes and skin and is essential for vision. Consumption of natural vegetables rich in flavonoids helps to protect from lung and oral cavity cancers.
- The root is also rich source of B-complex vitamins such as niacin (B-3), pantothenic acid (B-5), pyridoxine (B-6) and minerals such as iron, manganese, copper, and magnesium.
- Further, the root indeed has very good levels of potassium. 100 g fresh root has 325 mg of potassium or 7 per cent of daily requirements. Potassium lowers heart rate and regulates metabolism inside the cells by countering detrimental effects of sodium.

52

Bell Pepper

Bell pepper or sweet pepper is the most popular of the chili peppers in the *Capsicum annum* family. It is a fruit pod of small perennial shrub in the nightshade or *Solanaceae* family, of the genus, *capsicum*. Scientific name: *Capsicum annum*. Unlike their fellow capsicum members, sweet peppers have characteristic bell shape with crunchy, thick fleshy texture. On comparison to other members, bell (sweet) peppers feature characteristically least pungency that ranges from zero to very minimal hotness. For the same reasons, they generally treated like vegetables instead of spice.

Peppers are native to Mexico and other Central American region from where they spread to the rest of the world by Spanish and Portuguese explorers during 16th and 17th centuries and now grown widely in many parts of the world as an important commercial crop. As in other chili pepper varieties, bell peppers also have several cultivar types. However, the plant type and fruit pod (with 3-5 lobes) are common features in almost all cultivars. In structure, sweet pepper features blocky, cube like outer flesh enclosing numerous tiny, white, or cream coloured, circular and flat seeds. The seeds are actually clinging to the central core (placenta). To harvest, peppers are handpicked at different stages of maturity depending up on the local market preferences. All varieties of unripe peppers feature green colour pods, irrespective of their final destine colour. As the fruit matures, it gradually acquires its true genetic colour such as orange, red, purple, yellow, etc.

The hotness of peppers is measured in "Scoville heat units" (SHU). On the Scoville scale, a sweet bell pepper scores 0, while a jalapeno pepper around 2,500-4,000 and a Mexican habaneros 200,000 to 500,000 units.

Health Benefits of Bell Pepper

- ☆ Bell pepper contains an impressive list of plant nutrients that are known to have disease preventing and health promoting properties. Unlike other chili peppers, it is very low in calories and fats. 100 g provide just 31 calories.

Bell Peppers (*Capsicum annuum* var. *annuum*), Red, Raw, Nutrition Value per 100 g

Principle	*Nutrient Value*	*Percentage of RDA*
Energy	31 Kcal	1.5 per cent
Carbohydrates	6.03 g	4 per cent
Protein	0.99 g	2 per cent
Total Fat	0.30 g	1 per cent
Cholesterol	0 mg	0 per cent
Dietary Fiber	2.1 g	5.5 per cent
Vitamins		
Folates	46 µg	12 per cent
Niacin	0.979 mg	6 per cent
Pyridoxine	0.291 mg	22 per cent
Riboflavin	0.085 mg	6.5 per cent
Thiamin	0.054 mg	4.5 per cent
Vitamin A	3131 IU	101 per cent
Vitamin C	127.7 mg	213 per cent
Vitamin E	1.58 mg	11 per cent
Vitamin K	4.9 µg	4 per cent
Electrolytes		
Sodium	4 mg	<1 per cent
Potassium	211 mg	4.5 per cent
Minerals		
Calcium	7 mg	1 per cent
Copper	0.017 mg	2 per cent
Iron	0.43 mg	5 per cent
Magnesium	12 mg	3 per cent
Manganese	0.112 mg	5 per cent
Phosphorus	26 mg	4 per cent
Selenium	0.1 µg	<1 per cent
Zinc	0.25 mg	2 per cent
Phyto-nutrients		
Carotene-β	1624 µg	—
Carotene-α	20 µg	—
Cryptoxanthin-β	490 µg	—
Lutein-zeaxanthin	51 µg	—

☆ Sweet (bell) pepper contains small levels of health benefiting an alkaloid compound capsaicin. Early laboratory studies on experimental mammals suggest that capsaicin has anti-bacterial, anti-carcinogenic, analgesic and anti-diabetic properties. When used judiciously, it also found to reduce triglycerides and LDL cholesterol levels in obese individuals.

- ✰ Fresh bell peppers, red or green, are rich source of vitamin-C. This vitamin is especially concentrated in red peppers at highest levels. 100 g red pepper provides about 127.7 µg or about 213 per cent of RDA of vitamin C. Vitamin-C is a potent water soluble antioxidant. It is required for the collagen synthesis in the body. Collagen is the main structural protein in the body required for maintaining the integrity of blood vessels, skin, organs, and bones. Regular consumption of foods rich in this vitamin helps the human body protect from scurvy; develop resistance against infectious agents (boosts immunity) and scavenge harmful, pro-inflammatory free radicals from the body.
- ✰ It also contains good levels of vitamin-A. 100 g of sweet pepper has 3131 IU or 101 per cent of vitamin A. In addition, it contains anti-oxidant flavonoids such as α and β carotenes, lutein, zeaxanthin, and cryptoxanthin. Together, these antioxidant substances in capsicum help to protect the body from injurious effects of free radicals generated during stress and disease conditions.
- ✰ Bell pepper has adequate levels of essential minerals. Some of the main minerals in it are iron, copper, zinc, potassium, manganese, magnesium, and selenium. Manganese is used by the body as a co-factor for the antioxidant enzyme, superoxide dismutase. Selenium is an anti-oxidant micro-mineral that acts as a co-factor for enzyme, *superoxide dismutase*.
- ✰ Further, capsicum (sweet pepper) is also good in B-complex group of vitamins such as niacin, pyridoxine (vitamin B-6), riboflavin, and thiamin (vitamin B-1). These vitamins are essential in the sense that body requires them from external sources to replenish. B-complex vitamins facilitate cellular metabolism through various enzymatic functions.

53

Bitter Gourd

Bitter gourd (melon) is perhaps the secret vegetable of the Okinawa Islander longevity! Although the bitterness of Bitter-melon might turn some people away from, it can really sweeten your health because of its disease preventing and health promoting phyto chemical compounds. Botanically it belongs to the family of Cucurbitaceae, of the genus: Momordica and is a member of the same family of squash, watermelon, cantaloupes, cucumber, etc. Scientific name: *Momordica charantia.* Some of the related varieties are *balsam pear, cundeamor, la-kwa,* etc. Bitter melon is the immature pod vegetable, popular in many Asian countries. This widely grown as edible pod is, in fact, among the most bitter of all culinary vegetables. Bitter melon is a temperate / tropical vegetable probably originated in South-East Asia. Like other members of the Cucurbitaceae family, this plant is a fast-growing, trailing or climbing vine with thin stems and tendrils and requires trellis to support the climbing vine. The pods are characterized by smooth lengthwise ridges and uneven pebbly surface. Depending upon the cultivar type, immature pods are light to dark green and have oblong or oval shapes with a pointed tip at the blossom end. Internally, the flesh is white with rough edged seeds, somewhat similar to ridge gourd seeds. As the fruits begin to mature, they gradually turn yellow or orange.

Health Benefits of Bitter Gourd

- ☆ The vegetable is very low in calories, providing just 17 calories per 100g. Nevertheless, its pods are rich in phytonutrients like dietary fiber, minerals, vitamins and anti-oxidants.
- ☆ Bitter melon notably contains phyto-nutrient, polypeptide-P; a plant insulin known to lower blood sugar levels. In addition, it composes hypoglycemic agent called charantin. *Charantin* increases glucose uptake and glycogen synthesis in the cells of liver, muscle and adipose tissue. Together, these compounds are thought to be responsible for reduction of blood sugar levels in the treatment of type-2 diabetes.

Bitter Gourd or Bitter Melon (*Momordica charantia*), Fresh, Raw, Nutritive value per 100 g

Principle	*Nutrient Value*	*Percentage of RDA*
Energy	17 Kcal	<1 per cent
Carbohydrates	3.70 g	3 per cent
Protein	1.00 g	2 per cent
Total Fat	0.17 g	0.5 per cent
Cholesterol	0 mg	0 per cent
Dietary Fiber	2.80 g	7 per cent
Vitamins		
Folates	72 µg	18 per cent
Niacin	0.400 mg	2.5 per cent
Pantothenic acid	0.212 mg	4 per cent
Pyridoxine	0.043 mg	3 per cent
Riboflavin	0.040 mg	3 per cent
Thiamin	0.040 mg	3.5 per cent
Vitamin A	471 IU	16 per cent
Vitamin C	84 mg	140 per cent
Electrolytes		
Sodium	5 mg	<1 per cent
Potassium	296 mg	6 per cent
Minerals		
Calcium	19 mg	2 per cent
Copper	0.034 mg	4 per cent
Iron	0.43 mg	5 per cent
Magnesium	17 mg	4 per cent
Manganese	0.089 mg	4 per cent
Zinc	0.80 mg	7 per cent
Phyto-nutrients		
Carotene-β	190 µg	—
Carotene-α	185 µg	
Lutein-zeaxanthin	170 µg	—

☆ Fresh pods are an excellent source of folates; contain about 72 µg/100g (Provides 18 per cent of RDA). Folate helps reduce the incidence of neural tube defects in the newborns when taken by mothers during early pregnancy.

☆ Fresh bitter melon is an excellent source of vitamin-C (100 g of raw pod provides 84 mg or about 140 per cent of RDI). Vitamin-C, one of the powerful natural antioxidants, helps the body scavenge deleterious free radicals one of the reasons for cancer development.

- It is an excellent source of health benefiting flavonoids such as β-carotene, α-carotene, lutein, and zeaxanthin. It also contains a good amount of vitamin A. Together; these compounds help act as protective scavengers against oxygen-derived free radicals and reactive oxygen species (ROS) that play a role in ageing, cancers and various disease processes.
- Bitter melon stimulates easy digestion and peristalsis of food through the bowel until it is excreted from the body. Thus, helps in relieving indigestion and constipation problems.
- In addition, the vegetable is an also good source of niacin (vitamin B-3), pantothenic acid (vitamin B-5), pyridoxine (vitamin B-6) and minerals such as iron, zinc, potassium, manganese and magnesium.
- Early laboratory tests suggest that compounds in bitter melon might be effective for treating HIV infection.

54

Bok Choy

Bok choy or *leafy Chinese cabbage* is one of the popular mainland crops in China, Philippines, Vietnam and other oriental regions. Nonetheless, this humble Brassica family vegetable has gained popularity even in the western world for its sweet, succulent nutritious stalks. It is also named as *pe-tsai, pak choi, petsay, white-celery mustard, Chinese white cabbage* etc. Scientific name: *Brassica campestris* L. (Chinensis group). In structure, bok choy resembles collards and could be described as a non-heading cabbage (Acephala group). It is basically a small plant which grows upright from the ground with smooth white romaine lettuce like stalks, which spread at the end to fine, glossy green, oval or round leaves. Full grown-up plant may reach about 12-18 inches in length. Brassica campestris group can be further categorized according to the colour of the petiole in its leaves. White petiole varieties include *joi choi, pak-choy white, prize choi, lei choi, taisai, canton pak choi* etc. Green petiole types are *Chinese pak choi green, mei qing choi* etc.

Health Benefits of Bok Choy

- ☆ Bok choy is one of the popular leafy-vegetables very low in calories. Nonetheless, it is very rich source of many vital phyto-nutrients, vitamins, minerals and health-benefiting anti-oxidants.
- ☆ 100 g of bok choy contains just 13 calories. It is one of the recommended vegetables in the zero calorie or negative calorie category of foods which when eaten would add no extra weight to the body but in-turn facilitate calorie burns and reduction of weight.
- ☆ As in other Brassica family vegetables, bok choy too contains certain anti-oxidant plant chemicals like*thiocyanates, indole-3-carbinol, lutein, zeaxanthin, sulforaphane and isothiocyanates*. Along with dietary fiber, vitamins these compounds help to protect against breast, colon, and prostate cancers and help reduce LDL or "bad cholesterol" levels in the blood.

Bok Choy [(*Brassica campestris* (*Chinensis group*)], Nutrition Value per 100 g

Principle	*Nutrient Value*	*Percentage of RDA*
Energy	13 kcal	<1 per cent
Carbohydrates	2.18 g	1.5 per cent
Protein	1.5 g	3 per cent
Total Fat	0.20 g	1 per cent
Cholesterol	0 mg	0 per cent
Dietary Fiber	1 mg	2.5 per cent
Vitamins		
Folates	66 µg	16 per cent
Niacin	0.500 mg	3 per cent
Pantothenic acid	0.088 mg	1.5 per cent
Pyridoxine	0.194 mg	15 per cent
Riboflavin	0.070 mg	5 per cent
Thiamin	0.040 mg	3.5 per cent
Vitamin A	4468 IU	149 per cent
Vitamin C	45 mg	75 per cent
Vitamin K	45.5 µg	38 per cent
Electrolytes		
Sodium	65 mg	4 per cent
Potassium	252 mg	5 per cent
Minerals		
Calcium	105 mg	10.5 per cent
Iron	0.80 mg	10 per cent
Magnesium	19 mg	5 per cent
Manganese	0.159 mg	7 per cent
Phosphorus	37 mg	5 per cent
Zinc	0.19 mg	1.5 per cent
Phyto-nutrients		
Carotene-α	1 µg	—
Carotene-β	2681 µg	—
Lutein-zeaxanthin	40 µg	—

- ✫ Fresh pak choi is an excellent source of water-soluble antioxidant, vitamin-C (ascorbic acid). 100 g provides 45 mg or 75 per cent of daily requirements of vitamin C. Regular consumption of foods rich in vitamin C help the body develops resistance against infectious agents and scavenges harmful, pro-inflammatory free radicals.
- ✫ Bok-choy has more vitamin A, carotenes, and other flavonoid polyphenolic anti-oxidants than cabbage, cauliflower, etc. Just 100 g of fresh vegetable provides 4468 IU or 149 per cent of daily-required levels vitamin A.

- ☆ Pak choi is a very good source of vitamin K, provides about 38 per cent of RDA levels. Vitamin-K has a potential role in bone metabolism by promoting osteotrophic activity in bone cells. Therefore, enough vitamin K in the diet makes your bone stronger, healthier and delay osteoporosis. Further, vitamin-K also has established role in curing Alzheimer's disease patients by limiting neuronal damage in their brain.
- ☆ Fresh bok choy has many vital B-complex vitamins such as pyridoxine (vitamin B6), riboflavin, pantothenic acid (vitamin B5), pyridoxine, and thiamin (vitamin B-1). These vitamins are essential in the sense that our body requires them from external sources to replenish.
- ☆ Further, this leafy vegetable is a moderate source of minerals, particularly calcium, phosphorous, potassium, manganese, iron and magnesium. Potassium is an important electrolyte in the cell and body fluids that help regulate heart rate and blood pressure. Manganese is used by the body as a co-factor for the antioxidant enzyme, superoxide dismutase. Iron is required for the red blood cell formation.

55

Broccoli

Broccoli heads are rich source of phyto-nutrients that help protect from prostate cancer and stroke risks. It is actually a flower vegetable and known for its notable and unique nutrients that are found to have disease prevention and health promoting properties. Botanically, the vegetable is the member of large *cruciferous* (*Brassica*) family of vegetables, which also include cauliflower, brussel sprouts, cabbage, arugula, etc. Scientific name: *Brassica oleracea* var. italica.

Broccoli is a cool-season crop and demands fertile rich and well-drained soil to flourish. Technically; broccoli is categorized into two main types according to their appearance; heading and sprouting. Heading variety forms a large, solid head, whereas sprouting types forms many smaller heads or florets. Mature plant bears about 4-10 inches wide, dark green to purple colour flower-head depending on the cultivar type. Its central thick stalk measures about 6-10 inches in length. Both stalk and fleshy flower heads are edible. Many different hybrid-mix developed with other cruciferous family members such as *broccoflower* (hybrid of broccoli and cauliflower), *broccolini* (broccoli and Chinese kale) etc.

Health Benefits of Broccoli

- ☆ Broccoli is one of the very low calorie vegetables; provide just 34 calories per 100 g. Nevertheless, it is rich in dietary fiber, minerals, vitamins, and anti-oxidants that have proven health benefits. Total antioxidant strength measured in terms of *oxygen radical absorbance capacity* (ORAC) of broccoli is 1632 µmol TE/100 g.
- ☆ Fresh Broccoli is a storehouse of many phyto-nutrients such as *thiocyanates, indoles, sulforaphane, isothiocyanates* and *flavonoids like beta-carotene cryptoxanthin, lutein, and zeaxanthin*. Studies have shown that these compounds by modifying positive signaling at molecular receptor levels help protect from prostate, colon, urinary bladder, pancreatic, and breast cancers.

Broccoli (*Brassica oleracea* var. *italica*), Fresh, Raw, Nutrition Value per 100 g

Principle	*Nutrient Value*	*Percentage of RDA*
Energy	34 Kcal	1.5 per cent
Carbohydrates	6.64 g	5 per cent
Protein	2.82 g	5 per cent
Total Fat	0.37 g	1 per cent
Cholesterol	0 mg	0 per cent
Dietary Fiber	2.60 g	7 per cent
Vitamins		
Folates	63 μg	16 per cent
Niacin	0.639 mg	4 per cent
Pantothenic acid	0.573 mg	12 per cent
Pyridoxine	0.175 mg	13 per cent
Riboflavin	0.117 mg	9 per cent
Thiamin	0.071 mg	6 per cent
Vitamin A	623 IU	21 per cent
Vitamin C	89.2 mg	149 per cent
Vitamin E	0.17 mg	1.5 per cent
Vitamin K	101.6 μg	85 per cent
Electrolytes		
Sodium	33 mg	2 per cent
Potassium	316 mg	7 per cent
Minerals		
Calcium	47 mg	5 per cent
Copper	0.049 mg	5.5 per cent
Iron	0.73 mg	9 per cent
Magnesium	21 mg	5 per cent
Manganese	0.210 mg	9 per cent
Selenium	2.5 μg	5 per cent
Zinc	0.41 mg	4 per cent
Phyto-nutrients		
Carotene-β	361 μg	—
Crypto-xanthin-β	1 μg	—
Lutein-zeaxanthin	1403 μg	—

☆ Fresh vegetable is exceptionally rich source of vitamin-C. Provides 89.2 mg or about 150 per cent of RDA per 100 g. Vitamin-C is a powerful natural anti-oxidant and immune modulator, helps fight against flu causing viruses.

- ☆ Further, it contains very good amounts of another anti-oxidant vitamin, vitamin-A. 100 g fresh head provides 623 IU or 21 per cent of recommended daily levels. Together with other pro-vitamins like beta-carotene, alpha-carotene, and zeaxanthin, vitamin A helps maintain integrity of skin and mucus membranes. Vitamin A is essential for healthy eye-sight and helps prevent from macular degeneration of the retina in the elderly population.
- ☆ Broccoli leaves (green tops) are an excellent source of carotenoids and vitamin A; (provide 16000 IU of vitamin A per 100 g) contain these compounds several times greater than that in the flower-head.
- ☆ Fresh broccoli heads are an excellent source of folates; contain about 63 µg/100 g (Provides 16 per cent of RDA). Studies have shown that consumption of fresh vegetables and fruits rich in folates during pre-conception, and pregnancy helps prevent neural tube defects in the offspring.
- ☆ This flower vegetable is rich source of vitamin-K; and B-complex group of vitamins like niacin (vit B-3), pantothenic acid (vit.B-5), pyridoxine (vit.B-6), and riboflavin. The flower heads also have some amount of omega-3 fatty acids.
- ☆ It is also a good source of minerals like calcium, manganese, iron, magnesium, selenium, zinc and phosphorus.

56

Brussel Sprouts

Brussel sprouts are small leafy green buds resemble like miniature cabbages in appearance. The buds are exceptionally rich in protein, dietary fiber, vitamins, minerals, and antioxidants, which work wonders to get rid of many health troubles. In fact, a renewed interest is growing about health benefits these sprouts have to offer. Botanically, the sprouts belong to the same Brassica family, which also includes cabbage, collard greens,broccoli, and kale. Scientific name: *Brassica oleracea* (Gemmifera Group). Brussels sprouts are winter crops, flourishes well in cool weather and light frost conditions. Well-grown plant reaches about 90 cm in height. The sprouts are produced all along the stalk, starting from the base and moving upward. Each sprout, in general, features similarity in appearance and structure to cabbage, but only very small, measuring about 1-1.5 inches in diameter. In structure, each head consists of clusters of stiff leaves superimposed in compact layers giving it a round or globular shape as in cabbages. In order to get uniform sprouts, the tip of the stalk is cut as soon as sprouts at the bottom start to develop. In addition, sprouts exposed to hot weather do not form into compact buds. Sprouts are one of the most popular vegetables in the United States and Mediterranean Europe.

Health Benefits of Brussel Sprouts

- The sprouts are one of the low-glycemic nutritious vegetables that should be considered in weight reduction programs. 100 g brussel sprouts provide just 45 calories, nonetheless, contain 3.38 g of protein, 3.80 g of dietary fiber (10 per cent of RDA) and zero cholesterol.
- In fact, Brussels sprouts are a storehouse of several flavonoid anti-oxidants like *thiocyanates, indoles, lutein, zeaxanthin, sulforaphane* and *isothiocyanates*. Together, these phytochemicals offer protection from prostate, colon, and endometrial cancers.

Brussel Sprouts (*Brassica oleracea* Gemmifera Group), Fresh, Nutrition Value per 100 g

Principle	*Nutrient Value*	*Percentage of RDA*
Energy	43 Kcal	2 per cent
Carbohydrates	8.95 g	7 per cent
Protein	3.38 g	6 per cent
Total Fat	0.30 g	1 per cent
Cholesterol	0 mg	0 per cent
Dietary Fiber	3.80 g	10 per cent
Vitamins		
Folates	61 µg	15 per cent
Niacin	0.745 mg	4.5 per cent
Pantothenic acid	0.309 mg	6 per cent
Pyridoxine	0.219 mg	17 per cent
Riboflavin	0.90 mg	7 per cent
Thiamin	0.139 mg	13 per cent
Vitamin A	754 IU	25 per cent
Vitamin C	85 mg	142 per cent
Vitamin K	177 µg	147 per cent
Electrolytes		
Sodium	25 mg	1.5 per cent
Potassium	389 mg	8 per cent
Minerals		
Calcium	42 mg	4 per cent
Copper	0.70 mg	8 per cent
Iron	1.40 mg	17.5 per cent
Magnesium	23 mg	6 per cent
Manganese	0.337 mg	15 per cent
Phosphorus	69 mg	10 per cent
Selenium	1.6 µg	3 per cent
Zinc	0.42 mg	4 per cent
Phyto-nutrients		
Carotene-α	6 µg	—
Carotene-β	450 µg	—
Crypto-xanthin-β	0 µg	—
Lutein-zeaxanthin	1590 µg	—

- ☆ Di-indolyl-methane (DIM), a metabolite of *indole-3-carbinol* is found to be an effective immune modulator, anti-bacterial and anti-viral agent through its action of potentiating "Interferon-γ" receptors.

- ☆ In addition, brussel sprouts contain glucoside, sinigrin. Early laboratory studies suggest that *sinigrin* help protect from colon cancers by destroying pre-cancerous cells.
- ☆ Brussel sprouts are an excellent source of vitamin C; 100 g sprouts provide about 85 mg or 142 per cent of RDA. Together with other antioxidant vitamins such as vitamin A and E, it helps protect the body by trapping harmful free radicals.
- ☆ Zeaxanthin, an important dietary carotenoid in sprouts, is selectively absorbed into the retinal macula-lutea in the eyes where it is thought to provide anti-oxidant and protective light-filtering functions from UV rays. Thus, it helps prevent retinal damage, "age-related macular degeneration related macular degeneration disease" (ARMD), in the elderly.
- ☆ Sprouts are the good source of another anti-oxidant vitamin A, provides about 754 IU per 100g. Vitamin A is required for maintaining healthy mucus membranes and skin and is essential for acuity of vision. Foods rich in this vitamin have been found to offer protection against lung and oral cavity cancers.
- ☆ It is one of the excellent vegetable sources for vitamin-K; 100 g provides about 177 µg or about 147 per cent of RDA. Vitamin K has potential role bone health by promoting osteotrophic (bone formation and strengthening) activity. Adequate vitamin-K levels in the diet help limiting neuronal damage in the brain and thereby, preventing or at least, delay the onset of Alzheimer's disease.
- ☆ Further, the sprouts are notably good in many B-complex groups of vitamins such as niacin, vitamin B-6 (pyridoxine), thiamin, pantothenic acid, etc., that are essential for substrate metabolism inside the human body.
- ☆ They are also rich source of minerals like copper, calcium, potassium, iron, manganese, and phosphorus. 100 g fresh sprouts provide 25 mg (1.5 per cent of RDA) sodium and 389 mg (8 per cent of RDA) potassium. Potassium is an important component of cell and body fluids that helps controlling heart rate and blood pressure by countering effects of sodium. Manganese is used by the body as a co-factor for the antioxidant enzyme, *superoxide dismutase*. Iron is required for cellular oxidation and red blood cell formation.

Brussels sprouts are incredibly nutritious vegetable that offers protection from vitamin A deficiency, bone loss, iron-deficiency anemia, and believed to protect from cardiovascular diseases and colon and prostate cancers.

57

Butternut Squash

Butternut squash is the most popular among winter squash varieties. Oftentimes, the squash is recognized as a large pear shaped golden-yellow pumpkin fruit, which is put for sale in the markets. Butternuts are annual long trailing vines. They usually cultivated in warmer climates of South and Central American regions for their edible fruits, flowers, as well as seeds. Botanically, the vegetable belongs to the *Cucurbitaceae* family of field pumpkins; probably originated in the Central American region. Scientific name: *Cucurbita morschata*.

The butternut plant is monoecious as in pumpkins, and feature different male and female flowers that require honeybees for effective pollination. Butternut, in-fact, is the most common among winter squash fruits. Externally, the squash is better described as large-sized fruit featuring thick neck attached to a pear-shaped bottom and has smooth, ribbed skin. However, the fruit varies widely in its shape and size; with individual fruit may weigh up to 15 kg. Interiorly, its flesh is yellow to orange in colour. Cross-section of lower bulb part feature central hollow cavity containing mesh-like mucilaginous fibers interspersed with large, flat, elliptical seeds similar to Pepita (pumpkin seeds). The fruit's unique golden-yellow colour comes from yellow-orange pigments in their skin and pulp. Butternut squash seeds are used as nutritious snack food since they contain 35-40 per cent oil and 30 per cent protein. In Argentina, the fruit is also used to feed livestock.

Health Benefits of Butternut Squash

☆ Butternut squash contains many vital poly-phenolic anti-oxidants and vitamins. Similar to other Cucurbitaceae members, this too is one of the low-calorie vegetables, which provides just 45 calories per 100 g. It contains no saturated fats or cholesterol; however, is rich source of dietary fiber and phyto-nutrients. Squash is one of the common vegetables that often recommended by dieticians in the cholesterol controlling and weight reduction programs.

Butternut Squash (*Cucurbita moschata*), Fresh, Nutritive Value per 100 g

Principle	*Nutrient Value*	*Percentage of RDA*
Energy	45 Kcal	2 per cent
Carbohydrates	11.69 g	9 per cent
Protein	1.0 g	2 per cent
Total Fat	0.1 g	0.5 per cent
Cholesterol	0 mg	0 per cent
Dietary Fiber	2 g	5 per cent
Vitamins		
Folates	27 µg	7 per cent
Niacin	1.200 mg	8 per cent
Pantothenic acid	0.400 mg	8 per cent
Pyridoxine	0.154 mg	12 per cent
Riboflavin	0.020 mg	2 per cent
Thiamin	0.100 mg	8 per cent
Vitamin A	10630 IU	354 per cent
Vitamin C	21 mg	35 per cent
Vitamin E	1.44 mg	10 per cent
Vitamin K	1.1 µg	1 per cent
Electrolytes		
Sodium	4 mg	0.5 per cent
Potassium	352 mg	7 per cent
Minerals		
Calcium	48 mg	5 per cent
Copper	0.072 mg	8 per cent
Iron	0.70 mg	9 per cent
Magnesium	34 mg	9 per cent
Manganese	0.202 mg	1 per cent
Phosphorus	33 mg	5 per cent
Selenium	0.5 µg	<1 per cent
Zinc	0.15 mg	1 per cent
Phyto-nutrients		
Carotene-α	834 µg	—
Carotene-β	4226 µg	—
Crypto-xanthin-β	3471 µg	—
Lutein-zeaxanthin	0 µg	—

✰ It has more vitamin A than that in pumpkin. At 10630 IU per 100 g, it is perhaps the single vegetable source in the Cucurbitaceae family with the highest levels of vitamin-A, providing about 354 per cent of RDA. Vitamin A is a powerful natural anti-oxidant and is required by the body for maintaining the integrity of skin and mucus membranes. It is also an essential vitamin for good eye-sight. Research studies suggest that natural foods rich in vitamin A help the body protected against lung and oral cavity cancers.

- Furthermore, butternut squash has plenty of natural poly-phenolic flavonoid compounds like α *and* β*-carotenes, cryptoxanthin-*β*, and lutein*. These compounds convert to vitamin A inside the body and deliver same protective functions of vitamin A on the body.
- It is rich in B-complex group of vitamins like folates, riboflavin, niacin, vitamin B-6 (pyridoxine), thiamin, and pantothenic acid.
- It has similar mineral profile as that in pumpkin, containing adequate levels of minerals like iron, zinc, copper, calcium, potassium, and phosphorus.
- Butternut squash seeds are a good source of dietary fiber and mono-unsaturated fatty acids that benefit for heart health. In addition, they are rich in protein, minerals, and numerous health-benefiting vitamins. The seeds are an excellent source of health promoting amino acid, tryptophan. Tryptophan converts to health benefiting GABA neuro-chemical in the brain.

58

Cabbage

Rich in phyto-nutrients, this cool season leafy vegetable belongs to the "Brassica" family of vegetables, which also include brussels sprouts, cauliflower, bok choy, kale, and broccoli. It is one of the widely cultivated crops around the world. Scientific name: *Brassica oleracea* (capitata group).

Cabbage structurally consists of clusters of stiff leaves superimposed one over the other in compact layers, giving it a round or globular shape. Several varieties are cultivated worldwide including green, purple, red, and savoy (loose-wrinkled leaves). Bok-choy or "Chinese-cabbage" features similar appearance, but is derived from different species of the same Brassica genus vegetables. Bok-choy characteristically has a vigorous growth pattern. The whole plant has a long cylindrical shape, comprising of short, compact leaves. Napa cabbage is another Chinese vegetable variety in the Brassica family. The plant grows to oblong shaped head consisting of tightly arranged crinkly, thick, light-green colour leaves with prominent white veins.

Health Benefits of Cabbage

- ☆ Fresh, dark green-leafy cabbage is incredibly nutritious; however, very low in fat and calories. 100 g of leaves provide just 25 calories.
- ☆ The vegetable is the storehouse of phyto-chemicals like *thiocyanates, indole-3-carbinol, lutein, zeaxanthin, sulforaphane, and isothiocyanates.* These compounds are powerful antioxidants and known to help protect against breast, colon, and prostate cancers and help reduce LDL or "bad cholesterol" levels in the blood.
- ☆ Fresh cabbage is an excellent source of natural antioxidant, vitamin C. Provides 36.6 mg or about 61 per cent of RDA per 100 g. Regular consumption of foods rich in vitamin C helps the body develop resistance against infectious agents and scavenge harmful, pro-inflammatory free radicals.

Cabbage (*Brassica oleracea*, Capitata Group), Nutrition Value per 100 g

Principle	*Nutrient Value*	*Percentage of RDA*
Energy	25 kcal	1 per cent
Carbohydrates	5.8 g	4 per cent
Protein	1.3 g	2 per cent
Total Fat	0.1 g	0.5 per cent
Cholesterol	0 mg	0 per cent
Dietary Fiber	2.50 mg	6 per cent
Vitamins		
Folates	53 µg	13 per cent
Niacin	0.234 mg	1.5 per cent
Pantothenic acid	0.212 mg	4 per cent
Pyridoxine	0.124 mg	10 per cent
Riboflavin	0.040 mg	3 per cent
Thiamin	0.061 mg	5 per cent
Vitamin A	98 IU	3 per cent
Vitamin C	36.6 mg	61 per cent
Vitamin K	76 µg	63 per cent
Electrolytes		
Sodium	18 mg	1 per cent
Potassium	170 mg	3.5 per cent
Minerals		
Calcium	40 mg	4 per cent
Iron	0.47 mg	6 per cent
Magnesium	12 mg	3 per cent
Manganese	0.160 mg	7 per cent
Phosphorus	26 mg	3.5 per cent
Zinc	0.18 mg	1.5 per cent
Phyto-nutrients		
Carotene-α	33 µg	—
Carotene-β	42 µg	—
Lutein-zeaxanthin	30 µg	—

- ☆ Total antioxidant strength measured in terms of oxygen radical absorbance capacity (ORAC value) is 508 µmol TE/100 g. Red cabbages contain more antioxidant value, 2252 µmol TE/100 g.
- ☆ It is also rich in essential vitamins such as pantothenic acid (vitamin B-5), pyridoxine (vitamin B-6) and thiamin (vitamin B-1). These vitamins are essential in the sense that our body requires them from external sources to replenish.

- ☆ It also contains a adequate amount of minerals like potassium, manganese, iron, and magnesium. Potassium is an important component of cell and body fluids that helps controlling heart rate and blood pressure. Manganese is used by the body as a co-factor for the antioxidant enzyme, *superoxide dismutase.* Iron is required for the red blood cell formation.
- ☆ Cabbage is a very good source of vitamin K, provides about 63 per cent of RDA levels. Vitamin-K has the potential role in bone metabolism by promoting osteotrophic activity in them. So enough vitamin K in the diet gives you healthy bones. In addition, vitamin-K also has established role in curing Alzheimer's disease patients by limiting neuronal damage in their brain.

59

Carrots

Naturally sweet, delicious and crunchy, carrots are healthy additions you can make to the vegetable list in your diet. Indeed, these root vegetables come with wholesome health benefiting compounds such as beta-carotenes, vitamin A, minerals and anti-oxidants in ample amounts. Botanically these taproots belong to the *Apiaceae* or umbelliferous family of the genus; *Daucus* and known scientifically as *Daucus carota*. The other close Apiaceae members include parsnips, parsley, dill, cumin, etc.

Carrot plant is cultivated across the world for its prized taproot. The plant is biennial and bears flowers during second year of life. However, in general, the whole plant is harvested prematurely when the root reaches about an inch in diameter, tender and juicy. Carrots vary widely in colour and shape depending on the cultivar types. Generally, oriental taproots are long, flat upper ends with tapering, tail like, lower ends. They are winter season crops in many parts of Asia. European carrots, on the other hand, have more rounded ends with almost cylindrical body. In addition, European-variety feature bright orange colour in contrast to saffron coloured Asian cultivars.

Health Benefits of Carrots

- ☆ Sweet and succulent carrots are notably rich in anti-oxidants, vitamins and dietary fiber; however, they provide only 41 calories per 100 g, negligible amount of fat and no cholesterol.
- ☆ They are exceptionally rich source of carotenes and vitamin-A. 100 g fresh carrot contains 8285 µg of beta-carotene and 16706 IU of vitamin A. Studies have found that flavonoid compounds in carrots help protect from skin, lung and oral cavity cancers.
- ☆ Carotenes are converted into vitamin A in the liver. Beta-carotene is the major carotene that is present in these roots. Beta carotene is one of the powerful natural anti-oxidant helps protect body from harmful oxygen-

free radical injury. In addition, it also has all the functions of vitamin A such as vision, reproduction (sperm production), maintenance of epithelial integrity, growth and development.

Carrots (*Daucus carota*), Fresh, Raw, Nutrition Value per 100 g. Total-ORAC Value 666 umol TE/100 g

Principle	*Nutrient Value*	*Percentage of RDA*
Energy	41 Kcal	2 per cent
Carbohydrates	9.58 g	7 per cent
Protein	0.93 g	1.5 per cent
Total Fat	0.24 g	1 per cent
Cholesterol	0 mg	0 per cent
Dietary Fiber	2.8 g	7 per cent
Vitamins		
Folates	19 µg	5 per cent
Niacin	0.983 mg	6 per cent
Pantothenic acid	0.273 mg	5.5 per cent
Pyridoxine	0.138 mg	10 per cent
Riboflavin	0.058 mg	4 per cent
Thiamin	0.066 mg	6 per cent
Vitamin A	16706 IU	557 per cent
Vitamin C	5.9 mg	10 per cent
Vitamin K	13.2 µg	11 per cent
Electrolytes		
Sodium	69 mg	4.5 per cent
Potassium	320 mg	6.5 per cent
Minerals		
Calcium	33 mg	3 per cent
Copper	0.045 mg	5 per cent
Iron	0.30 mg	4 per cent
Magnesium	12 mg	3 per cent
Manganese	0.143 mg	6 per cent
Phosphorus	35 mg	5 per cent
Selenium	0.1 µg	<1 per cent
Zinc	0.24 mg	2 per cent
Phyto-nutrients		
Carotene-α	3427 µg	—
Carotene-β	8285 µg	—
Crypto-xanthin-β	0 µg	—
Lutein-zeaxanthin	256 µg	—

- Carrots are rich in poly-acetylene antioxidant falcarinol. Research study conducted by scientists at University of Newcastle on laboratory animals has found that *falcarinol* in carrots may help fight against cancers by destroying pre-cancerous cells in the tumors.
- Fresh roots are also good in vitamin C; provide about 9 per cent of RDA. Vitamin C is water soluble anti-oxidant. It helps the body maintain healthy connective tissue, teeth and gum. Its anti-oxidant property helps the body protect from diseases and cancers by scavenging harmful free radicals.
- In addition, this root vegetable is especially rich in many B-complex groups of vitamins such as folic acid, vitamin B-6 (pyridoxine), thiamin, pantothenic acid, etc., that acts as co-factors to enzymes during substrate metabolism in the body.
- Further, They also compose healthy levels of minerals like copper, calcium, potassium, manganese and phosphorus. Potassium is an important component of cell and body fluids that helps controlling heart rate and blood pressure by countering effects of sodium. Manganese is used by the body as a co-factor for the antioxidant enzyme, *superoxide dismutase*.

60

Cassava

Cassava or yuca is a nutty flavoured, starchy tuber of the spurge family (Euphorbiaceae) from the South-American origin. Its sweet crunchy underground tuber is a popular edible root since centuries in many parts of Africa, Asia and South American indigenous people. Together with other tropical roots and starch-rich foods like yam, taro, plantains, potato, etc., it too is an indispensable part of carbohydrate diet of millions of inhabitants living in these regions. Some of the common names include manioc, or mandioca in Brazil, manihot, tapioca and yuca. Scientific name: *Manihot esculenta* (Crantz). The cassava is a perennial plant that grows best under tropical, moist, fertile, well-drained soils. Completely grown plant reaches to a height of about 2-4 m. Under the cultivation fields; cut-stem sections are planted just as in sugarcanes. After about 8-10 months of plantation, long, globular roots or tubers radiate from the stem just underneath the soil surface up to a depth of 2-4 feet. Each tuber weighs one to several pounds depending up on the cultivar type and feature gray-brown, rough woody textured skin. Its flesh features white colour starch rich sweet-flavoured meat that should be eaten only after cooking.

Health Benefits of Cassava

- ☆ Cassava has nearly twice the calories than potatoes, perhaps highest for any tropical starch rich tubers and roots. 100 g root provides 160 calories. Their calorie mainly comes from sucrose forming the bulk of the sugars in tubers, accounting for more than 69 per cent of the total sugars. Complex sugar amylose is another major carbohydrate source (16-17 per cent).
- ☆ Cassava is very low in fats and protein than in cereals and pulses. Nonetheless, it has more protein than that of other tropical food sources like yam, potato, plantains, etc.
- ☆ As in other roots and tubers, cassava too is free from gluten. Gluten-free starch is used in special food preparations for celiac disease patients.

Cassava Root (*Manihot esculenta* (L.) Crantz), Raw, Nutrition Value per 100 g

Principle	*Nutrient Value*	*Percentage of RDA*
Energy	160 Kcal	8 per cent
Carbohydrates	38.06 g	29 per cent
Protein	1.36 g	2.5 per cent
Total Fat	0.28 g	1 per cent
Cholesterol	0 mg	0 per cent
Dietary Fiber	1.8 g	4 per cent
Vitamins		
Folates	27 µg	7 per cent
Niacin	0.854 mg	5 per cent
Pyridoxine	0.088 mg	7 per cent
Riboflavin	0.048 mg	4 per cent
Thiamin	0.087 mg	7 per cent
Vitamin A	13 IU	<1 per cent
Vitamin C	20.6 mg	34 per cent
Vitamin E	0.19 mg	1 per cent
Vitamin K	1.9 µg	1.5 per cent
Electrolytes		
Sodium	14 mg	1 per cent
Potassium	271 mg	6 per cent
Minerals		
Calcium	16 mg	1.6 per cent
Iron	0.27 mg	3 per cent
Magnesium	21 mg	5 per cent
Manganese	0.383 mg	1.5 per cent
Phosphorus	27 µg	4 per cent
Zinc	0.34 mg	3 per cent

- ✰ Young tender cassava (yuca) leaves are a good source of dietary proteins and vitamin K. Vitamin-K has a potential role in bone mass building by promoting osteotrophic activity in the bones. It also has established role in the treatment of Alzheimer's disease patients by limiting neuronal damage in the brain.
- ✰ Cassava is a moderate source of some of the valuable B-complex group of vitamins such as folates, thiamin, pyridoxine (vitamin B-6), riboflavin, and pantothenic acid.
- ✰ The root is the chief source of some important minerals like zinc, magnesium, copper, iron, and manganese for many inhabitants in the tropical belts. In addition, it has adequate amounts of potassium (271 mg per 100g or 6 per cent of RDA). Potassium is an important component of cell and body fluids that help regulate heart rate and blood pressure.

61

Cauliflower

Packed with rich nutrients, cauliflower or *cabbage flower* is one of the commonly used flower-vegetable. The flower heads contain numerous health benefiting phtyo-nutrients such as indole-3-carbinol, sulforaphane etc., that help prevent prostate, ovarian and cervical cancers. Botanically, it is a member of the cruciferous or brassicaceae family. It has got a similar nutritional and phyto-chemistry profile with broccoli and cabbage. Several cultivars exist other than common snow-white variety, including green, orange, purple, and romanesco heads. Like broccoli, cauliflower too is made up of tightly clustered florets that are begun to form but stopped at bud stage. This cool-season vegetable prefers fertile rich adequate moisture in the soil to flourish. To keep the flower heads creamy white, they should be protected from sunlight. This is done by tying the close-by leaves together over the heads when the heads are the size of a quarter. Over-maturity makes the heads get loose and grainy surfaced, and lose much of their flavour and tenderness.

Health Benefits of Cauliflower

- ☆ It is very low in calories. 100 g of the fresh cauliflower head provides only 26 calories. Nevertheless, it comprises of several health-benefiting antioxidants and vitamins in addition to be very low in fat and contains no cholesterol.
- ☆ Its florets contain about 2 g of dietary fiber per 100 g; providing about 5 per cent of recommended value.
- ☆ Cauliflower contains several anti-cancer phyto-chemicals like sulforaphane and plant sterols such as indole-3-carbinol, which appears to function as an anti-estrogen agent. Together these compounds have proven benefits against prostate, breast, cervical, colon, ovarian cancers by virtue of their cancer-cell growth inhibition, cytotoxic effects on cancer cells.

Cauliflower (*Brassica oleracea* (Botrytis Group), Raw, Nutritive Value per 100 g

Principle	*Nutrient Value*	*Percentage of RDA*
Energy	25 Kcal	1 per cent
Carbohydrates	4.97 g	4 per cent
Protein	1.92 g	4 per cent
Total Fat	0.28 g	1 per cent
Cholesterol	0 mg	0 per cent
Dietary Fiber	2.0 g	5 per cent
Vitamins		
Folates	57 μg	14 per cent
Niacin	0.507 mg	3 per cent
Pantothenic acid	0.667 mg	13 per cent
Pyridoxine	0.184 mg	14 per cent
Riboflavin	0.060 mg	4.5 per cent
Thiamin	0.050 mg	4 per cent
Vitamin A	0 IU	0 per cent
Vitamin C	48.2 mg	80 per cent
Vitamin E	0.08 mg	0.5 per cent
Vitamin K	15.5 μg	13 per cent
Electrolytes		
Sodium	30 mg	2 per cent
Potassium	299 mg	6 per cent
Minerals		
Calcium	22 mg	2 per cent
Copper	0.039 mg	4.5 per cent
Iron	0.42 mg	5 per cent
Magnesium	15 mg	3.5 per cent
Manganese	0.155 mg	7 per cent
Zinc	0.27 mg	2.5 per cent
Phyto-nutrients		
Carotene-β	0 μg	—
Lutein-zeaxanthin	1 μg	—

☆ Furthermore, Di-indolyl-methane (DIM), a lipid soluble compound present abundantly in Brassica group of vegetables has found effective as immune modulator, anti-bacterial and anti-viral compound by synthesis and potentiating Interferon-Gamma receptors. DIM has currently been found application in the treatment of recurring respiratory papillomatosis caused by the Human Papilloma Virus (HPV) and is in Phase III clinical trials for cervical dysplasia.

- Fresh cauliflower is an excellent source of vitamin C; 100 g provides about 48.2 mg or 80 per cent of daily recommended value. Vitamin-C is a proven antioxidant helps fight against harmful free radicals, boosts immunity and prevents from infections and cancers.
- It contains good amounts of many vital B-complex groups of vitamins such as folates, pantothenic acid (vitamin B5), pyridoxine (vitamin B6) and thiamin (vitamin B1), niacin (B3) as well as vitamin K. These vitamins are essential in the sense that body requires them from external sources to replenish and required for fat, protein and carbohydrate metabolism.
- Further, It is an also good source of minerals such as manganese, copper, iron, calcium and potassium. Manganese is used in the body as a co-factor for the antioxidant enzyme; *superoxide dismutase.Potassium* is an important intracellular electrolyte helps counter the hypertension effects of sodium.

62

Collard Greens

Collard greens are highly nutritious staple green "cabbage-like leaves" vegetable. Collards are one of the most popular members of the *Brassica* family, closely related to kale and cabbage and could be described as a non-heading (acephalous) cabbage. Botanically, the greens are named as *Brassica oleracea* L. (Acephala group).

Collards are probably originated in the eastern European or Asia Minor region and now days grown almost all parts of the cooler temperate regions. The plant grows up to 3-4 feet in height and bears dark-green leaves arranged in a rosette fashion around an upright, stocky main stem.

Several cultivar types of collard greens are grown around the planet depending on the soil type, climate, etc.

1. Blue Max: It has very attractive savoy- like blue-green leaves.
2. Georgia: It is also known as Georgia LS or Georgia Southern. It has blue-green and slightly savoy-leaves.
3. Vates: Plant is compact and leaves are smooth and dark green.
4. Champion: Low growing plant, featuring smooth, dark-green leaves with short internodes.
5. Flash: It is a very uniform Vates type with smooth, dark-green leaves.
6. Heavy-Crop: It has very large, slightly savoy like, blue-green leaves. Leaves have close internodes spacing so bunching can be more difficult.

Health Benefits of Collard Greens

- ☆ Wonderfully nutritious collard leaves are very low in calories (provide only 30 calories per 100 g) and contain no cholesterol. However, its green leaves contain a very good amount of soluble and insoluble dietary fiber that helps control LDL cholesterol levels and offer protection against hemorrhoids, constipation as well as colon cancer diseases.

☆ Widely considered to be wholesome foods, collards are rich in invaluable sources of phyto-nutrients with potent anti-cancer properties, such as di-indolyl-methane (DIM) and sulforaphane that have proven benefits against *prostate, breast, cervical, colon, ovarian* cancers by virtue of their cancer-cell growth inhibition and cytotoxic effects on cancer cells.

Collard Greens (*Brassica oleracea* L. Acephala Group), Fresh, Raw, Nutrition Value per 100 g

Principle	*Nutrient Value*	*Percentage of RDA*
Energy	30 Kcal	1.5 per cent
Carbohydrates	5.69 g	4 per cent
Protein	2.45 g	4 per cent
Total Fat	0.42 g	1.5 per cent
Cholesterol	0 mg	0 per cent
Dietary Fiber	3.60 g	9 per cent
Vitamins		
Folates	166 µg	41.5 per cent
Niacin	0.742 mg	5 per cent
Pantothenic acid	0.267 mg	5 per cent
Pyridoxine	0.165 mg	13 per cent
Riboflavin	0.130 mg	10 per cent
Thiamin	0.054 mg	4.5 per cent
Vitamin A	6668 IU	222 per cent
Vitamin C	35.3 mg	59 per cent
Vitamin E	2.26 mg	15 per cent
Vitamin K	510.8 µg	426 per cent
Electrolytes		
Sodium	20 mg	1 per cent
Potassium	169 mg	3.5 per cent
Minerals		
Calcium	145 mg	14.5 per cent
Copper	0.039 mg	4.5 per cent
Iron	0.19 mg	2.5 per cent
Magnesium	9 mg	2 per cent
Manganese	0.276 mg	12 per cent
Selenium	1.3 µg	2 per cent
Zinc	0.13 mg	1 per cent
Phyto-nutrients		
Carotene-β	3842 µg	—
Crypto-xanthin-β	80 µg	—
Lutein-zeaxanthin	8932 µg	—

- Di-indolyl-methane has also found to be effective immune modulator, anti-bacterial and anti-viral properties by potentiating Interferon-gamma receptors.
- The leaves are also an excellent source of folates, provides about 166 µg *or* 41.5 per cent of RDA. Folates are important in DNA synthesis and when given during the peri-conception period can prevent neural tube defects in the baby.
- Fresh collard leaves are also rich in vitamin-C, provides about 59 per cent of RDA per 100 g. Vitamin-C is a powerful natural anti-oxidant that offers protection against free radical injury and flu-like viral infections.
- Collard greens are an excellent source of vitamin-A (222 per cent of RDA per 100 g) and carotenoid anti-oxidants such as *lutein, carotenes, zeaxanthin, crypto-xanthin,* etc. These compounds are scientifically found to have antioxidant properties. Vitamin A also required maintaining healthy mucus membranes and skin and is also essential for healthy vision. Consumption of natural fruits rich in flavonoids helps to protect from lung and oral cavity cancers.
- This leafy vegetable contains amazingly high levels of vitamin-K, provides staggering 426 per cent of recommended daily levels per 100 leaves. Vitamin K has a potential role in the increase of bone mass by promoting osteotrophic activity in the bone. It also has the beneficial effect in Alzheimer's disease patients by limiting neuronal damage in their brain.
- Collards are rich in many vital B-complex groups of minerals such as niacin (vitamin B-3), pantothenic acid (vitamin B-5), pyridoxine (vitamin B-6) and riboflavin.
- Further, the leaves and stems are good in minerals like iron, calcium, copper, manganese, selenium and zinc.

63

Cucumber

Ever wonder how to beat the scorching summer heat. Just remember your humble crunchy cucumber. Nonetheless, this wonderful low calorie vegetable indeed has more nutrients to offer than just water and electrolytes. It is one of the oldest cultivated crops and believed to be originating in the northern plains of Indian subcontinent. The plant is a creeper (vine) like other members of Cucurbita family, for example, gourds, squashes, melons, zucchini, etc. Botanically; it belongs to the *Cucurbitaceae* family; and is known scientifically as *Cucumis sativus.* Cucumber is easy to grow. Varieties, varying in size, shape, and colour, are cultivated all around the world. In general, the fruit features dark-green skin, crispy moisture rich flesh, and small edible seeds concentrated at its core. As in other squash members, cucumbers too are best-harvested young, tender and just short of reaching maturity, at the stage when they taste sweet, have crunchy texture, and unique flavour. If left uninterrupted, the fruit continues to grow in size, its skin becomes tougher and turns yellow, and seeds become hard and inedible. Fresh cucumbers are available throughout the season and can be eaten raw as is or in vegetable salads or juicing.

Armenian cucumbers (*Cucumis melo var. flexuosus*) are long, crispy, and thin-ribbed, curved, and have light green colour. Although grouped botanically in the melon family, they appear and taste just like cucumbers.

Miniature varieties such as *gherkins,* American-*dills,* and French-*cornichons* are very small indeed and usually preferred in pickling.

Dosakayi is a yellow Indian curry cucumber, has sweet taste and neutral flavour. It is used extensively in the preparation of stews and curries, particularly during the summer season in southern parts of India and Sri Lanka.

Health Benefits of Cucumber

☆ It is one of the very low calorie vegetables; provide just 15 calories per 100 g. It contains no saturated fats or cholesterol. Cucumber peel is a good source of dietary fiber that helps reduce constipation, and offer some protection against colon cancers by eliminating toxic compounds from the gut.

Cucumber (*Cucumis sativus*), Raw, Nutritive Value per 100 g

Principle	*Nutrient Value*	*Percentage of RDA*
Energy	15 Kcal	<1 per cent
Carbohydrates	3.63 g	3 per cent
Protein	0.65 g	1 per cent
Total Fat	0.11 g	0.5 per cent
Cholesterol	0 mg	0 per cent
Dietary Fiber	0.5 g	1 per cent
Vitamins		
Folates	7 µg	2 per cent
Niacin	0.098 mg	<1 per cent
Pantothenic acid	0.259 mg	5 per cent
Pyridoxine	0.040 mg	3 per cent
Riboflavin	0.033 mg	3 per cent
Thiamin	0.027 mg	2 per cent
Vitamin A	105 IU	3.5 per cent
Vitamin C	2.8 mg	4.5 per cent
Vitamin E	0.03 mg	0 per cent
Vitamin K	16.4 µg	13.6 per cent
Electrolytes		
Sodium	2 mg	0 per cent
Potassium	147 mg	3 per cent
Minerals		
Calcium	16 mg	1.6 per cent
Iron	0.28 mg	3.5 per cent
Magnesium	13 mg	3 per cent
Manganese	0.079 mg	3.5 per cent
Phosphorus	24 mg	3 per cent
Phosphorus		
Zinc	0.20 mg	2 per cent
Phyto-nutrients		
Carotene-β	45 µg	—
Crypto-xanthin-β	26 µg	—
Lutein-zeaxanthin	23 µg	—

- It is a very good source of potassium, an important intracellular electrolyte. 100 g of cucumber provides 147 mg of potassium but only 2 mg of sodium. Potassium is a heart friendly electrolyte helps bring a reduction in total blood pressure and heart rates by countering effects of sodium.
- Cucumbers contain unique anti-oxidants in moderate ratios such as β-carotene and α-carotene, vitamin-C, vitamin-A, *zeaxanthin* and *lutein*. These compounds help act as protective scavengers against oxygen-derived free radicals and reactive oxygen species (ROS) that play a role in ageing and various disease processes. Their total antioxidant strength, measured in terms of *oxygen radical absorbance capacity* (ORAC value), is 214 µmol TE/ 100 g.
- Cucumbers have mild diuretic property, which perhaps attributed to their free-water, and potassium and low sodium content. This helps in checking weight gain and high blood pressure.
- They surprisingly have a high amount of vitamin K, provides about 17 µg of this vitamin per 100 g. Vitamin-K has been found to have a potential role in bone strength by promoting osteotrophic (bone mass building) activity. It also has established role in the treatment of Alzheimer's disease patients by limiting neuronal damage in their brain.

64

Eggplant/Brinjal

Eggplant or Brinjal, is a very low calorie vegetable and has healthy nutrition profile; good news for weight watchers. The veggie is popularly known as aubergine in the western world. Botanically, it belongs to *Solanaceae* family and named as *Solanum melongena*. This perennial plant is native to Indian subcontinent and now grown in many tropical and semitropical regions. Several varieties of aubergines are grown all around the world. Depending on the cultivar type, they vary greatly in size, shape, and colour. Generally, these veggies fall into two broad categories, either oval shaped, or thin and elongated. The plant reaches about 3-4 feet tall in quick time and bears many bright fruits. Each fruit has smooth, glossy skin. Internally, it features off-white colour pulp with numerous centrally arranged small, soft seeds. Fruits are generally harvested when they reach maturity but short of full stage ripeness.

Health Benefits of Eggplant

- ☆ Eggplant is very low in calories and fats but rich in soluble fiber content. 100 g provides just 24 calories but contributes about 9 per cent of RDA (recommended daily allowance) of fiber.
- ☆ Research studies conducted at the Institute of Biology of *São Paulo State University; Brazil* suggested that eggplant is effective to control high blood cholesterol.
- ☆ The peel or skin (deep blue/purple varieties) of aubergine has significant amounts of phenolic flavonoid phyto-chemicals called anthocyanins. Scientific studies have shown that these anti-oxidants have potential health effects against cancer, ageing, inflammation, and neurological diseases.
- ☆ Total antioxidant strength measured in terms of *oxygen radical absorbance capacity* (ORAC) of aubergines is 993 µmol TE/100 g. At value 15; they are one of the *low glycemic index* (GI) vegetables.

Brinjal (*Solanum melongena*), Raw, Nutritive Value per 100 g

Principle	*Nutrient Value*	*Percentage of RDA*
Energy	24 Kcal	1 per cent
Carbohydrates	5.7 g	4 per cent
Protein	1 g	2 per cent
Total Fat	0.19 g	1 per cent
Cholesterol	0 mg	0 per cent
Dietary Fiber	3.40 g	9 per cent
Vitamins		
Folates	22 µg	5.5 per cent
Niacin	0.649 mg	4 per cent
Pantothenic acid	0.281 mg	6 per cent
Pyridoxine	0.084 mg	6.5 per cent
Riboflavin	0.037 mg	3 per cent
Thiamin	0.039 mg	3 per cent
Vitamin A	27 IU	1 per cent
Vitamin C	2.2 mg	3.5 per cent
Vitamin E	0.30 mg	2 per cent
Vitamin K	3.5 µg	3 per cent
Electrolytes		
Sodium	2 mg	0 per cent
Potassium	230 mg	5 per cent
Minerals		
Calcium	9 mg	1 per cent
Copper	0.082 mg	9 per cent
Iron	0.24 mg	3 per cent
Magnesium	14 mg	3.5 per cent
Manganese	0.250 mg	11 per cent
Zinc	0.16 mg	1 per cent

- ✰ It contains good amounts of many essential B-complex groups of vitamins such as pantothenic acid (vitamin B5), pyridoxine (vitamin B6) and thiamin (vitamin B1), niacin (B3). These vitamins are essential in the sense that body requires them from external sources to replenish and required for fat, protein and carbohydrate metabolism.
- ✰ Further, this vegetable is an also good source of minerals like manganese, copper, iron and potassium. Manganese is used as a co-factor for the antioxidant enzyme, *superoxide dismutase.* Potassium is an important intracellular electrolyte helps counter the hypertension effects of sodium.

65

Endive

Endive, commonly popular as escarole, is a green leafy-vegetable with a hint of bitter flavour. Nevertheless, this well-known salad plant is much more than just a leafy green, packed with numerous health benefiting plant nutrients such as vitamin C, vitamin A, etc. Botanically this perennial herbaceous leafy plant belongs to the Asteraceae (daisy) family, of the genus Cichorium and is closely related to chicory, radicchio and Belgian endive (witloof). Its scientific name: *Cichorium endivia.* Endive is native to Asia Minor region. This cool-season crop requires well-drained fertile soil to flourish. There are two main cultivar varieties exist: curly-endive (frisee, cichorium endivia, var crispum) with curly narrow leaves and Escarole or scarole (cichorium endivia, var latifolia) with broad leaves. Escarole leaves have spine like-dentate margins (dandelion or lettuce like) with thick stalks. Its leaves are less bitter than narrow, curly, intensely bitter-taste of "frisee". *Belgian endive* or witloof is a popular winter season vegetable in Europe. It is a type of vegetable with smooth cream-coloured leaves compressed into a compact 10 to 12 cm long heads.

Health Benefits of Endive

- ✰ Endive is one of the very low calorie leafy vegetable. 100 g fresh leaves provide just 17 calories; however, it contributes about 8 per cent of daily-required intake (DRI) of fiber.
- ✰ Current research studies suggest that high inulin and fiber content in escarole help reduce glucose and LDL cholesterol levels in diabetes and obese patients.
- ✰ Endive is enriched with good amount Vitamin A and β-carotene. Both compounds are known to have antioxidant properties. Carotenes convert to vitamin A in the body. Furthermore, vitamin A is required for maintaining healthy mucus membranes and skin. In addition, it is also essential vitamin for good eye-sight. Consumption of natural vegetables/greens rich in vitamin A helps to protect from lung and oral cavity cancers.

Endive (*Cichorium endivia*), Raw, Nutritive Value per 100 g

Principle	*Nutrient Value*	*Percentage of RDA*
Energy	17 Kcal	<1 per cent
Carbohydrates	3.35 g	2.5 per cent
Protein	1.25 g	2 per cent
Total Fat	0.20 g	1 per cent
Cholesterol	0 mg	0 per cent
Dietary Fiber	3.10 g	8 per cent
Vitamins		
Folates	142 µg	36 per cent
Niacin	0.400 mg	2.5 per cent
Pantothenic acid	0.900 mg	18 per cent
Pyridoxine	0.020 mg	1.5 per cent
Riboflavin	0.075 mg	6 per cent
Thiamin	0.080 mg	7 per cent
Vitamin A	2167 IU	72 per cent
Vitamin C	6.5 mg	11 per cent
Vitamin E	0.44 mg	3 per cent
Vitamin K	231 µg	192 per cent
Electrolytes		
Sodium	22 mg	1.5 per cent
Potassium	314 mg	7 per cent
Minerals		
Calcium	52 mg	5 per cent
Copper	0.099 mg	11 per cent
Iron	0.83 mg	10 per cent
Magnesium	15 mg	4 per cent
Manganese	0.420 mg	18 per cent
Phosphorus	28 mg	4 per cent
Selenium	0.2 mcg	<1 per cent
Zinc	0.79 mg	7 per cent
Phyto-nutrients		
Carotene-β	1300 µg	—
Carotene-α	0 µg	—
Cryptoxanthin-β	0 µg	—
Lutein-zeaxanthin	0 µg	—

☆ Further, it contains good amounts of many essential B-complex groups of vitamins such as folic acid, pantothenic acid (vitamin B5), pyridoxine (vitamin B6) and thiamin (vitamin B1), niacin (B3). These vitamins are essential in the sense that body requires them from external sources to replenish and required for fat, protein, and carbohydrate metabolism.

- ☆ Additionally, escarole is a good source of minerals like manganese, copper, iron, and potassium. Manganese is used as a co-factor for the antioxidant enzyme, *superoxide dismutase.* Potassium is an important intracellular electrolyte helps counter the hypertension effects of sodium.

66

Fennel Bulb

Delicate, sweet-flavoured fennel bulb commonly features in many Mediterranean cuisines. Its succulent enlarged bulb imparts special "anise like" sweet flavour to the recipes. Bulb fennel is cultivated as a vegetable for its beautiful, squatted stems in many regions of the southern Europe, especially in the Italian plains. It is also known as *Florence fennel, finocchio, sweet fennel*, etc. Fennel is a member of the Apiaceae (parsley family) and is related to carrots, caraway, anise, cumin, dill, etc. Scientific name: *Foeniculum vulgare* var. azoricum.

Bulb fennel is a cool season perennial herb but grown as annual vegetable crop. Unlike seed fennel, the vegetable fennel is a small plant, growing up to only 2 feet in height. As the plant grows, its thickened lower leaves overlap one above the other to form a swollen, bulblike structure just above the ground. At maturity, its bulb measures about 3-5 inches wide and about 3 inches in length. As the plant grows, oftentimes, surrounding soil is pulled around the stem base to create a mound to obtain long blanched fronds.

Health Benefits of Fennel Bulb

- ☆ Fennel bulb is a versatile vegetable, used since ancient times for its nutritional and medicinal properties. This winter season has some noteworthy essential oils, flavonoid anti-oxidants, minerals, and vitamins that have known health benefits.
- ☆ Bulb fennel is one of very low calorie vegetables. 100 g bulb provides just 31 calories. Further, it contains generous amounts of fiber (3.1 g/100 g or 8 per cent of RDI), very little fat and zero cholesterol.
- ☆ Fresh bulbs give sweet anise-like flavour. Much of it is due to high concentration of aromatic essential oils like *anethole, estragole, and fenchone (fenchyl acetate)*. Anethole has been found to have anti-fungal and anti-bacterial properties.

Fennel Bulb (*Foeniculum vulgare* var. *azoricum*), Nutrient Value per 100 g

Principle	*Nutrient Value*	*Percentage of RDA*
Energy	31 Kcal	1 per cent
Carbohydrates	7.29 g	6 per cent
Protein	1.24 g	2 per cent
Total Fat	0.20 g	1 per cent
Cholesterol	0 mg	0 per cent
Dietary Fiber	3.1 g	8 per cent
Vitamins		
Folates	27 µg	7 per cent
Niacin	0.640 mg	4 per cent
Pantothenic acid	0.232 mg	5 per cent
Pyridoxine	0.047 mg	4 per cent
Riboflavin	0.032 mg	2.5 per cent
Thiamin	0.010 mg	1 per cent
Vitamin A	134 IU	4.5 per cent
Vitamin C	12 mg	20 per cent
Electrolytes		
Sodium	52 mg	3 per cent
Potassium	414 mg	9 per cent
Minerals		
Calcium	49 mg	5 per cent
Copper	0.066 mg	7 per cent
Iron	0.73 mg	9 per cent
Magnesium	17 mg	4 per cent
Manganese	0.191 mg	1 per cent
Phosphorus	50 mg	9 per cent
Selenium	0.7 µg	1 per cent
Zinc	0.20 mg	2 per cent

- ☆ The bulbs have moderate amounts of minerals and vitamins that are essential for optimum health. Their juicy fronds indeed contain several vital vitamins such as pantothenic acid, pyridoxine (vitamin B-6), folic acid, niacin, riboflavin, and thiamin in small but healthy proportions. 100 g fresh bulbs provide 27 µg of folates. Folic acid is essential for DNA synthesis and cell division. Their adequate levels in the diet during pregnancy can help prevent neural tube defects in the newborn babies.
- ☆ In addition, fennel bulb contains an average amount of water-soluble vitamins, vitamin-C. 100 g of fresh bulbs provide 12 mg or 20 per cent of vitamin C. Vitamin C helps the body develop resistance against infectious

agents and scavenge harmful, pro-inflammatory free radicals. Further, it has small amounts of vitamin A.

- ☆ The bulbs have very good levels of heart-friendly electrolyte potassium. 100 g provides 414 mg or 9 per cent of daily-recommended levels. It is an important electrolyte inside the cell. Potassium helps reduce blood pressure and rate of heartbeats by countering effects of sodium. Fennel also contains small amounts of minerals such as copper, iron, calcium, magnesium, manganese, zinc, and selenium.

67

Fiddlehead Ferns

Fiddlehead ferns are young tender tightly furled new-growth shoots of fern family plant, usually of ostrich fern. The curly ferns are so named after their unique resemblance to fiddle (violin) head. The fronds are very popular among the inhabitants of Maine, Vermont in US and New Brunswick provinces in Canada, where their short season in the later part of springs attracts many enthusiasts. Ostrich variety fern belongs to the unique flowerless plant species of the *Onocleaceae*, a small family of terrestrial ferns and has the scientific name: *Matteuccia struthioreris*. Ostrich ferns are the most common edible fiddle ferns in the North America. The fern is a clump forming (like an ostrich plume), deciduous plant, which typically grows in well-drained, moisture rich under shady environments. During each spring season, several fiddlehead fronds erupt along the length of rhizome spreads of big fern plant. Their harvesting season is very short and should be done before the fronds unfurl.

Each fiddlehead is a tightly curled, deep green stalk measuring about 4 cm in diameter, reaching to the height of about 10-12 cm off the ground. The tender shoots are covered with brown scales, which have to be scraped off before use in cooking. Young and tender fronds taste similar to that of asparagus, or green beans with crunchy texture of their own. Vegetable fern (*Diplazium esculentum*), known locally as lungru, are found in the hilly areas of North India and Nepal eaten as local delicacy. Lungru season lasts very briefly from May until June.

Health Benefits of Fiddlehead Ferns

- ✰ Fiddlehead ferns are unique in their appearance, taste, and nutrition profile. The curly young shoots contain only 34 calories per 100 g. Nonetheless; their high-quality plant-nutrition profile consists of health benefiting antioxidants, vitamins, and omega-3 and omega-6 essential fatty acids.
- ✰ Fresh fronds are very high in antioxidant vitamin-A, and carotenes. 100 g of fiddleheads contain 3617 IU of or 120 per cent of recommended daily

requirements of vitamin A. Vitamin A is a powerful natural anti-oxidant and is required by the body for maintaining the integrity of skin and mucus membranes. It is also an essential vitamin for vision. Research studies suggest that natural foods rich in vitamin A help the body protects against lung and oral cavity cancers.

Fiddlehead Ferns (*Matteuccia struthioreris*), Fresh, Raw, Nutrition Value per 100 g

Principle	*Nutrient Value*	*Percentage of RDA*
Energy	34 Kcal	1.7 per cent
Carbohydrates	5.54 g	4 per cent
Protein	4.55 g	8 per cent
Total Fat	0.40 g	2 per cent
Cholesterol	0 mg	0 per cent
Vitamins		
Niacin	4.980 mg	31 per cent
Riboflavin	0.210 mg	16 per cent
Thiamin	0.020 mg	1.5 per cent
Vitamin A	3617 IU	120.5 per cent
Vitamin C	26.6 mg	44 per cent
Electrolytes		
Sodium	1 mg	<1 per cent
Potassium	370 mg	8 per cent
Minerals		
Calcium	32 mg	3 per cent
Copper	0.320 mg	35.5 per cent
Iron	1.31 mg	16 per cent
Magnesium	34 mg	8.5 per cent
Manganese	0.510 mg	22 per cent
Selenium	0.7 mcg	1 per cent
Zinc	0.83 mg	7.5 per cent
Phyto-nutrients		
Carotene-β	2040 µg	—
Carotene-α	261µg	—

- ✰ They are an excellent source of many natural poly-phenolic flavonoid compounds such as α and β-carotenes. Carotenes convert to vitamin A inside the body.
- ✰ Their unique sweet taste comes from their richness in vitamin C. 100 g of fresh fronds contains 26.6 mg or 44 per cent of daily-required levels. Vitamin C is a moderately potential water soluble anti-oxidant. Together with flavonoid compound like carotenes, it helps scavenge harmful free radicals, and offer protection from cancers, inflammation, and viral cough and cold.

- Fern shoots are a very good source of minerals and electrolytes, especially potassium, iron, manganese and copper. 100 g of fresh shoots contains 370 mg or 7 per cent of daily-required levels of potassium. Potassium is a heart friendly electrolyte, which helps reduces blood pressure and heart rate by countering sodium effects.
- Further, they contain small to moderate levels of some of the valuable B-complex group of vitamins such as niacin, riboflavin, and thiamin.

68

Green Beans

Tender, flexible green beans are a delight of vegetarian lovers for their wholesome nutritional qualities. They are unripe or immature pods obtained from the bean plant belonging to common *Fabaceae* family, and known scientifically as *Phaseolus vulgaris*. Broadly, tender beans are classified depending on their growth habits as "bush beans" which stand erect without support and "Pole Beans" that need climb supports. The other varieties of unripe fruits of bean family include shell beans, Pinto or mottled beans, White beans, Red or kidney beans, Black beans, Pink beans and Yellow wax beans. All most all the varieties are available year around across the world. Fava or broad beans (*Vicia faba*) are large, flattened light green pods usually eaten shelled for their delicious beans. Lima beans (*Phaseolus lunatus*) are large, plump, pale-green pods with kidney-shaped seeds.

Health Benefits of Green beans

☆ Fresh green beans are very low in calories (31 kcal per 100 g of raw beans) and contain no saturated fat. Nevertheless, the lean vegetables are a very good source of vitamins, minerals, and plant derived micro-nutrients.

☆ They are very rich source of dietary fiber (9 per cent per100g RDA) which acts as a bulk laxative that helps to protect the mucous membrane of the colon by decreasing its exposure time to toxic substances as well as by binding to cancer-causing chemicals in the colon. Adequate amount of fiber has also been shown to reduce blood cholesterol levels by decreasing reabsorption of cholesterol-binding bile acids in the colon.

☆ Green beans contain excellent levels of vitamin A, and health promoting flavonoid poly phenolic antioxidants such as lutein, zeaxanthin and β-carotene in good amounts. These compounds help act as protective scavengers against oxygen-derived free radicals and reactive oxygen species (ROS) that play a role in ageing and various disease processes.

Green Beans (*Phaseolus vulgaris*), Raw, Nutritive Value per 100 g

Principle	*Nutrient Value*	*Percentage of RDA*
Energy	31 Kcal	1.5 per cent
Carbohydrates	7.13 g	5.5 per cent
Protein	1.82 g	3 per cent
Total Fat	0.34 g	1 per cent
Cholesterol	0 mg	0 per cent
Dietary Fiber	3.4 g	9 per cent
Vitamins		
Folates	37 µg	9 per cent
Niacin	0.752 mg	5 per cent
Pantothenic acid	0.094 mg	2 per cent
Pyridoxine	0.074 mg	5.5 per cent
Riboflavin	0.105 mg	8 per cent
Thiamin	0.084 mg	7 per cent
Vitamin A	690 IU	23 per cent
Vitamin C	16.3 mg	27 per cent
Vitamin K	14.4 µg	12 per cent
Electrolytes		
Sodium	6 mg	0.4 per cent
Potassium	209 mg	5.5 per cent
Minerals		
Calcium	37 mg	3.7 per cent
Iron	1.04 mg	13 per cent
Magnesium	25 mg	6 per cent
Manganese	0.214 mg	9 per cent
Phosphorus	38 mg	6 per cent
Zinc	0.24 mg	2 per cent
Phyto-nutrients		
Carotene-β	379 µg	—
Carotene-α	69 µg	—
Lutein-zeaxanthin	640 µg	—

- ✰ Zeaxanthin, an important dietary carotenoid in the beans, selectively absorbed into the retinal macula lutea in the eyes where it thought to provide antioxidant and protective UV light filtering functions. It is, therefore, green beans offer some protection in preventing *age-related macular disease* (ARMD) in the elderly.
- ✰ Snap beans are a good source of *folates*. 100 g fresh beans provide 37 µg or 9 per cent of folates. Folate along with vitamin B-12 is one of the essential

components of DNA synthesis and cell division. Good folate diet when given during preconception periods and during pregnancy helps prevent neural-tube defects among the offspring.

- They also contain good amounts of vitamin-B6 (pyridoxine), thiamin (vitamin B-1), and vitamin-C. Consumption of foods rich in vitamin C helps the body develop resistance against infectious agents and scavenge harmful oxygen-free radicals.
- In addition, beans contain healthy amounts of minerals like iron, calcium, magnesium, manganese, and potassium, which are very essential for body metabolism. Manganese is a co-factor for the antioxidant enzyme, *superoxide dismutase,* which is a very powerful free radical scavenger. Potassium is an important component of cell and body fluids that helps controlling heart rate and blood pressure.

69

Jicama

Jicama, also known as yam bean, is a round, fleshy taproot vegetable of bean family plant. Its underground starchy root is one of the popular edible root vegetables grown in many parts of Central American, South Asian, Caribbean, and some Andean South American regions. It's refreshing; crispy, ice-white fruit-like pulp can be eaten raw or cooked in a variety of sweet and savory dishes worldwide. Some of the common names of this tuber are *Mexican water chestnut, Mexican turnip, sengkwang, yacon*...etc. It is pronounced as *hecama*. Scientific name: *Pachyrhizus erosus*. Jicama is a perennial vine plant growing vigorously under semitropical and tropical climates. It has similar growth characteristics as that of lima bean or any other bean species plant. The most distinguishing feature, however, is that it bears globular, fleshy, turnip-like starchy edible root below the ground surface. Unlike other starch roots like potato, and sweet-potato wherein the peel may be eaten; jicama features thick dust-brown colour inedible skin. Inside, its white starchy flesh has crisp texture and fruit-like succulent, sweet-starchy taste. Each tuber weighs about 250 to 1200 g. There exist at least five cultivar types of *Pachyrhizus* species; however, the three popular cultivated varieties include *Pachyrhizus erosus* (*Mexican yam bean*), *Pachyuhizus ahipa* (*Andean yam bean*), and *Pachyrhizus tuberosus* (*Amazonian yam bean, jíquima*). *P. erosus* ('jícama de leche') or Mexican yam bean is the popular variety imported in the USA. Another cultivar, *P. palmatilobus*, locally known as 'jícama de leche', has deeply lobed leaflets, a milky sap and less agreeable taste.

Health Benefits of Jicama

☆ Jicama is very low calorie root vegetables; contains only 35 calories per 100 g. However, its high quality phyto-nutrition profile comprises of dietary fiber, and anti-oxidants, in addition to small proportions of minerals, and vitamins.

Jicama (*Pachyrhizus erosus*), Fresh, Raw, Nutrition Value per 100 g

Principle	*Nutrient Value*	*Percentage of RDA*
Energy	38 Kcal	2 per cent
Carbohydrates	8.82 g	7 per cent
Protein	0.72 g	1 per cent
Total Fat	0.19 g	<1 per cent
Cholesterol	0 mg	0 per cent
Dietary Fiber	4.9 g	13 per cent
Vitamins		
Folates	12 µg	3 per cent
Niacin	0.200 mg	1.5 per cent
Pantothenic acid	0.135 mg	3 per cent
Pyridoxine	0.042 mg	3 per cent
Riboflavin	0.029 mg	2 per cent
Thiamin	0.020 mg	2 per cent
Vitamin A	21 IU	1 per cent
Vitamin C	20.2 mg	34 per cent
Vitamin E	0.46 mg	3 per cent
Vitamin K	0.3 µg	<1 per cent
Electrolytes		
Sodium	4 mg	<1 per cent
Potassium	150 mg	3 per cent
Minerals		
Calcium	12 mg	1 per cent
Copper	0.048 mg	5 per cent
Iron	0.60 mg	7 per cent
Magnesium	12 mg	3 per cent
Manganese	0.60 mg	3 per cent
Zinc	0.16 mg	1 per cent
Phyto-nutrients		
Carotene-β	13 µg	—
Carotene-α	0 µg	—
Lutein-zeaxanthin	0 µg	—

☆ It is one of the finest source dietary fiber and excellent source of oligofructose inulin, a soluble dietary fiber. The root pulp provides 4.9 mg or 13 per cent of fiber. Inulin is a zero calorie, sweet inert carbohydrate and does not metabolize in the human body, which make the root an ideal sweet snack for diabetics and dieters.

- As in turnips, fresh yam bean tubers are rich in vitamin C; provide about 20.2 mg or 34 per cent of DRA of vitamin C per 100 g. Vitamin-C is a powerful water-soluble anti-oxidant that helps body scavenge harmful free radicals, thereby offers protection from cancers, inflammation and viral cough and cold.
- It also contains small levels of some of valuable B-complex group of vitamins such as folates, riboflavin, pyridoxine, pantothenic acid and thiamin.
- Further, it provides healthy amounts of some important minerals like magnesium, copper, iron and manganese.

70

Kale

Kale or *borecole* is rich in numerous health benefiting polyphenolic flavonoid compounds such as lutein, zeaxanthin, and beta-carotene, and vitamins than found in any other green leafy vegetables. It is widely cultivated across Europe, Japan, and the United States for its "frilly"leaves. Botanically, the plant belongs to the "cabbage" (Brassica) family subgroup of *Brassica oleracea* (*acephala group*), characteristic of headless, leafy greens. It features close similarity in growth and appearance to collard greens. Other common vegetables closely related being: broccoli, cauliflower, brussels sprouts, etc.

Kale is an annual plant, flourishes well in rich organic soil and prefers cool climate and light frost conditions. Its succulent, curly leaves appear "rosette" like and may have dark green to blue-green colour depending on the cultivar type. It is grown mainly for autumn and winter harvest, because cool weather further enhances its sweet taste quality. Some of the important cultivars grown around the globe are Scottish curly leaf (*Brassica napus* (*Pabularia Group*)), Red Russian, Blue curled, Winterbor cultivars. Tuscan kale, also known as *cavalo nero*, is popular winter-season green in the Northern parts of Italy. It features distinctive very long, curly, blue-green leaves with embossed surface resembling dinosaur skin, giving its name as dinosaur kale.

Health Benefits of Kale

- ☆ Kale is a very versatile and nutritious green leafy vegetable. It is a widely popular vegetable since ancient Greek and Roman times for its low fat, no cholesterol but health benefiting anti-oxidant properties.
- ☆ Kale, like other members of the Brassica family, contains health-promoting phytochemicals, sulforaphane and indole-3-carbinol that appear to protect against prostate and colon cancers.

Kale (*Brassica oleracea.* Acephala Group), Fresh, Raw Leaves, Nutrition Value per 100 g

Principle	*Nutrient Value*	*Percentage of RDA*
Energy	50 Kcal	2.5 per cent
Carbohydrates	10.01 g	8 per cent
Protein	3.30 g	6 per cent
Total Fat	0.70 g	3 per cent
Cholesterol	0 mg	0 per cent
Dietary Fiber	2.0 g	5 per cent
Vitamins		
Folates	29 µg	7 per cent
Niacin	1.000 mg	6 per cent
Pantothenic acid	0.091 mg	1.5 per cent
Pyridoxine	0.271 mg	21 per cent
Riboflavin	0.130 mg	10 per cent
Thiamin	0.110 mg	9 per cent
Vitamin A	15376 IU	512 per cent
Vitamin C	120 mg	200 per cent
Vitamin K	817 µg	681 per cent
Electrolytes		
Sodium	43 mg	3 per cent
Potassium	447 mg	9.5 per cent
Minerals		
Calcium	135 mg	13.5 per cent
Copper	0.290 mg	32 per cent
Iron	1.70 mg	21 per cent
Magnesium	34 mg	8.5 per cent
Manganese	0.774 mg	34 per cent
Phosphorus	56 mg	8 per cent
Selenium	0.9 µg	1.5 per cent
Zinc	0.44 mg	4 per cent
Phyto-nutrients		
Carotene-β	9226 µg	—
Crypto-xanthin-β	0 µg	—
Lutein-zeaxanthin	39550 µg	—

☆ Di-indolyl-methane (DIM), a metabolite of *indole-3-carbinol* is an effective immune modulator, anti-bacterial and anti-viral agent through its action of potentiating "Interferon-Gamma" receptors.

- Borecole is very rich source of β-carotene, lutein and zeaxanthin. These flavonoids have strong anti-oxidant and anti-cancer activities. Beta-carotene is converted to vitamin A in the body.
- Zeaxanthin, an important dietary carotenoid, is selectively absorbed into the retinal macula lutea in the eyes where it is thought to provide antioxidant and protective light-filtering functions. Thus, it helps prevent retinal detachment and offer protection against "age-related macular degeneration related macular degeneration disease" (ARMD) in the elderly.
- It is very rich in vitamin A, 100 g leaves provide 512 per cent of RDA. Vitamin A is required for maintaining healthy mucus membranes and skin and is essential for vision. Foods rich in this vitamin are known to offer protection against lung and oral cavity cancers.
- It is one of the excellent vegetable sources for vitamin-K; 100 g provides about 700 per cent of recommended intake. Vitamin K has potential role bone health by promoting osteotrophic (bone formation and strengthening) activity. Adequate vitamin-K levels in the diet help limiting neuronal damage in the brain; thus, has established role in the treatment of patients suffering from Alzheimer's disease.
- 100 g of fresh leaves contain 120 mg or 200 per cent of daily-recommended levels of vitamin C. Scottish curly leaf variety yet has more of this vitamin, 130 mg/100g. Vitamin C is a powerful antioxidant, which helps the body develop resistance against infectious agents and scavenge harmful oxygen-free radicals.
- This leafy vegetable is notably good in many B-complex groups of vitamins such as niacin, vitamin B-6 (pyridoxine), thiamin, pantothenic acid, etc., that are essential for substrate metabolism in the body.
- It is also rich source of minerals like copper, calcium, sodium, potassium, iron, manganese, and phosphorus. Potassium is an important component of cell and body fluids that helps controlling heart rate and blood pressure by countering effects of sodium. Manganese is used by the body as a co-factor for the antioxidant enzyme, *superoxide dismutase*. Iron is required for cellular oxidation and red blood cell formation.

Kale provides rich nutrition ingredients that offer protection from vitamin A deficiency, osteoporosis, iron-deficiency anemia, and believed to protect from cardiovascular diseases and colon and prostate cancers.

71

Knol-Khol

Kohlrabi or *knol-khol or German Turnip* is a stout, round, tuberous vegetable of the Brassica family; same as that of cabbage, broccoli, cauliflower, kale, collard greens, and brussels sprouts belongs to. This stem vegetable is native to Europe. Scientific name: *Brassica oleracea* (Gongylodes Group). Knol knol is a perennial, cool season vegetable. It is grown all over the temperate climates for its succulent round shaped modified stem as well for its turnip-flavoured top greens. The plant requires well-drained fertile soil and average sun shine to flourish. Its edible stem attains maturity and ready for harvest in 55-60 days after sowing. Approximate weight is 150 g. The tuber is tolerant of cracking and has a good standing ability for up to 30 days after maturity.

Knol-knols have similar taste and texture as that of a broccoli stem or cabbage, but milder and sweeter. The younger stems have crispy, pleasant taste and rich flavour. Two main varieties of knol-knol exist, white and purple. However, internally, both the "white" (actually light green colour) as well as purple types have similar cream-yellow colour edible flesh. Some of the popular cultivars of kohlrabi grown worldwide are: *white Vienna, white Danube, purple vienna,* and *grand duke.*

Health Benefits of Kohlrabi (Knol-khol)

- Mildly sweet, succulent kohlrabi is notably rich in vitamins and dietary fiber; however, it has only 27 calories per 100 g, a negligible amount of fat, and zero cholesterol.
- Fresh kohlrabi stem is rich source of *vitamin-C;* provide 62 mg per 100 g weight that is about 102 per cent of RDA. Vitamin C (ascorbic acid) is a water-soluble vitamin and powerful anti-oxidant. It helps the body maintain healthy connective tissue, teeth, and gum. Its anti-oxidant property helps the human body protect from diseases and cancers by scavenging harmful free radicals from the body.

Knol-knol (*Brassica oleracea*. Gongylodes Group), Fresh, Raw, Nutrition Value per 100 g

Principle	*Nutrient Value*	*Percentage of RDA*
Energy	27 Kcal	1.5 per cent
Carbohydrates	6.20 g	5 per cent
Protein	1.70 g	3 per cent
Total Fat	0.10 g	<1 per cent
Cholesterol	0 mg	0 per cent
Dietary Fiber	3.6 g	10 per cent
Vitamins		
Folates	16 μg	4 per cent
Niacin	0.400 mg	2.5 per cent
Pantothenic acid	0.165 mg	3 per cent
Pyridoxine	0.150 mg	11.5 per cent
Riboflavin	0.020 mg	1.5 per cent
Thiamin	0.050 mg	4 per cent
Vitamin A	36 IU	1 per cent
Vitamin C	62 mg	102 per cent
Vitamin K	0.1 μg	<1 per cent
Electrolytes		
Sodium	20 mg	1 per cent
Potassium	350 mg	7 per cent
Minerals		
Calcium	24 mg	2.5 per cent
Copper	0.129 mg	14 per cent
Iron	0.40 mg	5 per cent
Magnesium	19 mg	5 per cent
Manganese	0.139 mg	6 per cent
Phosphorus	46 mg	6.5 per cent
Selenium	0.7 μg	1 per cent
Zinc	0.03 mg	<1 per cent
Phyto-nutrients		
Carotene-β	22 μg	—
Crypto-xanthin-β	0 μg	—
Lutein-zeaxanthin	0 μg	—

☆ Kohlrabi, like other members of the Brassica family, contains health-promoting phytochemicals such as isothiocyanates, sulforaphane, and indole-3-carbinol that are supposed to protect against prostate and colon cancers.

- It especially contains good amounts of many B-complex groups of vitamins such as niacin, vitamin B-6 (pyridoxine), thiamin, pantothenic acid, etc., that acts as co-factors to enzymes during various metabolism inside the body.
- Knol-knol notably has good levels of minerals; copper, calcium, potassium; manganese, iron, and phosphorus are especially available in the stem. Potassium is an important component of cell and body fluids that helps controlling heart rate and blood pressure by countering effects of sodium. Manganese is used by the body as a co-factor for the antioxidant enzyme, *superoxide dismutase*.
- In addition, its creamy colour flesh contains small amounts of vitamin A and carotenes.
- Kohlrabi leaves or tops, like turnip greens, are also very nutritious greens abundant in carotenes, vitamin A, vitamin K, minerals, and B-complex group of vitamins.

72

Leeks

Delicate flavoured leeks are cylindrical, leafy stalks obtained from the onion-like plant in the *Allium* family. They are biennial, tall, slender plants with long cylindrical stem composed of overlapping leaves. They are commonly used as vegetables in many parts of Europe, America, and Asia. Botanically, they belong to the *Alliaceae* family of bulbous plants, of the genus: *Allium*. However, unlike their fellow allium members such as onion, shallots, garlic etc., they do not form bulbs. Scientific name: *Allium ampeloprasum* var. porrum. The plant requires well-drained fertile soil to flourish. In general, it is cultivated as annual crop in many parts of Europe and Asia. Planting can be done by either sowing seeds or transplanting seedlings that take about 100-120 days to harvest. In general, leeks are planted in deep trenches to deprive sun light exposure which otherwise would turn their stems green (chlorophyll pigmentation) due to photosynthesis. As the plant grow, fill-in the trench by pulling surrounding earth to create a mound around it. This method is applied in order to obtain long, blanched stem.

Health Benefits of Leeks

- ✰ Leeks contain many noteworthy flavonoid anti-oxidants, minerals, and vitamins that have proven health benefits.
- ✰ Leeks are low in calories. 100 g fresh stalks contain 61 calories. Further, their elongated stalks provide good amounts of soluble and insoluble fiber.
- ✰ Though leeks contain proportionately less thio-sulfinites than that in garlic, they still possess significant amounts of these anti-oxidants such as *diallyl disulfide, diallyl trisulfide* and *allyl propyl disulfide*. The compounds convert to allicin by enzymatic reaction when the leek-stalk disturbed (crushing, cutting, etc). The total measured anti-oxidant strength (ORAC value) of 100 g leek is 490 TE (Trolex Equivalents).

Leeks (*Allium ampeloprasum* var. *porrum*), Nutrient Value per 100 g

Principle	*Nutrient Value*	*Percentage of RDA*
Energy	61 Kcal	2 per cent
Carbohydrates	14.15 g	11 per cent
Protein	1.50 g	3 per cent
Total Fat	0.30 g	1 per cent
Cholesterol	0 mg	0 per cent
Dietary Fiber	1.8 g	5 per cent
Vitamins		
Folates	64 µg	16 per cent
Niacin	0.400 mg	2.5 per cent
Pantothenic acid	0.140 mg	3 per cent
Pyridoxine	0.233 mg	18 per cent
Riboflavin	0.030 mg	2 per cent
Thiamin	0.060 mg	5 per cent
Vitamin A	1667 IU	55 per cent
Vitamin C	12 mg	20 per cent
Vitamin E	0.92 mg	6 per cent
Vitamin K	47 µg	39 per cent
Electrolytes		
Sodium	20 mg	1 per cent
Potassium	180 mg	4 per cent
Minerals		
Calcium	59 mg	6 per cent
Copper	0.120 mg	13 per cent
Iron	2.10 mg	26 per cent
Magnesium	28 mg	7 per cent
Manganese	0.481 mg	2 per cent
Phosphorus	35 mg	5 per cent
Selenium	1 µg	2 per cent
Zinc	1.2 mg	11 per cent
Phyto-nutrients		
Carotene-β	1000 µg	—
Crypto-xanthin-β	0 µg	—
Lutein-zeaxanthin	1900 µg	—

☆ Laboratory studies show that *allicin* reduces cholesterol production by inhibiting the *HMG-CoA reductase*enzyme in the liver cells. Further, it also found to have anti-bacterial, anti-viral and anti-fungal activities.

- ☆ Additionally, allicin decreases blood vessel stiffness by release of nitric oxide (NO) and, thereby bring a reduction in the total blood pressure. It also blocks platelet clot formation and has fibrinolytic action in the blood vessels, which helps decrease an overall risk of coronary artery disease (CAD), peripheral vascular diseases (PVD), and stroke.
- ☆ Leeks are a great source of minerals and vitamins that are essential for optimum health. Their leafy stems indeed contain several vital vitamins such as pyridoxine, folic acid, niacin, riboflavin, and thiamin in healthy proportions. 100 g fresh stalks provide 64 µg of folates. Folic acid is essential for DNA synthesis and cell division. Their adequate levels in the diet during pregnancy can help prevent neural tube defects in the newborn babies.
- ☆ In addition, leeks are one of the good sources of vitamin-A (1667 IU or 55 per cent of RDA per 100 g) and other flavonoid phenolic anti-oxidants such as carotenes, xanthin, and lutein. They also have some other essential vitamins such as vitamin C, K, and vitamin E. Vitamin C helps the human body develop resistance against infectious agents and scavenge harmful, pro-inflammatory free radicals.
- ☆ Further, its stalks have small amounts of minerals such as potassium, iron, calcium, magnesium, manganese, zinc, and selenium.

73

Lettuce

Crispy, green/crimson-red coloured lettuce is one of the incredible leafy vegetables packed with essential nutrients that benefit overall general health. Indeed, it is among the most sought-after greens, be it in *your* crunchy, green salads or healthy vegetable sandwiches. Botanically this marvelous, nutrition rich leafy green belongs to the daisy family of *Asteraceae*. Scientific name: *Lactuca sativa*. *Lactuca sativa* is a small size annual plant flourishes in well-draining, humus soil. Its leaves exude milk like fluid (sap) when cut, and hence their name derived from Latin, *lactuca* for milk. There are about six varieties of cultivars exist based upon head formation and leaf structure. The leaf varieties with more bitter taste are rather rich in antioxidants.

Here are some popular varieties grown around the globe:

1. *Butter-head*, with loose heads; it has a buttery texture. Butter head cultivars are most popular and widely grown in Europe.
2. *Chinese variety or celtuce*, generally have long, tapering, non-head forming, strong-flavoured leaves unlike its Western counterparts. They are, therefore, used preferred in stir fried dishes and stews.
3. *Crisp-head* variety forms tight, dense heads that resemble cabbage. They are generally the mildest form, valued more for their crunchy texture than flavour. Cultivars of the crisp head are the most familiar type used in the USA.
4. *Loose-leaf* variety features tender, delicate and fully flavoured leaves with a loose bunch. This group includes green oak leaf, red oak leaf, valeria and lolla-rosa-types.
5. *Romaine-lettuce* grows in a long head of sturdy leaves with a firm rib almost reaching to the tip of the leaf. Cultivars of Romaine are also the most popular types in the USA.
6. *Summer Crisp* variety forms moderately dense heads with a crunchy texture; this type is intermediate between crisp-head and loose-leaf types.

Health Benefits of Lettuce

☆ Lettuce leaves are one of the very low calorie green-vegetables. 100 g fresh greens provide just 15 calories. Nonetheless, they are the store house of many phyto-nutrients that have health promoting and disease prevention properties.

Lettuce (*Lactuca sativa* var. *crispa*), Raw, Green-Leaf, Nutritive Value per 100 g

Principle	*Nutrient Value*	*Percentage of RDA*
Energy	15 Kcal	1 per cent
Carbohydrates	2.79 g	2 per cent
Protein	1.36 g	2 per cent
Total Fat	0.15 g	0.5 per cent
Cholesterol	0 mg	0 per cent
Dietary Fiber	1.3 g	3 per cent
Vitamins		
Folates	38 μg	9.5 per cent
Niacin	0.375 mg	2 per cent
Pantothenic acid	0.134 mg	2.5 per cent
Pyridoxine	0.090 mg	7 per cent
Riboflavin	0.080 mg	6 per cent
Thiamin	0.070 mg	6 per cent
Vitamin A	7405 IU	247 per cent
Vitamin C	9.2 mg	15 per cent
Vitamin E-α	0.29 mg	2 per cent
Vitamin K	126.3 μg	105 per cent
Electrolytes		
Sodium	28 mg	2 per cent
Potassium	194 mg	4 per cent
Minerals		
Calcium	36 mg	3.5 per cent
Copper	0.029 mg	3 per cent
Iron	0.86 mg	10 per cent
Magnesium	13 mg	3 per cent
Manganese	0.250 mg	11 per cent
Phosphorus	29 mg	4 per cent
Zinc	0.18 mg	1.5 per cent
Phyto-nutrients		
Carotene-β	4443 μg	—
Crypto-xanthin-β	0 μg	—
Lutein-zeaxanthin	1730 μg	—

- Vitamins in lettuce are plentiful. Fresh leaves are an excellent source of several Vitamin A and beta carotenes. Just 100 g of fresh, raw-lettuce provides 247 per cent of daily vitamin A, and 4443 µg of beta-carotene (Carotenes convert to vitamin A in the body; 2 µg of carotene is considered equivalent to 1 IU of vitamin A). These compounds have antioxidant properties. Vitamin A is required for maintaining healthy mucus membranes and skin, and is also essential for vision. Consumption of natural fruits and vegetables rich in flavonoids helps to protect the body from lung and oral cavity cancers.
- It is a rich source of vitamin K. Vitamin K has a potential role in the bone metabolism where it thought to increase bone mass by promoting osteotrophic activity in the bone cells. It also has established role in Alzheimer's disease patients by limiting neuronal damage in the brain.
- Fresh leaves contain good amounts folates and vitamin C. Folates require for DNA synthesis and therefore, vital in prevention of the neural tube defects in-utero fetus during pregnancy. Vitamin C is a powerful natural antioxidant; regular consumption of foods rich in vitamin C helps the body develop resistance against infectious agents and scavenge harmful, pro-inflammatory free radicals.
- Zeaxanthin (1730 µg per 100 g), an important dietary carotenoid in lettuce, is selectively absorbed into the retinal macula lutea, where it thought to provide antioxidant and filter UV rays falling on the retina. Diet rich in xanthin and carotenes is thought to offer some protection against *age-related macular disease* (*ARMD*) in the elderly.
- It also contains good amounts of minerals like iron, calcium, magnesium, and potassium, which are very essential for body metabolism. Potassium is an important component of cell and body fluids that helps controlling heart rate and blood pressure. Manganese is used by the body as a co-factor for the antioxidant enzyme, *superoxide dismutase.* Copper is required in the production of red blood cells. Iron is essential for red blood cell formation.
- It is rich in B-complex group of vitamins like thiamin, vitamin B-6 (pyridoxine), riboflavins.

Regular inclusion of lettuce in salads is known to prevent osteoporosis, iron-deficiency anemia and believed to protect from cardiovascular diseases, ARMD, Alzheimer's disease and cancers.

74

Lotus Root

Crunchy, delicate flavoured lotus root is an under-water edible rhizome of lotus plant. Since centuries, the lotus rhizome has held high esteem in oriental regions, especially in Chinese and Japanese cultures. Almost all the parts of the plant: root, young flower stalks, and seeds are being used in the cuisine. Lotus plant is an herbaceous, perennial aquatic plant belonging to *Nelumbonaceae* family. Scientific name: *Nelumbo nucifera.*

Lotus root is grown as annual root vegetable crop in customized ponds. Although lotus can be raised from the seeds, commercially, the rhizomes with meristems (growing points) are preferred for cultivation since it takes less time for crop production. Rhizome activity in the plant usually coincides with the appearance of large floating leaves about 5-6 months after implantation. The rhizomes grow underwater in mud. They are actually modified tubers storing energy in the form of starch. The rhizome develops into sausage-like three to five jointed nodes of about 2-4 feet length. Each rhizome segment features smooth, grey-white colour and measure about 10-20 cm in length, 6-10 cm in diameter. Internally, the root has white, crunchy flesh with mild sweet, water chestnut like flavour. The cut sections reveal visually appealing display of symmetrically arranged air canals (holes) traversing along the length of the root. The lotus fruit is an enlarged receptacle akin to sunflower head where in numerous edible seeds embedded in its head.

Health Benefits of Lotus root

- Lotus root is one of the moderate calorie root vegetables. 100 g root-stem provides about 74 calories. Nevertheless, it composed of several health benefiting phyto-nutrients, minerals, and vitamins.
- Lotus rhizome is very good source dietary fibers; 100 g flesh provides 4.9 g or 13 per cent of daily-requirement of fiber. The fiber, together with slow digesting complex carbohydrates in the root help reduce blood cholesterol, sugar, body weight and constipation conditions.

Lotus Root (*Nelumbo nucifera*), Raw, Nutrition Value per 100 g

Principle	*Nutrient Value*	*Percentage of RDA*
Energy	74 Kcal	3.7 per cent
Carbohydrates	17.23 g	13 per cent
Protein	2.60 g	5 per cent
Total Fat	0.10 g	0.5 per cent
Cholesterol	0 mg	0 per cent
Dietary Fiber	4.9 g	13 per cent
Vitamins		
Folates	13 µg	3 per cent
Niacin	0.400 mg	2.5 per cent
Pantothenic acid	0.377 mg	7.5 per cent
Pyridoxine	0.258 mg	20 per cent
Riboflavin	0.220 mg	17 per cent
Thiamin	0.160 mg	13 per cent
Vitamin A	0 IU	0 per cent
Vitamin C	44 mg	73 per cent
Electrolytes		
Sodium	40 mg	3 per cent
Potassium	556 mg	12 per cent
Minerals		
Calcium	45 mg	4.5 per cent
Copper	0.257 mg	29 per cent
Iron	1.16 mg	14 per cent
Magnesium	23 mg	6 per cent
Manganese	0.261 mg	1 per cent
Selenium	0.7 µg	1 per cent
Zinc	0.39 mg	3.5 per cent
Phyto-nutrients		
Carotene-β	0 µg	—
Carotene-α	0 µg	—
Cryptoxanthin-β	0 µg	—
Lutein-zeaxanthin	0 µg	—

☆ Lotus root is one of the excellent sources of vitamin C. 100 g root provides 44 mg or 73 per cent of daily-recommended values. Vitamin C is a powerful water soluble anti-oxidant. It is required for the collagen synthesis in the body. Collagen is the main structural protein in the body required for maintaining the integrity of blood vessels, skin, organs, and bones. Regular consumption of foods rich in vitamin C helps the body protect from scurvy, develop resistance against viral infection, boosts immunity, hasten wound healing and remove cancer causing harmful free radicals from the body.

- In addition, it contains moderate levels of some of valuable B-complex group of vitamins such as pyridoxine (vitamin B-6), folates, niacin, riboflavin, pantothenic acid, and thiamin. Pyridoxine (vitamin B-6) acts as a coenzyme in the neuro-chemical synthesis in the brain which influences mood. Adequate pyridoxine levels help control nervous irritability, headache, and tension. It also protects heart-attack risk by controlling harmful homocysteine levels in the blood.
- Further, the root provides healthy amounts of some important minerals like copper, iron, zinc, magnesium, and manganese. Copper is a cofactor for many vital enzymes, including *cytochrome c-oxidase* and *superoxide dismutase* (other minerals function as cofactors for this enzyme are manganese and zinc). Along with iron, it is also required in the production of red blood cells.
- The crunchy, sweet yet delicate flavour of root lotus is because of its optimum electrolyte balance. It composes agreeable ratio of sodium to potassium at the value 1:4. While sodium gives the sweet taste to the root, potassium acts to counter negative effects of sodium by regulating heart rate and blood pressure.

75

Moringa/Drumstick

Moringa oleifera, known popularly as *drumstick tree*, is an herbaceous plant grown for its nutritious greens, flowers, and mineral-rich pods. It is a well-recognized member in the *Moringaceae* family of trees and thought to be originated in the sub-Himalayan ranges of Indian subcontinent. The plant possesses horseradish like root and, hence, known to the western world as horseradish tree. The seed pods are popular as *murnga* in Tamil, and *malunggay* in Philippines. Scientific name: *Moringa oleifera*. Moringa is a drought tolerant, medium-sized, evergreen tree that prefers warm, frost-free climates to flourish. Its tender leaves and twigs can be harvested from a well-established, 1.5 to 2 meters height plant. Taller plants bear cream-white colour small size flowers in clusters throughout the season, which subsequently develop to long slender dark-green colour, three-sided, edible seedpods with tapering ends. Each pod measures about 6-18 inches in length, with constrictions at the seed nodes giving them a typical drumstick appearance. Inside, each pod features fleshy pulp and round pea-sized seeds encased in a wing-shaped coat (hull).

Health Benefits of moringa

- ☆ Moringa plant possesses unique nutritional qualities that hold promise to millions of impoverished communities around the world those lacks in many nutritional supplements such as protein, minerals, and vitamins.
- ☆ Leaves are an excellent source of protein that can be rarely found in any other herbs and green leafy vegetables. 100 g of fresh raw leaves provide 9.8 g of protein or about 17.5 per cent of daily-required levels. Dry, powdered leaves indeed are a much-concentrated source of many quality amino acids.
- ☆ Fresh pods and seeds are a good source of oleic acid, a health-benefiting monounsaturated fat. Moringa as high-quality oilseed crop can be grown alternatively to improve nutrition levels of populations in many drought-prone regions of Africa and Asia.

Moringa (drumstick) pods and leaves, raw, Nutritive Value per 100 g. Percentages of daily-recommended values are marked in brackets.

Principle	*Nutrient Value-Pods*	*Nutrient Value-Leaves*
Energy	37 Kcal (2 per cent)	64 Kcal (3 per cent)
Carbohydrates	8.53 g (6.5 per cent)	8.28 per cent (6 per cent)
Protein	2.10 g (4 per cent)	9.40 g (17 per cent)
Total Fat	0.20 g (1 per cent)	1.40 per cent (7 per cent)
Cholesterol	0 mg (0 per cent)	0 mg (0 per cent)
Dietary Fiber	3.2 g (8 per cent)	2.0 g (5 per cent)
Vitamins		
Folates	44 µg (11 per cent)	40 µg (10 per cent)
Niacin	0.680 mg (4 per cent)	2.220 mg (14 per cent)
Pyridoxine	0.120 mg (9 per cent)	1.200 mg (92 per cent)
Riboflavin	0.074 mg (6 per cent)	0.660 mg (51 per cent)
Thiamin	0.053 mg (4.5 per cent)	0.257 mg (21.5 per cent)
Vitamin A	74 IU (2.5 per cent)	7564 IU (252 per cent)
Vitamin C	141mg (235 per cent)	51.7 mg (86 per cent)
Electrolytes		
Sodium	42 mg (3 per cent)	9 mg (0.5 per cent)
Potassium	461 mg (10 per cent)	337 mg (7 per cent)
Minerals		
Calcium	30 mg (3 per cent)	185 mg (18.5 per cent)
Iron	0.36 mg (4.5 per cent)	4.00 mg (50 per cent)
Magnesium	45 mg (11 per cent)	147 mg (37 per cent)
Phosphorus	50 mg (9 per cent)	112 mg (20 per cent)
Selenium	8.2 µg (15 per cent)	0.9 µg (1.5 per cent)
Zinc	0.45 mg (4 per cent)	0.60 mg (5 per cent)

- ☆ Fresh leaves and growing tips of moringa are the richest source of vitamin A. *100 g of fresh leaves 7564 IU or 252 per cent of daily-required levels.* Vitamin A is one of the fat-soluble anti-oxidant offering several benefits, including mucus membrane repair, maintenance of skin integrity, vision, and immunity.
- ☆ Fresh moringa pods and leaves are excellent sources of vitamin-C. 100 g of pods contain 145 µg or 235 per cent of daily-required levels of vitamin C. 100 g of greens provide 51.7 µg or 86 per cent of daily-recommended intake values of this vitamin. Research studies have shown that consumption of fruits/vegetables rich in vitamin C helps the body develop immunity against infectious agents, and scavenge harmful oxygen-free radicals from the body.
- ☆ The greens as well as pods also contain good amounts of many vital B-complex vitamins such as folates, vitamin-B6 (pyridoxine), thiamin (vitamin

B-1), riboflavin, pantothenic acid, and niacin. Much of these vitamin functions as co-enzymes in carbohydrate, protein, and fat metabolism.

- Furthermore, its leaves are one of the fine sources of minerals like calcium, iron, copper, manganese, zinc, selenium, and magnesium. Iron alleviates anemia. Calcium is required for bone strengthening. Zinc plays a vital role in hair-growth, spermatogenesis, and skin health.

76

Mustard Greens

Spicy, crunchy mustard greens, also known as leaf mustard, are indeed one of the most nutritious green-leafy vegetables. The lovely top greens have actually more vitamin A, carotenes, vitamin K, and flavonoid anti-oxidants than some of commonly consumed fruits and vegetables. This crispy leafy vegetable obtained from mustard plant belonging to *Brassica* family, which also includes cabbage, broccoli, Brussels sprouts etc. Scientific name: *Brassica juncea*. The mustard plant is native to sub-Himalayan plains of Indian sub-continent commonly cultivated for its winter season leaves and oil seeds since ancient times. Several cultivars exist. Mustards are winter crops when the leaves are more flavourful from November until March.

In general, its young tender green leaves that is harvested when the plant reaches about 2 feet in height and used as a green leafy vegetable. Completely grown plant reaches about 4-5 feet in height and bears golden yellow coloured flowers. Fresh mustards feature dark green coloured broad leaves with flat surface and may have either toothed, frilled or lacy edges depending up on the cultivar type. Its light-green stem branches out extensively into many laterals. Mustard seeds used as spice are being used extensively in cooking as well as in oil production n many parts of South-Asian region.

Health Benefits of Mustard Greens

- ☆ Mustard greens, like spinach, are the storehouse of many phyto-nutrients that have health promotional and disease prevention properties.
- ☆ Mustards are very low in calories (26 calories per 100 g raw leaves) and fats. However, its dark-green leaves contain a very good amount of fiber that helps control cholesterol level by interfering with its absorption in the gut. Additionally, adequate fiber in the food aids in smooth bowel movements and thereby offers protection from hemorrhoids, constipation as well as colon cancer diseases.

Mustard Greens (*Brassica juncea*), Fresh, Raw, Nutrition Value per 100 g

Principle	*Nutrient Value*	*Percentage of RDA*
Energy	26 Kcal	1 per cent
Carbohydrates	4.9 g	4 per cent
Protein	2.70 g	5 per cent
Total Fat	0.20 g	1 per cent
Cholesterol	0 mg	0 per cent
Dietary Fiber	3.30 g	9 per cent
Vitamins		
Folates	187 μg	47 per cent
Niacin	0.800 mg	5 per cent
Pantothenic acid	0.210 mg	5 per cent
Pyridoxine	0.180 mg	14 per cent
Riboflavin	0.110 mg	8 per cent
Thiamin	0.080 mg	7 per cent
Vitamin A	10500 IU	350 per cent
Vitamin C	70 mg	117 per cent
Vitamin E	2.01 mg	13 per cent
Vitamin K	497.3 μg	414 per cent
Electrolytes		
Sodium	25 mg	2 per cent
Potassium	354 mg	7.5 per cent
Minerals		
Calcium	103 mg	10 per cent
Copper	0.147 mg	16 per cent
Iron	1.46 mg	18 per cent
Magnesium	32 mg	8 per cent
Manganese	0.480 mg	21 per cent
Selenium	0.9 μg	1.5 per cent
Zinc	0.20 mg	2 per cent
Phyto-nutrients		
Carotene-β	6300 μg	—
Crypto-xanthin-β	0 μg	—
Lutein-zeaxanthin	9900 μg	—

☆ The greens are supposed to be one of the highest among leafy vegetables, which provide vitamin K. 100 g of fresh leaves contains about 497 μg or about 500 per cent of daily requirement of vitamin K-1 (phylloquinone). Vitamin K has found to have a potential role in bone mass building function by promoting osteo-trophic activity in the bone. It also has established role in Alzheimer's disease patients by limiting neuronal damage in the brain.

- Fresh leaves are also a very good source of folic acid. 100 g provide about 187 µg (about 47 per cent of RDA) of folic acid. This water-soluble vitamin has an important role in DNA synthesis and when given before, and early pregnancy may help prevent neural tube defects in the newborn baby.
- Mustard greens are rich source of anti-oxidants flavonoids, indoles, sulforaphane, carotenes, lutein and zeaxanthin. Indoles, mainly *di-indolyl-methane* (*DIM*) and *sulforaphane* have proven benefits against prostate, breast, colon and ovarian cancers by virtue of their cancer-cell growth inhibition, cytotoxic effects on cancer cells.
- Fresh mustard leaves are an excellent source of vitamin-C. Provides 70 µg or about 117 per cent of RDA per 100 g. Vitamin-C (ascorbic acid) is a powerful natural anti-oxidant that offers protection against free radical injury and flu-like viral infections.
- Mustard leaves are also incredible sources of vitamin-A (provide 10500 IU or 350 per cent of RDA per 100 g). Vitamin A is an essential nutrient for maintaining healthy mucus membranes and skin and is required for good eye-sight. Consumption of natural fruits rich in flavonoids helps to protect from lung and oral cavity cancers.
- Fresh mustard greens are an excellent source of several essential minerals such as calcium, iron, magnesium, potassium, zinc, selenium, and manganese.
- Regular consumption of mustard greens in the diet is known to prevent arthritis, osteoporosis, iron deficiency anaemia and believed to protect from cardiovascular diseases, asthma and colon and prostate cancers.

77

Okra

Okra, also known as *"lady finger"* is one of the highly nutritious vegetables, usually eaten while the pod is green, tender, and immature. Botanically, this perennial flowering plant belongs to the Malvaceae (*mallows*) family and named scientifically as Abelmoschus esculentus. The plant is cultivated throughout the tropical and warm temperate regions around the world for its fibrous fruits or "pods." It grows best in well-drained and manure rich soil. The plant bears numerous dark green coloured pods measuring about 5-15 cm in length. It takes about 45-60 days to get ready-to-harvest fruits. Internally, the pods feature small, round, mucilaginous white coloured seeds arranged in vertical rows. The pods are handpicked while just short of reaching maturity and eaten as a vegetable.

Health Benefits of Okra

- ✰ The pods are among the very low calorie vegetables. They provide just 30 calories per 100 g besides containing no saturated fats or cholesterol. Nonetheless, they are rich sources of dietary fiber, minerals, and vitamins; often recommended by nutritionists in cholesterol controlling and weight reduction programs.
- ✰ The rich fiber and *mucilaginous* content in okra pods help in smooth peristalsis of digested food particles and relieve constipation condition.
- ✰ The pods contain healthy amounts of vitamin A, and flavonoid anti-oxidants such as beta carotenes, xanthin and lutein. It is one of the *green* vegetables with highest levels of these anti-oxidants. These compounds are known to have antioxidant properties and are essential for vision. Vitamin A is also required for maintaining healthy mucus membranes and skin. Consumption of natural vegetables and fruits rich in flavonoids helps to protect from lung and oral cavity cancers.

Okra (*Abelmoschus esculentus*), Fresh, Raw Pods, Nutrition Value per 100 g

Principle	*Nutrient Value*	*Percentage of RDA*
Energy	1.5 per cent	31 Kcal
Carbohydrates	7.03 g	5.4 per cent
Protein	2.0 g	4 per cent
Total Fat	0.1 g	0.5 per cent
Cholesterol	0 mg	0 per cent
Dietary Fiber	9 per cent	3.2 g
Vitamins		
Folates	88 µg	22 per cent
Niacin	1.000 mg	6 per cent
Pantothenic acid	0.245 mg	5 per cent
Pyridoxine	0.215 mg	16.5 per cent
Riboflavin	0.060 mg	4.5 per cent
Thiamin	0.200 mg	17 per cent
Vitamin C	21.1 mg	36 per cent
Vitamin A	375 IU	12.5 per cent
Vitamin E	0.36 mg	2.5 per cent
Vitamin K	53 µg	44 per cent
Electrolytes		
Sodium	8 mg	0.5 per cent
Potassium	303 mg	6 per cent
Minerals		
Calcium	81 mg	8 per cent
Copper	0.094 mg	10 per cent
Iron	0.80 mg	10 per cent
Magnesium	57 mg	14 per cent
Manganese	0.990 mg	43 per cent
Phosphorus	63 mg	9 per cent
Selenium	0.7 µg	1 per cent
Zinc	0.60 mg	5.5 per cent
Phyto-nutrients		
Carotene-β	225 µg	—
Crypto-xanthin-β	0 µg	—
Lutein-zeaxanthin	516 µg	—

✰ Fresh pods are the good source of folates; provide about 22 per cent of RDA per 100 g. Consumption of foods rich in folates, especially during the pre-conception period helps decrease the incidence of neural tube defects in the offspring.

- The gumbo pods are also an excellent source of anti-oxidant vitamin, vitamin-C, providing about 36 per cent of daily-recommended levels. Research suggests that consumption of foods rich in vitamin-C helps the body develops immunity against infectious agents, reduce episodes of cold and cough and protect the body from harmful free radicals.
- The veggies are rich in B-complex group of vitamins like niacin, vitamin B-6 (pyridoxine), thiamin and pantothenic acid. The pods also contain good amounts of vitamin K. Vitamin K is a co-factor for blood clotting enzymes and is required for strengthening of bones.
- The pods are an also good source of many important minerals such as iron, calcium, manganese and magnesium.

78

Onion

Can you imagine a recipe without the Onions? This wonderful bulb vegetable, one of the oldest edible sources known to humankind, is found in a bewildering array of recipes and preparations, be it your favourite salad, or mouth-watering gravy or curries. It has also been in used in traditional medicines since ancient times for its health promoting and curative properties. Botanically, the vegetable belongs to the *Alliaceae* family of the genus, *Allium* and known scientifically as: *Allium cepa*. The plant grows about 2 feet tall and bears an underground globular stem, which consists of modified leaves arranged in whorls. There are many cultivar onions grown around the world. The average crop takes about three to four months time. Top greens or scallions and flower heads are also eaten all around the world.

The sharp, pungent smell of onions is due to its sulfur compound, *allyl propyl disulphide*. Spanish red onions are generally less strongly flavoured than white or brown varieties, which makes them ideal for use in raw salads. Shallot (*Allium cepa* L. var. *aggregatum*) is a variety of the onion that produces a cluster of small-elongated bulbs from a single planted bulb. Shallots are relatively smaller and tastes sweeter than onions.

Health Benefits of Onions

- ☆ Onions are very low in calories (just 40 calories per 100 g) and fats; however, rich in soluble dietary fiber.
- ☆ Phyto-chemical compounds *allium* and *Allyl disulphide* in the onion convert to *allicin* by enzymatic reaction when its modified leaves are distorted (crushing, cutting, etc.). Studies have shown that these compounds have anti-mutagenic (protects from cancers) and anti-diabetic properties (helps lower blood sugar levels in diabetics).
- ☆ Laboratory studies show that *allicin* reduces cholesterol production by inhibiting the *HMG-CoA reductase* enzyme in the liver cells. Further, it also found to have anti-bacterial, anti-viral, and anti-fungal activities.

Onoin (*Allium cepa*), Raw, Nutrition Value per 100 g

Principle	*Nutrient Value*	*Percentage of RDA*
Energy	40 Kcal	2 per cent
Carbohydrates	9.34 g	7 per cent
Protein	1.10 g	2 per cent
Total Fat	0.10 g	0.5 per cent
Cholesterol	0 mg	0 per cent
Dietary Fiber	1.7 g	4.5 per cent
Vitamins		
Folates	19 µg	5 per cent
Niacin	0.116 mg	1 per cent
Pantothenic acid	0.123 mg	2.5 per cent
Pyridoxine	0.120 mg	9 per cent
Riboflavin	0.027 mg	2 per cent
Thiamin	0.046 mg	4 per cent
Vitamin A	2 IU	0 per cent
Vitamin C	7.4 mg	12 per cent
Vitamin E	0.02 mg	0 per cent
Electrolytes		
Sodium	4 mg	0 per cent
Potassium	146 mg	3 per cent
Minerals		
Calcium	23 mg	2 per cent
Copper	0.039 mg	4 per cent
Iron	0.0.21 mg	3 per cent
Magnesium	10 mg	2.5 per cent
Manganese	0.129 mg	5.5 per cent
Phosphorus	29 mg	4 per cent
Zinc	0.17 mg	1.5 per cent
Phyto-nutrients		
Carotene-β	1 µg	—
Cryptoxanthin-β	0 µg	—
Lutein-zeaxanthin	4 µg	—

☆ In addition, *Allicin* also decreases blood vessel stiffness by releasing nitric oxide (NO) and thereby bring a reduction in the total blood pressure. Further, it blocks platelet-clot formation and has fibrinolytic action in the blood vessels. Altogether, it helps decrease an overall risk of coronary artery disease (CAD), peripheral vascular diseases (PVD), and stroke.

- ✰ Onions are rich source of chromium, the trace mineral that helps tissue cells respond appropriately to insulin levels in the blood. It thus helps facilitate insulin action and control sugar levels in diabetes.
- ✰ They are an also good source of antioxidant flavonoid *quercetin,* which is found to have anti-carcinogenic, anti-inflammatory, and anti-diabetic functions.
- ✰ They are also good in antioxidant vitamin, vitamin-C and mineral manganese. Manganese is required as a co-factor for anti-oxidant enzyme, *superoxide dismutase.* In addition, *isothiocyanate* anti-oxidants in them help provide relief from cold and flu by exerting anti-inflammatory actions.
- ✰ Onions are also good in B-complex group of vitamins like pantothenic acid, pyridoxine, folates and thiamin. Pyridoxine or vitamin B-6 helps keep up GABA levels in the brain, which works against neurotic conditions.

79

Parsnips

Parsnips are closely related to carrot family of vegetables; grown for their sweet, succulent underground taproots. The root vegetable has similar appearance and growth characteristics as other *Apiaceae* family members like carrots, parsley, celery, cumin, dill, etc. Botanically, they belong to the *Umbelliferae* (Apiaceae) family of the genus: *Pastinaca*. Scientific name: *Pastinaca sativa. Pastinaca sativa* is a biennial cool season crop native to Mediterranean region. In the first year, it grows about 1-1.5 meter tall and bears underground taproots, which are generally harvested after the first-frost of the season. If left undisturbed, the plant develops umbrella-shaped clusters of small yellow flowers and seeds during the next season. Its fleshy, stout roots appear like that of carrots, but are white or cream in colour and sweeter than that of carrots. Good winter frost is essential for better crop, as frost converts much of the starch to sugars and helps develop long, narrow, and firm parsnips. The roots are generally harvested when they reach about six to ten inches in length, by pulling the entire plant along with its root (uprooting) as in carrots.

Health Benefits of Parsnips

- ☆ Generally, parsnip contains more sugar than carrots, radish, and turnips. It has calories (100 g provide 75 calories) comparable to that of some fruits like banana, and grapes. Nonetheless, its sweet, juicy root is rich in several health-benefiting phyto-nutrients, vitamins, minerals, and fiber.
- ☆ It is one of the excellent sources of soluble and insoluble dietary fiber. 100 g root provides 4.9 mg or 13 per cent of fiber. Adequate fiber in the diet helps reduce blood cholesterol levels, obesity and constipation conditions.
- ☆ As in carrots and other members of apiaceae family vegetables, parsnip too contains many poly-acetylene anti-oxidants such as *falcarinol, falcarindiol, panaxydiol,* and *methyl-falcarindiol.*

Parsnips (*Pastinaca sativa*), Fresh, Raw, Nutrition Value per 100 g

Principle	*Nutrient Value*	*Percentage of RDA*
Energy	75 Kcal	4 per cent
Carbohydrates	17.99 g	14 per cent
Protein	1.20 g	2 per cent
Total Fat	0.30 g	1 per cent
Cholesterol	0 mg	0 per cent
Dietary Fiber	4.9 g	13 per cent
Vitamins		
Folates	67 µg	17 per cent
Niacin	0.700 mg	4 per cent
Pantothenic acid	0.600 mg	12 per cent
Pyridoxine	0.90 mg	7 per cent
Riboflavin	0.050 mg	4 per cent
Thiamin	0.090 mg	7.5 per cent
Vitamin A	0 IU	0 per cent
Vitamin C	17 mg	29 per cent
Vitamin K	22.5 µg	19 per cent
Electrolytes		
Sodium	10 mg	<1 per cent
Potassium	375 mg	8 per cent
Minerals		
Calcium	36 mg	3.5 per cent
Copper	0.120 mg	13 per cent
Iron	0.59 mg	7.5 per cent
Magnesium	29 mg	7 per cent
Manganese	0.560 mg	24 per cent
Phosphorus	71 mg	10 per cent
Selenium	1.8 µg	3 per cent
Zinc	0.59 mg	5 per cent
Phyto-nutrients		
Carotene-α	0 µg	—
Carotene-β	0 µg	—
Crypto-xanthin-β	0 µg	—
Lutein-zeaxanthin	0 µg	—

☆ Several research studies from scientists at University of Newcastle at Tyne found that these compounds have anti-inflammatory, anti-fungal, and anti-cancer function and offer protection from colon cancer and acute lymphoblastic leukemia (ALL).

- ☆ Fresh roots are also good in vitamin C; provide about 17 mg or 28 per cent of RDA. Vitamin C is a powerful water-soluble anti-oxidant, easily available to us from natural sources. It helps the body maintain healthy connective tissue, teeth, and gum. Its anti-oxidant property helps protect from diseases and cancers by scavenging harmful free radicals from the body.
- ☆ Further, the root is rich in many B-complex groups of vitamins such as folic acid, vitamin B-6 (pyridoxine), thiamin, and pantothenic acid as well as vitamin K and vitamin E.
- ☆ In addition, it also has healthy levels of minerals like iron, calcium, copper, potassium, manganese and phosphorus. Potassium is an important component of cell and body fluids that helps controlling heart rate and blood pressure by countering effects of sodium.

80

Green Peas

Starchy, and sweet green peas or garden peas are one of the ancient cultivated vegetables grown for their succulent nutritious green-pods. Peas are probably originated in the sub-Himalayan plains of north-west India. Now, this versatile legume is one of the major commercial crops grown all over the temperate and semi-tropical regions. Botanically, pea plant is an herbaceous vine. It belongs to the family of *Fabaceae* of the genus, *Pisum*. Scientific name: *Pisum sativum*. Some of the common names include *English peas, sweet peas, garden peas, pease* etc. Pea is a quick growing, an annual herbaceous vine that requires the trellis to support growth. It flourishes well in well-drained, sandy soil supplemented with adequate moisture and cool weather conditions. Short stalked green pods appear during late winter or spring. The pods measure about 2-3 inches long, swollen or compressed, straight or slightly curved, filled with single row of 2-10 light-green coloured, smooth edible seeds.

In general, the pods harvested while they are just short of reaching maturity, at the point when their seeds are green, soft, sweet and edible as raw. Allowing the pods to mature further would make the seeds less sugary and turn colour to light-green to yellow. Pea tendrils are also edible. They are delicate, tender top shoots of young pea plants, featuring flavour akin to peas. The tendrils and leafy-shoots are favoured in cooking as well in salads. Snow peas or sugar snap peas are different species of peas wherein the whole immature green pod, including its outer peel eaten like a vegetable.

Health Benefits of Green Peas

- ☆ Green peas are one of the most nutritious leguminous vegetables, rich in health benefiting phyto-nutrients, minerals, vitamins and anti-oxidants.
- ☆ Peas are relatively low in calories on comparison with beans, and cowpeas. 100 g of green peas provide only 81 calories, and no cholesterol. Nonetheless, the legumes are a good source of proteins, and soluble as well as insoluble fiber.

Green Peas (*Pisum sativum*), Fresh, Raw, Nutrition Value per 100 g

Principle	*Nutrient Value*	*Percentage of RDA*
Energy	81 Kcal	4 per cent
Carbohydrates	14.45 g	11 per cent
Protein	5.42 g	10 per cent
Total Fat	0.40 g	2 per cent
Cholesterol	0 mg	0 per cent
Dietary Fiber	5.1 g	13 per cent
Vitamins		
Folates	65 µg	16 per cent
Niacin	2.090 mg	13 per cent
Pantothenic acid	0.104 mg	2 per cent
Pyridoxine	0.169 mg	13 per cent
Riboflavin	0.132 mg	10 per cent
Thiamin	0.266 mg	22 per cent
Vitamin A	765 IU	25.5 per cent
Vitamin C	40 mg	67 per cent
Vitamin E	0.13 mg	1 per cent
Vitamin K	24.8 µg	21 per cent
Electrolytes		
Sodium	5 mg	<1 per cent
Potassium	244 mg	5 per cent
Minerals		
Calcium	25 mg	2.5 per cent
Copper	0.176 mg	20 per cent
Iron	1.47 mg	18 per cent
Magnesium	33 mg	8 per cent
Manganese	0.410 mg	18 per cent
Selenium	1.8 µg	3 per cent
Zinc	1.24 mg	11 per cent
Phyto-nutrients		
Carotene-β	449 µg	—
Crypto-xanthin-β	0 µg	—
Lutein-zeaxanthin	2477 µg	—

☆ Fresh pea pods are excellent source of folic acid. 100 g provides 65 µg or 16 per cent of recommended daily levels of folates. Folates are B-complex vitamins required for DNA synthesis inside the cell. Well established research studies suggest that adequate folate rich foods in expectant mothers would help prevent neural tube defects in the newborn babies.

- ☆ Fresh green peas are very good in ascorbic acid (vitamin C). Contain 40 mg/100 g or 67 per cent of daily requirement of vitamin C. Vitamin C is a powerful natural water-soluble anti-oxidant. Vegetables rich in this vitamin helps body develop resistance against infectious agents and scavenge harmful, pro-inflammatory free radicals from the body.
- ☆ Peas contain phytosterols especially β-sitosterol. Studies suggest that vegetables like legumes, fruits and cereals rich in plant sterols help lower cholesterol levels in the body.
- ☆ Garden peas are also good in vitamin K. 100 g of fresh leaves contains about 24.8 μg or about 21 per cent of daily requirement of vitamin K-1 (phylloquinone). Vitamin K has found to have a potential role in bone mass building function by promoting osteo-trophic activity in the bone. It also has established role in Alzheimer's disease patients by limiting neuronal damage in the brain.
- ☆ Fresh green peas also contain adequate amounts of anti-oxidants flavonoids such as carotenes, lutein and zeaxanthin as well as vitamin-A (provide 765 IU or 25.5 per cent of RDA per 100 g). Vitamin A is an essential nutrient required for maintaining health of mucus membranes, skin and eye-sight. Further, consumption of natural fruits rich in flavonoids helps to protect from lung and oral cavity cancers.
- ☆ In addition to folates, peas are also good in many other essential B-complex vitamins such as pantothenic acid, niacin, thiamin, and pyridoxine. Furthermore, they are rich source of many minerals such as calcium, iron, copper, zinc, and manganese.

81

Potato

Potato, nutritionally rich tuberous root vegetable, is a good source of starch, vitamins and fiber. The humble tuber is one of the most widely grown perennial crops and one of the cheap staple food items of the poor population all over the world. Botanically, it belongs to the various perennial subspecies of *Solanum tuberosum* of the *Solanaceae* family. The plant grows about 12 to 18 inches in length and bears many tubers underground. The tubers usually have round to oval *or* oblong shape and vary widely in size. Internally, the flesh features bright cream-white, rose red or russet colour depending on the variety with moist texture and have special buttery "potato" flavour.

Some of the popular cultivars are:

- ☆ White/yellow skin and flesh: Yukan gold, Yellow finn, Russian banana, Milva.
- ☆ Red skin and flesh: Ida rose, Norland, Cal red, French fingerling.
- ☆ Russet skin and flesh: Russet burbank, Ranger russet, Utamilla russet.

Health Benefits of Potato

- ☆ Potatoes are one of the richest sources of starch, vitamins, minerals and dietary fiber. 100 g provides 70 calories; however, they contain very little fat (just 0.1 g per100 g) and no cholesterol.
- ☆ They are very good natural sources of both soluble and insoluble fiber. The dietary fiber in them increases the bulk of the stool, thus, it helps prevent constipation, decrease absorption of dietary cholesterol and there by lower plasma LDL cholesterol. Additionally, the rich fiber content also helps protect from colon polyps and cancer.
- ☆ The fiber content aids in slow digestion starch and absorption of simple sugars in the gut. It thus helps keep blood sugar levels within the normal

range and avoid wide fluctuations. For the same reason, potato is considered as reliable source of carbohydrates in diabetics.

Potatoes (*Solanum tuberosum*), Nutritive Value per 100 g

Principle	*Nutrient Value*	*Percentage of RDA*
Energy	70 Kcal	3.5 per cent
Carbohydrates	15.90 g	12 per cent
Protein	1.89 g	3 per cent
Total Fat	0.10 g	0.5 per cent
Cholesterol	0 mg	0 per cent
Dietary Fiber	2.5 g	7 per cent
Vitamins		
Folates	18mcg	4.5 per cent
Niacin	1.149 mg	7 per cent
Pantothenic acid	0.279 mg	6 per cent
Pyridoxine	0.239 mg	18 per cent
Riboflavin	0.038 mg	3 per cent
Thiamin	0.081 mg	7 per cent
Vitamin A	7 IU	<1 per cent
Vitamin C	11.4 mg	20 per cent
Vitamin K	2.9 mcg	2.5 per cent
Electrolytes		
Sodium	6 mg	0.4 per cent
Potassium	455 mg	10 per cent
Minerals		
Calcium	10 mg	1 per cent
Iron	0.73 mg	9 per cent
Magnesium	22mg	5.5 per cent
Manganese	0.141mg	6 per cent
Phosphorus	61 mg	9 per cent
Zinc	0.33 mg	3 per cent
Phyto-nutrients		
Carotene-β	4 mcg	—
Crypto-xanthin-β	0 mcg	—
Lutein-zeaxanthin	21 mcg	—

☆ The tubers are one of the richest sources of B-complex group of vitamins such as pyridoxine (vitamin B6), thiamin, niacin, pantothenic acid and folates.

- Fresh potato along with its skin is good source of antioxidant vitamin; vitamin-C. 100 g of fresh tuber provides 11.4 mg or 20 per cent of daily required levels of this vitamin. Regular consumption of foods rich in vitamin-C helps body develop resistance against infectious agents and scavenge harmful, pro-inflammatory free radicals.
- They also contain adequate amounts of many essential minerals like Iron, manganese, magnesium, phosphorous, copper and potassium.
- Red and *russet potatoes* contain good amount vitamin A, and antioxidant flavonoids like carotenes and zeaxanthins.
- Recent studies at Agricultural research service (by plant genetics scientist Roy Navarre) suggests that flavonoid antioxidant, *quercetin* present in potatoes has anti-cancer and cardio-protective properties.

82

Pumpkin

Pumpkin fruit is one of the widely grown vegetables that are incredibly rich in vital antioxidants, and vitamins. This humble backyard low calorie vegetable contains vitamin A, flavonoid poly-phenolic antioxidants such as leutin, xanthin, and carotenes in abundance. The plant is a fast-growing vine that creeps on the surface in a similar fashion like that of other *Cucurbitaceae* family vegetables and fruits such as cucumber, squash, cantaloupes etc. It is one of the most popular field crops cultivated around the world, including the USA at the commercial scale for its fruit, and seeds.

Pumpkins vary greatly in shape, size and colours. Giant pumpkins generally weigh 4–6 kg with the largest capable of reaching a weight of over 25 kg. Golden-nugget pumpkins are flat, smaller in size and have sweet creamy orange colour flesh. Pumpkins in general feature orange or yellow colour; however, some varieties exhibit dark to pale green, brown, white, red and gray. The colour is largely influenced by yellow-orange pigments in their skin and pulp. The rind is smooth with light, vertical ribs. In structure, its fruit pulp features golden-yellow to orange colour depending up on the poly-phenolic pigments in it. The fruit has a hollow center, with numerous small, off-white coloured seeds interspersed in the net like structure. Pumpkin seeds are a great source of protein, minerals, vitamins and omega-3 fatty acids.

Health Benefits of Pumpkin

- ☆ It is one of the very low calorie vegetables. 100 g fruit provides just 26 calories and contains no saturated fats or cholesterol; however, it is rich in dietary fiber, anti-oxidants, minerals, vitamins. The vegetable is one of the food items recommended by dieticians in cholesterol controlling and weight reduction programmes.
- ☆ Pumpkin is a storehouse of many anti-oxidant vitamins such as vitamin-A, vitamin-C and vitamin-E.

Pumpkin (*Cucurbita spp.*), fresh, Nutritive Value per 100 g

Principle	*Nutrient Value*	*Percentage of RDA*
Energy	26 Kcal	1 per cent
Carbohydrates	6.50 g	5 per cent
Protein	1.0 g	2 per cent
Total Fat	0.1 g	0.5 per cent
Cholesterol	0 mg	0 per cent
Dietary Fiber	0.5 g	2 per cent
Vitamins		
Folates	16 mcg	4 per cent
Niacin	0.600 mg	4 per cent
Pantothenic acid	0.298 mg	6 per cent
Pyridoxine	0.061 mg	5 per cent
Riboflavin	0.110 mg	8.5 per cent
Thiamin	0.050 mg	4 per cent
Vitamin A	7384 IU	246 per cent
Vitamin C	9.0 mg	15 per cent
Vitamin E	1.06 mg	7 per cent
Vitamin K	1.1 mcg	1 per cent
Electrolytes		
Sodium	1 mg	0.5 per cent
Potassium	340 mg	7 per cent
Minerals		
Calcium	21 mg	2 per cent
Copper	0.127 mg	14 per cent
Iron	0.80 mg	10 per cent
Magnesium	12 mg	3 per cent
Manganese	0.125 mg	0.5 per cent
Phosphorus	44mg	5 per cent
Selenium	0.3 mcg	<0.5 per cent
Zinc	0.32 mg	3 per cent
Phyto-nutrients		
Carotene-α	515 mcg	—
Carotene-β	3100 mcg	—
Crypto-xanthin-β	2145 mcg	—
Lutein-zeaxanthin	1500 mcg	—

- ☆ With 7384 mg per 100 g, it is one of the vegetables in the Cucurbitaceae family featuring highest levels of vitamin-A, providing about 246 per cent of RDA. Vitamin A is a powerful natural anti-oxidant and is required by the body for maintaining the integrity of skin and mucus membranes. It is also an essential vitamin for good visual sight. Research studies suggest that natural foods rich in vitamin A help a body protects against lung and oral cavity cancers.

- ☆ It is also an excellent source of many natural poly-phenolic flavonoid compounds such as α, β carotenes, cryptoxanthin, lutein and zeaxanthin. Carotenes convert into vitamin A inside the body.
- ☆ Zeaxanthin is a natural anti-oxidant which has UV (ultra-violet) rays filtering actions in the macula lutea in retina of the eyes. Thus, it helps protect from "age-related macular disease" (ARMD) in the elderly.
- ☆ The fruit is a good source of B-complex group of vitamins like folates, niacin, vitamin B-6 (pyridoxine), thiamin and pantothenic acid.
- ☆ It is also rich source of minerals like copper, calcium, potassium and phosphorus.
- ☆ Pumpkin seeds indeed are an excellent source of dietary fiber and mono-unsaturated fatty acids, which are good for heart health. In addition, the seeds are concentrated sources of protein, minerals and health-benefiting vitamins. For instance, 100 g of pumpkin seeds provide 559 calories, 30 g of protein, 110 per cent RDA of iron, 4987 mg of niacin (31 per cent RDA), selenium (17 per cent of RDA), zinc (71 per cent) etc., but no cholesterol. Further, the seeds are an excellent source of health promoting amino acid tryptophan. Tryptophan is converted to GABA in the brain.

83

Purslane

Soft, succulent purslane leaves have more omega-3 fatty acids than in some of the fish oils. If you are a vegetarian and pledge to avoid all forms of animal products, then here is the answer! Go for this healthy dark-green leafy vegetable and soon you will forget fish. Botanically, this herbaceous leafy vegetable belongs to the family of *Portulacaceae* and scientifically known as *Portulaca oleracea.* Other common names in place for this green leafy vegetable are *pursley, pigweed, or verdolaga*.

Purslane is native to Indian sub-continent and now widely distributed across the continents actually as a wild weed. There exist varieties of pusley with variation in leaf size, thickness and leaf arrangement and pigment distribution. It is actually hard herb plant requiring comparatively less water and soil nutrients and grows well in sunny conditions. The plant grows up to 12-15 cm in height as a low-lying spread. Pursley is widely grown in many Asian and European regions as a staple leafy vegetable. Its leaves appear thick, contain mucilaginous substance, and have a slightly sour and salty taste. Leaves and tender stems have a slightly sour, and salty taste. In addition to succulent stems and leaves, its yellow flower buds are also edible. Purslane seeds, appear like black tea powder, are often used to make some herbal drinks.

Health Benefits of Purslane

- ☆ This wonderful green leafy vegetable is very low in calories (just 16 kcal/ 100g) and fats; nonetheless, it is rich in dietary fiber, vitamins, and minerals.
- ☆ Fresh leaves contain surprisingly more omega-3 fatty acids (α-linolenic acid) than any other leafy vegetable plant. 100 grams of fresh purslane leaves provide about 350 mg of α-linolenic acid. Research studies show that consumption of foods rich in ù-3 fatty acids may reduce the risk of coronary heart disease, stroke, and help prevent the development of ADHD, autism, and other developmental differences in children.

Purslane (*Portulaca oleracea*), Raw, Fresh, Nutritive Value per 100 g

Principle	*Nutrient Value*	*Percentage of RDA*
Energy	16 Kcal	1.5 per cent
Carbohydrates	3.4 g	3 per cent
Protein	1.30 g	2 per cent
Total Fat	0.1 g	0.5 per cent
Cholesterol	0 mg	0 per cent
Vitamins		
Folates	12 µg	3 per cent
Niacin	0.480 mg	3 per cent
Pantothenic acid	0.036 mg	1 per cent
Pyridoxine	0.073 mg	5.5 per cent
Riboflavin	0.112 mg	8.5 per cent
Thiamin	0.047 mg	4 per cent
Vitamin A	1320 IU	44 per cent
Vitamin C	21 mg	35 per cent
Electrolytes		
Sodium	45 mg	3 per cent
Potassium	494 mg	10.5 per cent
Minerals		
Calcium	65 mg	6.5 per cent
Copper	0.113 mg	12.5 per cent
Iron	1.99 mg	25 per cent
Magnesium	68 mg	17 per cent
Manganese	0.303 mg	13 per cent
Phosphorus	44 mg	6 per cent
Selenium	0.9 µg	2 per cent
Zinc	0.17 mg	1.5 per cent

- ✰ It is an excellent source of Vitamin A, (1320 IU/100 g, provides 44 per cent of RDA) one of the highest among green leafy vegetables. Vitamin A is a known powerful natural antioxidant and is essential for vision. This vitamin is also required to maintain healthy mucus membranes and skin. Consumption of natural vegetables and fruits rich in vitamin A is known to help to protect from lung and oral cavity cancers.
- ✰ Purslane is also a rich source of vitamin C, and some B-complex vitamins like riboflavin, niacin, pyridoxine and carotenoids, as well as dietary minerals, such as iron, magnesium, calcium, potassium, and manganese.
- ✰ Furthermore, present in purslane are two types of betalain alkaloid pigments, the reddish *beta-cyanins* and the yellow *beta-xanthins*. Both pigment types are potent anti-oxidants and have been found to have anti-mutagenic properties in laboratory studies.

84

Radicchio

Radicchio is a quick growing Mediterranean red coloured leafy vegetable. It is actually one of the varieties of leaf chicory used in salads in Veneto regions of Italy for centuries. Its wine-red succulent, bitter flavoured leaves hold several unique compounds like lactucopicrin (intybin), zeaxanthin, vitamin K and several other vitamins, minerals and antioxidants. Binomially, this beautiful leafy vegetable belongs to the *Asteracae* or daisy family, of the genus: *Cichorium*. Scientific name: *Cichorium intybus* L.

Radicchio is a perennial, small cabbage like plant. It prefers cool weather supplanted with well draining, fertile, moisture rich soil. The crop is ready to harvest after about 75-90 days after seedling. Hot weather and inadequate watering might results in small, dense, and bolting heads. Well-grown radicchio features compact wine-red colour leaves with prominent white veins about the size of a romaine leaves with prominent white veins about the size of romaine lettuce head. Different cultivars are grown generally by the name of Veneto provincial cities. **Chioggia** variety has compact, beet-red, bitter leaves. **Treviso** variety has been less pungent, long, conical, compactly arranged leaves. Radicchio variegate di Castlefranco is a hybrid between radicchio and endive (*Cichorium endiva*). Castlefranco has loose, mild flavour leaves. **Verona** is another non-heading type and has red colour, open leaves with broad white veins as in cabbage.

Health Benefits of Radicchio

- ☆ Radicchio, like other chicory class vegetables, is very low in calories. 100 g fresh leaves provide just 23 calories.
- ☆ The bitter principle in the radicchio is lactucopicrin (intybin), a sesquiterpene lactone. Lactucopicrin is a potent anti-malarial agent and has a sedative and analgesic (painkiller) effect.

Radicchio (*Cichorium intybus* L.), Raw, Nutritive Value per 100 g

Principle	*Nutrient Value*	*Percentage of RDA*
Energy	23 Kcal	1 per cent
Carbohydrates	4.48 g	3.5 per cent
Protein	1.43 g	3 per cent
Total Fat	0.25 g	1 per cent
Cholesterol	0 mg	0 per cent
Dietary Fiber	0.9 g	2 per cent
Vitamins		
Folates	60 μg	15 per cent
Niacin	0.255 mg	1.5 per cent
Pantothenic acid	0.269 mg	5 per cent
Pyridoxine	0.057 mg	4 per cent
Riboflavin	0.028 mg	2 per cent
Thiamin	0.016 mg	1 per cent
Vitamin A	27 IU	1 per cent
Vitamin C	8 mg	13 per cent
Vitamin E	2.26 mg	15 per cent
Vitamin K	255.2 μg	212 per cent
Electrolytes		
Sodium	22 mg	1.5 per cent
Potassium	302 mg	6 per cent
Minerals		
Calcium	19 mg	2 per cent
Copper	0.341 mg	38 per cent
Iron	0.57 mg	7 per cent
Magnesium	13 mg	3 per cent
Manganese	0.138 mg	6 per cent
Phosphorus	40 mg	5.5 per cent
Selenium	0.9 μg	1 per cent
Zinc	0.62 mg	6 per cent
Phyto-nutrients		
Carotene-β	16 μg	—
Carotene-α	0 μg	—
Cryptoxanthin-β	0 μg	—
Lutein-zeaxanthin	8832 μg	—

☆ Its leaves are an excellent source of phenolic flavonoid antioxidants such as *zeaxanthin* and *lutein*. 100 g leaves provide 8832 μg of these pigments.

Zeaxanthin is a xanthophyll category of flavonoid carotenoid (yellow pigment) which concentrates mainly in the central retina in the eyes. Together with lutein, it helps protect eyes from *age-related macular disease* (ARMD) by filtering harmful ultra-violet rays.

- Fresh leaves contain moderate amounts of essential B-complex groups of vitamins such as folic acid, pantothenic acid (vitamin B5), pyridoxine (vitamin B6) and thiamin (vitamin B1), niacin (B3). These vitamins are essential in the sense that body requires them from external sources to replenish and required for fat, protein and carbohydrate metabolism.
- Fresh radicchio is one of the excellent sources of vitamin K. 100 g provides about 255.2 µg or 212 per cent of daily-recommended values. Vitamin K has a potential role in bone health by promoting osteotrophic (bone formation and strengthening) activity. Further, adequate vitamin-K levels in the diet help limiting neuronal damage in the brain; thus, has established role in the treatment of patients suffering from Alzheimer's disease.
- Further, it is an also good source of minerals like manganese, copper, iron, zinc, and potassium. Manganese is used as a co-factor for the antioxidant enzyme, *superoxide dismutase.* Potassium is an important intracellular electrolyte helps counter the hypertension effects of sodium.

85

Radish

Radish is one of the nutritious root vegetables featured in both salads as well as in main recipes. This widely used root vegetable belongs to the family of Brassica. In China, it along with cabbage and soybean curd (tofu) is believed as healthy food. A popular Chinese proverb goes like this, "Eating pungent radish and drinking hot tea, let the starved doctors beg on their knees." They are thought to be originated from the mainland China centuries ago but now cultivated and consumed throughout the world. Botanically the veggie is named as *Raphanus sativus.* Radishes come in different forms varying in size, colour and duration of required cultivation time. They can be broadly categorized into four main types-summers, fall, winter, and spring while growers classify them by shapes, colours, and sizes, such as black or white coloured, with round or elongated roots. Their biting pungent flavour comes from "isothiocyanate" compound in them, which ranges from mild in case of white-icicles to be very hot in red globe and other pigmented radishes. The top greens are also eaten as leafy-vegetables in some parts of the world.

Daikon or Japanese radishes are native to Asia. They are generally grown during winter months and have elongated smooth, icy-white roots.

Black Spanish radishes are peppery and more flavourful than their white counterparts.

Green radish is native to Northern China region. Its outer peel near the top stem end features leafy-green colour which, gradually changes to white colour near the lower tip. Inside, its flesh has beautiful jade green colour, sweet and less pungent flavour.

Watermelon radishes have watermelon like flesh inside. However, they taste sweet and less peppery, something similar to that of white varieties.

When left to grow for longer than the usual root harvest period, all radish varieties bear small flowers, which subsequently develop into edible fruit pods. Podding or a *rat-tailed* radish is a type of seed pod variety grown exclusively for their long rat-tail like tapering pods. The pods feature mild radish-like flavour and spiciness.

Health Benefits of Tadish

☆ Since ancient times, Chinese believe that eating radish and other brassica group vegetables such as cabbage, cauliflower, and napa would immensely benefit overall health.

Radish (*Raphanus sativus*), Fresh, Raw, Nutrition Value per 100 g

Principle	*Nutrient Value*	*Percentage of RDA*
Energy	16 Kcal	1 per cent
Carbohydrates	3.40 g	3 per cent
Protein	0.68 g	1 per cent
Total Fat	0.10 g	<1 per cent
Cholesterol	0 mg	0 per cent
Dietary Fiber	1.6 g	4 per cent
Vitamins		
Folates	25 µg	6 per cent
Niacin	0.254 mg	1.5 per cent
Pyridoxine	0.071 mg	5.5 per cent
Riboflavin	0.039 mg	3 per cent
Vitamin A	7 IU	<1 per cent
Vitamin C	14.8 mg	25 per cent
Vitamin E	0 mg	9 per cent
Vitamin K	1.3 µg	1 per cent
Electrolytes		
Sodium	39 mg	2.5 per cent
Potassium	233 mg	5 per cent
Minerals		
Calcium	25 mg	2.5 per cent
Copper	0.050 mg	5 per cent
Iron	0.34 mg	4 per cent
Magnesium	10 mg	2.5 per cent
Manganese	0.069 mg	2.5 per cent
Zinc	0.28 mg	2 per cent
Phyto-nutrients		
Carotene-β	4 µg	—
Carotene-α	0 µg	—
Lutein-zeaxanthin	10 µg	—

- ✰ They are are one of very low calorie root vegetables. Fresh root provides just 16 calories per 100 g., nonetheless; they are a very good source of anti-oxidants, electrolytes, minerals, vitamins and dietary fiber.
- ✰ Radish, like other cruciferous and Brassica family vegetables, contains *isothiocyanate* anti-oxidant compound called sulforaphane. Studies suggest that sulforaphane has proven role against prostate, breast, colon and ovarian cancers by virtue of its cancer-cell growth inhibition, and cyto-toxic effects on cancer cells.
- ✰ Fresh roots are rich in vitamin C; provide about 15 mg or 25 per cent of DRI of vitamin C per 100 g. Vitamin C is a powerful water soluble anti-oxidant required by the body for synthesis of collagen. Vitamin C helps the body scavenge harmful free radicals, prevention from cancers, inflammation and help boost immunity.
- ✰ In addition, they contain adequate levels of folates, vitamin B-6, riboflavin, thiamin and minerals such as iron, magnesium, copper and calcium.
- ✰ Further, they contain many phytochemicals like *indoles* which are detoxifying agents and zeaxanthin, lutein and beta carotene, which are flavonoid antioxidants. Their total antioxidant strength, measured in terms of oxygen radical absorbance capacity (ORAC value), is 1736 µmol TE/100 g.

86

Rhubarb

Rhubarb is a perennial vegetable grown for its attractive succulent rose red colour edible leafy stalks. This cool-season herbaceous plant is native to Siberia and popular in many regions of Europe and North America as "pie plant." In its natural habitat, the plant expands over the ground surface as a large spread. Botanically, it belongs to the family of *Polygonaceae*, of the genus: *Rheum*, and known as *Rheum rhabarbarum*. Rhubarb is easy to grow and lives for many years (10-15 years) once established. The plant is usually propagated by dividing the old rhizomes (roots). Well grown plant features broad heart shaped dark-green leaves with 12 to 18 inches long leaf-petioles. It is these stalks, which are used, and their top greens discarded, as they are unfit for human consumption. Usually its stalks can be harvested from second year onwards after planting when the foliage spread and stalks reached sufficient girth of about one to two inches thick. Several cultivars exist. Some of the popular varieties grown in the USA are *Canada red, cherry red, Burgess-`Colossal', MacDonald, ruby, valentine e*tc. Generally, red variety preferred as they have more subtle stalks.

Health Benefits of Rhubarb

- ✰ Rhubarb is one of the least calorie vegetables. 100 g fresh petioles provide just 21 calories. Nonetheless, it contains some vital phyto-nutrients such as dietary fiber, poly-phenolic anti-oxidants, minerals, and vitamins. Further, its petioles contain no saturated fats or cholesterol.
- ✰ The stalks are rich in several B-complex vitamins such as folates, riboflavin, niacin, vitamin B-6 (pyridoxine), thiamin, and pantothenic acid.
- ✰ Red colour stalks contain more vitamin A than in the green varieties. Further, the stalks also contain small amounts of poly-phenolic flavonoid compounds like β-carotene, zeaxanthin, and lutein. These compounds convert to vitamin A inside the body and deliver same protective effects of

vitamin A on the body. Vitamin A is a powerful natural anti-oxidant and is required by the body for maintaining the integrity of skin and mucus membranes. It is also an essential vitamin for healthy eye-sight. Research studies suggest that natural foods rich in vitamin A help the body protects against lung and oral cavity cancers.

Rhubarb (*Rheum rhabarbarum*), Fresh, Nutritive Value per 100 g

Principle	*Nutrient Value*	*Percentage of RDA*
Energy	21 Kcal	1 per cent
Carbohydrates	4.54 g	3 per cent
Protein	0.90 g	2 per cent
Total Fat	0.20 g	1 per cent
Cholesterol	0 mg	0 per cent
Dietary Fiber	1.8 g	5 per cent
Vitamins		
Folates	7 µg	2 per cent
Niacin	0.300 mg	2 per cent
Pantothenic acid	0.085 mg	1.5 per cent
Pyridoxine	0.024 mg	2 per cent
Riboflavin	0.030 mg	2 per cent
Thiamin	0.020 mg	2 per cent
Vitamin A	102 IU	3.5 per cent
Vitamin C	8 mg	13 per cent
Vitamin E	0.27 mg	2 per cent
Vitamin K	29.3 µg	24 per cent
Electrolytes		
Sodium	4 mg	0.5 per cent
Potassium	288 mg	6 per cent
Minerals		
Calcium	86 mg	8.5 per cent
Copper	0.021 mg	2 per cent
Iron	0.22 mg	3 per cent
Magnesium	12 mg	3 per cent
Manganese	0.196 mg	8.5 per cent
Phosphorus	14 mg	2 per cent
Selenium	1.1 µg	2 per cent
Zinc	0.10 mg	1 per cent
Phyto-nutrients		
Carotene-α	0 µg	—
Carotene-β	61 µg	—
Crypto-xanthin-β	0 µg	—
Lutein-zeaxanthin	170 µg	—

- As in other greens like kale, spinach, etc., rhubarb stalks also provide good amounts of vitamin-K. 100 g of fresh stalks provide 29.3 µg or about 24 per cent of daily recommended intake of this vitamin. Vitamin K has a potential role in bone health by promoting osteotrophic (bone formation and strengthening) activity. Adequate vitamin-K levels in the diet help limiting neuronal damage in the brain; thus, has established role in the treatment of Alzheimer's disease.
- Its stalks also contain healthy levels of minerals like iron, copper, calcium, potassium, and phosphorus. However, most of these minerals do not absorb into the body as they are subject to chelating into insoluble complexes by oxalic acid, and excreted out from the body.

87

Shallots

Shallots or *eschalots* are long slender bulbs in the Allium family of root vegetables. In general, they differ from the onions in being smaller, and grow in clusters of bulbs from each plant's root system. The bulbs are characteristically less pungency than that of onions and garlic, which makes them one of the favourite ingredients of the chefs all around the world. This bulb vegetable is probably originated in the Central Asian region and today, cultivated as a major commercial crop in all the continents. Botanically, it too like its fellow bulb and stem (leeks) vegetables belong to the *Alliaceae* family of the genus *Allium*. Scientific name: *Allium cepa* var. aggregatum.

Shallots are cool season perennials; however, they have grown as annual crop like onions. The easiest way to grow them is from sets, usually available from nearby garden centers. The plant reaches about 50 cm in height and bears 1-5 cloves of bulblets just underneath the surface. It takes about 100 days when the bulbs are ready for harvest. Mature bulbs are then cured under dry air and stored in the same manner as onions. Like in other allium species, shallot's top greens and flower heads are also eaten in many parts of the world.

Health Benefits of Shallots

- ✰ Shallots have better nutrition profile than onions. On weight per weight basis, they have more anti-oxidants, minerals, and vitamins than onions.
- ✰ They are rich source of flavonoid anti-oxidants such as quercetin, kemferfol...etc. Further; they contain sulphur anti-oxidant compounds such as *diallyl disulfide, diallyl trisulfide* and *allyl propyl disulfide*. These compounds convert to allicin through enzymatic action following disruption of their cell surface while crushing, and chopping.
- ✰ Research studies show that *allicin* reduces cholesterol production by inhibiting the *HMG-CoA reductase* enzyme in the liver cells. Further, it also found to have anti-bacterial, anti-viral, and anti-fungal activities.

Shallots (*Allium cepa* var. *aggregatum*), Raw, Nutrition Value per 100 g

Principle	*Nutrient Value*	*Percentage of RDA*
Energy	72 Kcal	3.6 per cent
Carbohydrates	16.80 g	13 per cent
Protein	2.50 g	5 per cent
Total Fat	0.10 g	0.5 per cent
Cholesterol	0 mg	0 per cent
Vitamins		
Folates	34 µg	9 per cent
Niacin	0.200 mg	1.5 per cent
Pantothenic acid	0.290 mg	6 per cent
Pyridoxine	0.345 mg	26.5 per cent
Riboflavin	0.020 mg	2 per cent
Thiamin	0.060 mg	5 per cent
Vitamin A	1190 IU	35 per cent
Vitamin C	8 mg	13 per cent
Electrolytes		
Sodium	12 mg	1 per cent
Potassium	334 mg	7 per cent
Minerals		
Calcium	37 mg	4 per cent
Copper	0.088 mg	10 per cent
Iron	1.20 mg	15 per cent
Magnesium	21 mg	5 per cent
Manganese	0.292 mg	13 per cent
Phosphorus	60 mg	8.5 per cent
Selenium	1.2 µg	2 per cent
Zinc	0.40 mg	4 per cent

- ✰ Allicin also decreases blood vessel stiffness by releasing vasodilator chemical *nitric oxide* (NO) and thereby help bring a reduction in the total blood pressure. Further research studies suggest that allicin inhibit the platelet clot-formation in the blood vessels that helps decrease an overall risk of coronary artery disease (CAD), peripheral vascular diseases (PVD), and stroke.
- ✰ The phyto-chemical compounds *allium* and *Allyl disulfide* in onion have anti-mutagenic (protects from cancers) and anti-diabetic properties (helps lower blood sugar levels in diabetics).
- ✰ Shallots contain several fold more concentration of vitamins and minerals than in onions, especially vitamin A, pyridoxine, folates, thiamin, vitamin

C etc. Pyridoxine (B-6) helps raises GABA chemical levels in the brain that help sooth nervous irritability. In addition, 100 g fresh shallots have 1190 IU (35 per cent RDA) of vitamin A. Vitamin A is a powerful antioxidant that helps protect from lung and oral cavity cancers.

- ☆ Further, they are also good in minerals and electrolytes than onions; particularly iron, calcium, copper, potassium, and phosphorus.

88

Spinach

Spinach (*Spinacia oleracea*) is a wonderful green-leafy vegetable often recognized as one of the *functional foods* for its nutritional, antioxidants and anti-cancer constituents. Its tender, crispy, dark-green leaves are favourite ingredients of chefs all around the planet. Botanically it belongs to the *Amaranthaceae* family and its scientific name: *Spinacia oleracea.* Spinacia plant grows about 1 foot in height. Although, it can be grown year round, fresh greens are best available just after the winter season in the Northern hemisphere from March through May and from September until November, in the South of the equatorial line. At least, two varieties of spinach are cultivated for their edible leaves; Savoy type with dark-green *crinkle* (*wrinkled*) *leaves* and *flat-leaf* type with smooth surfaced leaves.

Health Benefits of Spinach

- ✰ Spinach is store house for many phyto-nutrients that have health promotional and disease prevention properties.
- ✰ Very low in calories and fats (100 g of raw leaves provide just 23 calories). It contains a good amount of soluble dietary fiber; no wonder green spinach is one of the finest vegetable sources recommended in cholesterol controlling and weight reduction programs!
- ✰ Fresh 100 g of spinach contains about 25 per cent of daily intake of iron; one of the richest among green leafy vegetables. Iron is an important trace element required by the body for red blood cell production and as a co-factor for oxidation-reduction enzyme, *cytochrome-oxidase* during the cellular metabolism.
- ✰ Fresh leaves are rich source of several vital anti-oxidant vitamins like vitamin A, vitamin C, and flavonoid poly phenolic antioxidants such as lutein, zeaxanthin and beta-carotene. Together these compounds help act as protective scavengers against oxygen-derived free radicals and reactive

oxygen species (ROS) that play a healing role in ageing and various disease processes.

Spinach (*Spinacia oleracea*), Raw, Nutritive Value per 100 g

Principle	*Nutrient Value*	*Percentage of RDA*
Energy	23 Kcal	1 per cent
Carbohydrates	3.63 g	3 per cent
Protein	2.86 g	5 per cent
Total Fat	0.39 g	1.5 per cent
Cholesterol	0 mg	0 per cent
Dietary Fiber	2.2 g	6 per cent
Vitamins		
Folates	194 µg	48.5 per cent
Niacin	0.724 mg	4.5 per cent
Pantothenic acid	0.065 mg	1 per cent
Pyridoxine	0.195 mg	15 per cent
Riboflavin	0.189 mg	14.5 per cent
Thiamin	0.078 mg	6.5 per cent
Vitamin A	9377 IU	312 per cent
Vitamin C	28.1 mg	47 per cent
Vitamin E	2.03 mg	13.5 per cent
Vitamin K	482.9 µg	402 per cent
Electrolytes		
Sodium	79 mg	5 per cent
Potassium	558 mg	12 per cent
Minerals		
Calcium	99 mg	10 per cent
Copper	0.130 mg	14 per cent
Iron	2.71 mg	34 per cent
Magnesium	79 mg	20 per cent
Manganese	0.897 mg	39 per cent
Zinc	0.53 mg	5 per cent
Phyto-nutrients		
Carotene-β	5626 µg	—
Crypto-xanthin-β	0 µg	—
Lutein-zeaxanthin	12198 µg	—

☆ *Zeaxanthin*, an important dietary carotenoid, is selectively absorbed into the retinal macula lutea in the eyes where it is thought to provide antioxidant and protective light-filtering functions. It thus helps protect from "age-related macular disease" (ARMD), especially in the elderly.

- In addition, vitamin A is required for maintaining healthy mucus membranes and skin and is essential for normal eye-sight. Consumption of natural vegetables and fruits rich in vitamin A and flavonoids also known to help the body protect from lung and oral cavity cancers.
- Spinach leaves are an excellent source of vitamin K. 100 g of fresh greens provides 402 per cent of daily vitamin-K requirements. Vitamin K plays a vital role in strengthening the bone mass by promoting osteotrophic (bone building) activity in the bone. Additionally, it also has established role in patients' with *Alzheimer's disease* by limiting neuronal damage in the brain.
- This green leafy vegetable also contains good amounts of many B-complex vitamins such as vitamin-B6 (pyridoxine), thiamin (vitamin B-1), riboflavin, folates and niacin. Folates help prevent neural tube defects in the offspring.
- 100 g of farm fresh spinach has 47 per cent of daily recommended levels of *vitamin* C. Vitamin C is a powerful antioxidant, which helps the body develop resistance against infectious agents and scavenge harmful oxygen-free radicals.
- Its leaves also contain a good amount of minerals like *potassium*, manganese, magnesium, copper and zinc. Potassium is an important component of cell and body fluids that helps controlling heart rate and blood pressure. Manganese and copper are used by the body as a co-factor for the antioxidant enzyme, *superoxide dismutase.* Copper is required in the production of red blood cells. Zinc is a co-factor in many enzymes that regulate growth and development, sperm generation, digestion and nucleic acid synthesis.
- It is also rich source of omega-3 fatty acids.

Regular consumption of spinach in the diet helps prevent osteoporosis (weakness of bones), iron-deficiency anemia. Besides, it is believed to protect the body from cardiovascular diseases and cancers of colon and prostate.

89

Sweet Corn

Sweet corn is a special maize variety in which it's tender, delicious kernels eaten as a vegetable. In contrast to traditional field corn, sugar corn varieties are harvested when the ears just reached milk stage and used fresh since the sugars in the kernel quickly convert to starch. Corn is native to Central American region, which later introduced to the rest of the world by Spanish explorers. Genetically, it differs from the field corn by mutation at the sugary (su) locus. The crop has achieved success as one of the important commercial cash crops in many tropical and semi-tropical countries. Scientific name: *Zea mays* var. saccharata.

Corn grows to about 7-10 feet in height. It grows quickly under sunny, fertile, and well-drained soil supplemented with good moisture conditions. Each plant bears about 2-6, long husked ears filled with rows of tooth like seeds surrounding a central woody core (cob). Optimum pollination is essential for full kernel development. Several different sweet corn cultivars with many variations in their sweetness, colour, and maturation time are grown according to the local and regional requirements. Depending upon the cultivar type, the crop may be ready for harvesting in 65-90 days. Harvest sweet corn when the silk end of the ear is completely filled out, the silk has turned brown and the kernels are firm but in the milk stage. Oftentimes, farmers check the kernels by pricking them with the thumbnail in order to ascertain timing of harvest. Baby corns are very young, miniature ears harvested when their kernels are still at incipient stage. The central core is sweet and tender enough to be eaten raw. Baby corns measure about 3-5 inches in length and weighs about 20-50 g.

Health Benefits of Sweet Corn

☆ At 86 calories per 100 g, sugar corn kernels are moderately high in calories on comparison to other vegetables. However, fresh kernels have been much lower in calories than field corn and other cereals like wheat, rice, etc. Their calorie mainly comes from simpler carbohydrates like glucose, sucrose than complex sugars like amylose and amylopectin as in cereals.

Sweet corn (*Zea mays* var. *saccharata.*), Raw, Nutrition Value per 100 g

Principle	*Nutrient Value*	*Percentage of RDA*
Energy	86 Kcal	4 per cent
Carbohydrates	18.70 g	14 per cent
Protein	3.27 g	6 per cent
Total Fat	1.35 g	7 per cent
Cholesterol	0 mg	0 per cent
Dietary Fiber	2.0 g	5 per cent
Vitamins		
Folates	42 µg	10.5 per cent
Niacin	1.770 mg	11 per cent
Pantothenic acid	0.717 mg	14 per cent
Pyridoxine	0.093 mg	7 per cent
Riboflavin	0.055 mg	4 per cent
Thiamin	0.155 mg	13 per cent
Vitamin A	187 IU	6 per cent
Vitamin C	6.8 mg	11 per cent
Vitamin E	0.07 mg	<1 per cent
Vitamin K	0.3 µg	2 per cent
Electrolytes		
Sodium	15 mg	1 per cent
Potassium	270 mg	6 per cent
Minerals		
Calcium	2 mg	<1 per cent
Copper	0.054 mg	6 per cent
Iron	0.52 mg	6.5 per cent
Magnesium	37 mg	9 per cent
Manganese	0.163 mg	7 per cent
Selenium	0.6 µg	1 per cent
Zinc	0.46 mg	4 per cent
Phyto-nutrients		
Carotene-β	47 µg	—
Carotene-α	16 µg	—
Cryptoxanthin-β	115 µg	—
Lutein-zeaxanthin	644 µg	—

- ✰ Sweet corn is gluten-free cereal and may be used safely much like rice, quinoa, etc., in celiac disease individuals.
- ✰ Corn features high-quality phyto-nutrition profile comprising of dietary fiber, vitamins, and antioxidants in addition to moderate proportions of

minerals. It is one of the finest source dietary fibers; 100 g kernels provide 2 g or 5 per cent of daily-requirement of dietary-fiber. Together with slow digesting complex carbohydrates; moderate amounts of fiber in the food regulate a gradual rise in blood sugar levels. However, corn, in line with rice, potato, etc., is one of the high glycemic index food items, limiting its role as the chief food ingredient in diabetes patients.

- ✰ Yellow variety corn has significant levels of phenolic flavonoid pigment antioxidants such as β-*carotenes,* and *lutein, xanthins and cryptoxanthin* pigments along with vitamin A. 100 g fresh kernels provide 187 IU or 6 per cent of daily-requirement of vitamin-A. Altogether; these compounds are required for maintaining healthy mucus membranes, skin and vision. Consumption of natural foods rich in flavonoids helps to protect from lung and oral cavity cancers.
- ✰ Corn is a good source of phenolic flavonoid antioxidant, ferulic acid. Several research studies suggest that ferulic acid plays vital role in preventing cancers, ageing, and inflammation in humans.
- ✰ It also contains good levels of some of the valuable B-complex group of vitamins such as thiamin, niacin, pantothenic acid, folates, riboflavin, and pyridoxine. Many of these vitamins function as co-factors to enzymes during substrate metabolism.
- ✰ Further, it contains healthy amounts of some important minerals like zinc, magnesium, copper, iron, and manganese.

90

Sweet Potato

Sweet potato, not only is just sweet to your taste buds but also good for your cardiovascular health. This starchy root vegetable is rich source of flavonoid anti-oxidants, vitamins, minerals, and dietary fiber that are essential for optimal health. Botanically, this starch-rich tuber belongs to the family of *Convolvulaceae*, and named botanically as *Ipomoea batatas.* It is grown throughout the tropical and warm temperate regions. The crop requires just sufficient water and attention for their cultivation. The tuberous root features oblong/elongated shape with tapering ends and has smooth outer skin whose colour ranges from red, purple, brown, and white, depending up on the variety.

Sweet potatoes should not be confused with yams, another starchy root commonly grown in Western Africa. Yams are indeed larger in size that can grow up to 120 pounds in weight and 2 meters in length. Yams are the tropical crops and never grow where the temperature dips below 68 degrees F. Important differentiating features that distinguish sweet potatoes from yams are:

- ☆ Sweet potatoes (*Ipomoea batatas*) are dicotyledonous, relatively smaller and possess very thin peel.
- ☆ Whereas, yams are monocotyledons, larger, features thick, rough, dark brown to pink skin depending up on the cultivar type.

Internally, it has starchy flesh which, depending upon the pigment concentration, ranges from white through yellow, orange, and purple.

Boniatos, also known as *Cuban sweet-potatoes,* feature dry, starchy flesh underneath the reddish-brown skin. They have mildly sweet flavour and coked in a similar way like potatoes.

Sweet potato leaves (top greens) are also edible. In fact, the greens contain more nutrients and dietary fiber than some of the green-leafy vegetables.

Health Benefits of Sweet potato

☆ Sweet potato is one of the average calories starch foods (provide just 90 calories/100 g vs. 70 calories/100 g of potato). The tuber, however, contains no saturated fats or cholesterol and is rich source of dietary fiber, antioxidants, vitamins, and minerals.

Sweet Potatoes (*Ipomoea batatas*), Raw, Nutrition Value per 100 g

Principle	*Nutrient Value*	*Percentage of RDA*
Energy	86 Kcal	4 per cent
Carbohydrates	20.12 g	15.5 per cent
Protein	1.6 g	3 per cent
Total Fat	0.05 g	<0.5 per cent
Cholesterol	0 mg	0 per cent
Dietary Fiber	3 g	8 per cent
Vitamins		
Folates	11 µg	3 per cent
Niacin	0.557 mg	3.5 per cent
Pantothenic acid	0.80 mg	16 per cent
Pyridoxine	0.209 mg	15 per cent
Riboflavin	0.061 mg	5.5 per cent
Thiamin	0.078 mg	6.5 per cent
Vitamin A	14187 IU	473 per cent
Vitamin C	2.4 mg	4 per cent
Vitamin E	0.26 mg	2 per cent
Vitamin K	1.8 µg	1.5 per cent
Electrolytes		
Sodium	55 mg	3.5 per cent
Potassium	337 mg	7 per cent
Minerals		
Calcium	30 mg	3 per cent
Iron	0.61 mg	7.5 per cent
Magnesium	25 mg	6 per cent
Manganese	0.258 mg	11 per cent
Phosphorus	47 mg	7 per cent
Zinc	0.30 mg	3 per cent
Phyto-nutrients		
Carotene-α	7 µg	—
Carotene-β	8509 µg	—
Crypto-xanthin-β	0 µg	—

- Its calorie content mainly comes from starch, a complex carbohydrate. Sweet potato has higher amylose to the amylopectin ratio than potato. Amylose raises the blood sugar levels *slowly* on comparison to simple sugars and therefore, recommended as a healthy food supplement even in diabetes.
- The tuber is an excellent source of *flavonoid phenolic compounds* such as beta-carotene and vitamin A (*100 g tuber provides 14187 IU of vitamin A and 8509 μg of β-carotene*). The value is one of the highest among the root-vegetables categories. These compounds are powerful natural antioxidants. Vitamin A is also required by the body to maintain integrity of healthy mucus membranes and skin. It is a vital nutrient for acuity of vision. Consumption of natural vegetables and fruits rich in flavonoids helps to protect from lung and oral cavity cancers.
- The total antioxidant strength of raw sweet potato measured in terms of *oxygen radical absorbance capacity* (*ORAC*) is 902 μmol TE/100 g.
- The tubers are packed with many essential vitamins such as pantothenic acid (vitamin B-5), pyridoxine (vitamin B-6), and thiamin (vitamin B-1), niacin, and riboflavin. These vitamins are essential in the sense that body requires them from external sources to replenish. These vitamins function as co-factors for various enzymes during metabolism.
- Sweet potato provides a good amount of vital minerals such as iron, calcium, magnesium, manganese, and potassium that are very essential for enzyme, protein, and carbohydrate metabolism.
- Sweet potato leaves are indeed more nutritious than the tuber itself. Weight per weight, 100 g of fresh leaves contain more iron, vitamin C, folates, vitamin K, and potassium but less sodium than the tuber.

91

Swiss Chard

Succulent Swiss chard, also known as *spinach-chard* or *silverbeet*, is one of the popular green leafy vegetables of European origin. Botanically, it belongs within the beet family (*Chenopodiaceae*) of vegetables, which also includes *table-beets, sugar-beets, garden-beets*, etc. Scientific name: *Beta vulgaris*, Cicla group. Chard is an annual crop widely grown greens around Mediterranean region and is available at its best during summer season from June until November months. Chard features distinctly large dark-green leaves with prominent petiole well-developed edible stalk. Generally, its leaves are harvested at various stages of maturity. While the whole plant with its tender young leaves can be harvested for salad preparation, individual large-size, mature leaves with slightly tough texture stem may be picked up for sautéing and cooking in dishes.

Swiss chard comes in variety of types based on their shiny, crunchy stalks and petiole:

- ☆ Green stalk: *Lucullus*.
- ☆ Red stalk: *Charlotte, Rhubarb Chard*.
- ☆ Multi-colour stalk: Bright lights (white, orange, yellow, purple, pink).

Health Benefits of Swiss chard

- ☆ Swiss chard, like spinach, is the store-house of many phytonutrients that have health promotional and disease prevention properties.
- ☆ Chard is very low in calories (19 kcal per 100 g fresh, raw leaves) and fats, recommended in cholesterol controlling and weight reduction programmes.
- ☆ Chard leaves are an excellent source of antioxidant vitamin, vitamin-C. Its fresh leaves provide about 33 per cent of recommended levels per 100 g. As a powerful water-soluble antioxidant, vitamin C helps to quench free radicals and reactive oxygen species (ROS) through its reduction potential

properties. Research studies suggest that regular consumption of foods rich in vitamin C help maintain normal connective tissue, prevent iron deficiency, and also help the human body develop resistance against infectious agents by boosting immunity.

Swiss Chard (*Beta vulgaris* sub sp. *vulagaris*) Fresh, Raw Leaves, Nutritiove Value per 100 g

Principle	*Nutrient Value*	*Percentage of RDA*
Energy	1 per cent	19 Kcal
Carbohydrates	3.74 g	3 per cent
Protein	3.27 g	6 per cent
Total Fat	0.20 g	1 per cent
Cholesterol	0 mg	0 per cent
Dietary Fiber	1.6 g	4 per cent
Vitamins		
Folates	14 µg	4.5 per cent
Niacin	0.400 mg	2 per cent
Pantothenic acid	0.172 mg	3 per cent
Pyridoxine	0.99 mg	7.5 per cent
Riboflavin	0.090 mg	7 per cent
Thiamin	0.040 mg	3 per cent
Vitamin A	6116 IU	204 per cent
Vitamin C	30 mg	50 per cent
Vitamin E	1.89 mg	12.5 per cent
Vitamin K	830 µg	692 per cent
Electrolytes		
Sodium	213 mg	14 per cent
Potassium	379 mg	8 per cent
Minerals		
Calcium	51 mg	5 per cent
Copper	0.179 mg	20 per cent
Iron	1.80 mg	22.5 per cent
Magnesium	81 mg	20 per cent
Manganese	0.366 mg	16 per cent
Phosphorus	46 mg	6 per cent
Selenium	0.9 µg	1.5 per cent
Zinc	0.39 mg	3 per cent
Phyto-nutrients		
Carotene-β	3647 µg	—
Carotene-α	45 µg	—
Lutein-zeaxanthin	11000 µg	—

- ✰ Chard is one of the excellent vegetable sources for vitamin-K; 100 g provides about 700 per cent of recommended intake. Vitamin K has potential role bone health by promoting osteotrophic (bone formation and strengthening) activity. Adequate vitamin-K levels in the diet help limiting neuronal damage in the brain; thus, has established role in the treatment of patients suffering from Alzheimer's disease.
- ✰ It is also rich source of omega-3 fatty acids; vitamin-A, and flavonoids anti-oxidants like β-carotene, α-carotene, lutein and zeaxanthin. Carotenes convert to vitamin A inside the body.
- ✰ It is also rich in B-complex group of vitamins such as folates, niacin, vitamin B-6 (pyridoxine), thiamin and pantothenic acid that are essential for optimum cellular metabolic functions.
- ✰ It is also rich source of minerals like copper, calcium, sodium, potassium, iron, manganese and phosphorus. Potassium is an important component of cell and body fluids that helps controlling heart rate and blood pressure by countering effects of sodium. Manganese is used by the body as a co-factor for the antioxidant enzyme, *superoxide dismutase*. Iron is required for cellular oxidation and red blood cell formation.

Regular inclusion of chard in the diet has been found to prevent *osteoporosis, iron-deficiency anemia,* and *vitamin-A deficiency*; and believed to protect from cardiovascular diseases and colon and prostate cancers.

92

Tomato

Tomato, a nutritious fruit commonly used as a vegetable, is another wonderful gift of Mayans. The humble vegetable has grabbed the attention of millions of health seekers for its incredible phyto-chemical properties. Interestingly, it has more health-benefiting properties than that in an apple. Botanically, it belongs to *Solanaceae* or nightshade family of common vegetables, which also includes chili peppers, potato, eggplant, etc. Its scientific name is *Lycopersicon esculentum.* This exotic vegetable of all seasons is native to the Central America and was cultivated by the Aztecs centuries before the Spanish explorers introduced it to all over the world. Many hundreds of cultivar varieties with different type and size, either organic, hybrid or genetically modified are being in use. Most cultivars produce red fruit; however, a number of cultivars with yellow, orange, pink, purple, green, or white colours are also common scenes.

Heirloom tomatoes are becoming increasingly popular, particularly among organic producers. Heirloom varieties tend to produce more natural and flavourful fruits. They are now grown worldwide as the front line vegetable in line with onions, chili, potato etc.

Cherry tomatoes (*Solanum lycopersicum* var. cerasiforme) are small, round, cherry-sized fruits of the same Solanaceae family. The fruits are popular in Mediterranean regions.

Health Benefits of Tomato

- ✰ Tomatoes are one of the low-calorie vegetables containing just 18 calories per 100 g. They are also very low in any fat contents and have zero cholesterol levels. Nonetheless, they are excellent sources of antioxidants, dietary fiber, minerals, and vitamins. Because of their all-round qualities, dieticians and nutritionists often recommend them to be included in cholesterol controlling and weight reduction programmes.

Tomato (*Lycopersicon esculentum*), Raw, Nutrition Value per 100 g

Principle	*Nutrient Value*	*Percentage of RDA*
Energy	18 Kcal	1 per cent
Carbohydrates	3.9 g	3 per cent
Protein	0.9 g	1.6 per cent
Total Fat	0.2 g	0.7 per cent
Cholesterol	0 mg	0 per cent
Dietary Fiber	1.2 g	3 per cent
Vitamins		
Folates	15 µg	4 per cent
Niacin	0.594 mg	4 per cent
Pyridoxine	0.080 mg	6 per cent
Thiamin	0.037 mg	3 per cent
Vitamin A	833 IU	28 per cent
Vitamin C	13 mg	21.5 per cent
Vitamin E	0.54 mg	4 per cent
Vitamin K	7.9 µg	6.5 per cent
Electrolytes		
Sodium	5 mg	>1 per cent
Potassium	237 mg	5 per cent
Minerals		
Calcium	10 mg	1 per cent
Iron	0.3 mg	4 per cent
Magnesium	11 mg	3 per cent
Manganese	0.15 mg	6.5 per cent
Phosphorus	24 mg	3 per cent
Zinc	0.17 mg	1.5 per cent
Phyto-nutrients		
Carotene-β	449 µg	—
Carotene-α	101 µg	—
Lutein-zeaxanthin	123 µg	—
Lycopene	2573 µg	—

- ✰ The antioxidants present in tomatoes are scientifically found to be protective of cancers, including colon, prostate, breast, endometrial, lung, and pancreatic tumors. Total -ORAC (Oxygen Radical Absorbance Capacity) in this vegetable is 367 µmol TE/100 g.
- ✰ Lycopene, a flavonoid antioxidant, is the unique phytochemical present in the tomatoes. Red varieties are especially concentrated in this antioxidant. Together with carotenoids, it can protect cells and other structures in the

body from harmful oxygen-free radicals. Studies have shown that *lycopene* prevents skin damage from ultra-violet (UV) rays and offers protection from skin cancer.

- ☆ Zeaxanthin is another flavonoid compound present abundantly in this vegetable. Zeaxanthin helps protect eyes from "age-related macular related macular disease" (ARMD) in the elderly persons by filtering harmful ultra-violet rays.
- ☆ The vegetable contains very good levels of vitamin A, and flavonoid anti-oxidants such as α and β-carotenes, xanthins and lutein. Altogether, these pigment compounds are found to have antioxidant properties and take part in vision, maintain healthy mucus membranes and skin, and bone health. Consumption of natural vegetables and fruits rich in flavonoids is known to help protect from lung and oral cavity cancers.
- ☆ Additionally, they are also good source of antioxidant vitamin-C (provide 21 per cent of recommended daily levels per 100 g); consumption of foods rich in vitamin C helps the body develop resistance against infectious agents and scavenge harmful free radicals.
- ☆ Fresh tomato is very rich in potassium. 100 g contain 237 mg of potassium and just 5 mg of sodium. Potassium is an important component of cell and body fluids that helps controlling heart rate and blood pressure caused by sodium.
- ☆ Further, they contain moderate levels of vital B-complex vitamins such as folates, thiamin, niacin, riboflavin as well some essential minerals like iron, calcium, manganese and other trace elements.

93

Turnip

Turnips are nutritious root vegetables sought after in a variety of cuisines across Europe, Asia, and Eastern American regions. It is one of the cool-season vegetables belonging within the broad *Brassicaceae* family, which also includes cabbage, kale, brussels-sprouts, etc. The roots have been cultivated as staple food during ancient Greek and Roman periods. Although, its bulbous root which is widely popular, it is its top fresh greens rather more nutritious, several times richer in vitamins, minerals, and antioxidants. Scientific name: *Brassica rapa* (Rapifera Group).

Small young turnips or "baby turnips" are called when the roots harvested early in the growing stage. They are delicate, sweeter and eaten raw in salads. However, as they advance in age and maturity, their flavour becomes more pronounced and texture firm and woody. Rutabaga, another root vegetable, is closely related to turnips. Rutabagas are larger, more round, mostly feature yellow colour flesh, and sweeter than turnips.

Health Benefits of Turnips

- ✰ Turnips are very low calorie root vegetables; contains only 28 calories per 100 g. However, they are very good source of anti-oxidants, minerals, vitamins and dietary fiber.
- ✰ Fresh roots are indeed one of the vegetables rich in vitamin C; provide about 21mg or 35 per cent of DRA of vitamin C per 100 g. Vitamin-C is a powerful water-soluble anti-oxidant required by the body for synthesis of collagen. It also helps the body scavenge harmful free radicals, prevents from cancers, inflammation, and helps boost immunity.
- ✰ Turnip greens indeed are the storehouse of many vital nutrients; contain certain minerals and vitamins several fold more than that in the roots. The greens are very rich in antioxidants like vitamin A, vitamin C, carotenoid, xanthin, and lutein. In addition, the leafy-tops are an excellent source of vitamin K.

Turnips Root [*Brassica rapa* (Rapifera Group)], Fresh, Raw, Nutrition Value per 100 g

Principle	*Nutrient Value*	*Percentage of RDA*
Energy	28 Kcal	1.5 per cent
Carbohydrates	6.43 g	5 per cent
Protein	0.90 g	1.5 per cent
Total Fat	0.10 g	<1 per cent
Cholesterol	0 mg	0 per cent
Dietary Fiber	1.8 g	5 per cent
Vitamins		
Folates	15 µg	4 per cent
Niacin	0.400 mg	2.5 per cent
Pantothenic acid	0.200 mg	4 per cent
Pyridoxine	0.090 mg	7 per cent
Riboflavin	0.030 mg	2.5 per cent
Thiamin	0.040 mg	4 per cent
Vitamin A	0 IU	0 per cent
Vitamin C	21 mg	35 per cent
Vitamin E	0.03 mg	<1 per cent
Vitamin K	0.1 µg	<1 per cent
Electrolytes		
Sodium	39 mg	2.5 per cent
Potassium	233 mg	5 per cent
Minerals		
Calcium	30 mg	3 per cent
Copper	0.085 mg	9 per cent
Iron	0.30 mg	4 per cent
Magnesium	11 mg	2.5 per cent
Manganese	0.134 mg	6 per cent
Zinc	0.27 mg	2 per cent
Phyto-nutrients		
Carotene-β	0 µg	—
Carotene-α	0 µg	—
Lutein-zeaxanthin	0 µg	—

- ✰ In addition, its top greens are also a very good source of B-complex group of vitamins such as folates, riboflavin, pyridoxine, pantothenic acid and thiamin.
- ✰ Furher, the fresh greens are also excellent sources of important minerals like calcium, copper, iron and manganese.

94

Watercress

Watercress is nutrient rich perennial herb found in abundance near springs and alongside slow running water ways. This peppery aquatic plant has been in cultivation since ancient times for its food and medicinal uses in Europe, Central Asia and Americas. Botanically this fast growing, leafy vegetable belongs to *Brassicaceae* family and closely related to mustard greens, garden cress, cabbage, salad rocket (arugula), etc. Scientific name: *Nasturtium officianale*. The plant features small, oval, deep green succulent leaves with high moisture content. The leaves feature sharp; peppery and slightly sour taste somewhat like mustard greens and garden cress (Lepidium sativum). Racemes of tiny white flowers appear in summer, which turns into small pods containing two rows of seeds. Its ripe seeds are also edible.

Health Benefits of Watercress

- ✰ Peppery and tangy flavoured cress is a storehouse of many natural phytonutrients like *isothiocyanates* that have health promotional and disease prevention properties.
- ✰ Cress is one of the very low-calorie green leafy vegetables (only 11 calories per 100 g raw leaves) and contains negligible amounts of fats. Being an antioxidant rich, fewer calories and low-fat vegetable it is often recommended in cholesterol controlling and weight reduction programmes.
- ✰ Cress leaves and stem contains gluconasturtiin, a glucosinolate compound that gives the peppery flavour. Research studies suggest that the hydrolysis product of gluconasturtiin, 2-phenethyl isothiocyanate (PEITC), is believed to be cancer preventing by inhibition of phase I enzymes (*mono-oxygenases* and *cytochrome P450s*).
- ✰ Fresh cress has more concentration of ascorbic acid (vitamin C) than some of the fruits and vegetables. 100 g of leaves provide 47 mg or 72 per cent of RDA of vitamin C. As an anti-oxidant, vitamin C helps to quench free

radicals and reactive oxygen species (ROS) through its reduction potential properties. Lab studies suggest that regular consumption of foods rich in vitamin C help maintain normal connective tissue, prevent iron deficiency, and also help the body develop resistance against infectious agents by boosting immunity.

Watercress (*Nasturtium officianale*), Fresh, Raw Leaves, Nutrition Value per 100 g

Principle	*Nutrient Value*	*Percentage of RDA*
Energy	11 kcal	<1 per cent
Carbohydrates	1.29 g	1 per cent
Protein	2.30 g	4 per cent
Total Fat	0.10 g	0.5 per cent
Cholesterol	0 mg	0 per cent
Dietary Fiber	0.5 g	1 per cent
Vitamins		
Folates	9 µg	2 per cent
Niacin	0.200 mg	1 per cent
Pantothenic acid	0.310 mg	6 per cent
Pyridoxine	0.129 mg	10 per cent
Riboflavin	0.120 mg	9 per cent
Thiamin	0.090 mg	7.5 per cent
Vitamin A	3191 IU	106 per cent
Vitamin C	43 mg	72 per cent
Vitamin E	1.0 mg	7 per cent
Vitamin K	250 µg	208 per cent
Electrolytes		
Sodium	41 mg	3 per cent
Potassium	330 mg	7 per cent
Minerals		
Calcium	120 mg	12 per cent
Copper	0.077 mg	8.5 per cent
Iron	0.20 mg	2.5 per cent
Magnesium	21 mg	5 per cent
Manganese	0.244 mg	11 per cent
Phosphorus	60 mg	8 per cent
Selenium	0.9 µg	1.5 per cent
Zinc	0.11 mg	1 per cent
Phyto-nutrients		
Carotene-β	1914 µg	—
Carotene-α	0 µg	—
Lutein-zeaxanthin	5767 µg	—

- ☆ It is one of the excellent vegetable sources for vitamin-K; 100 g provides over 200 per cent of daily recommended intake. Vitamin K has potential role bone health by promoting osteotrophic (bone formation and strengthening) activity. Adequate vitamin-K levels in the diet help limiting neuronal damage in the brain; thus, has established role in the treatment of patients suffering from Alzheimer's disease.
- ☆ Cress is also an excellent source of vitamin-A, and flavonoids anti-oxidants like β-carotene, lutein and zeaxanthin.
- ☆ It is also rich in B-complex group of vitamins such as riboflavin, niacin, vitamin B-6 (pyridoxine), thiamin and pantothenic acid that are essential for optimum cellular metabolic functions.
- ☆ Further, it is also rich source of minerals like copper, calcium, potassium, magnesium, manganese and phosphorus. Potassium is an important component of cell and body fluids that helps controlling heart rate and blood pressure by countering effects of sodium. Manganese is used by the body as a co-factor for the antioxidant enzyme, *superoxide dismutase*. Calcium is required as bone/teeth mineral and in the regulation of heart and skeletal muscle activity.

Regular inclusion of cress in the diet is found to prevent osteoporosis, anaemia, and vitamin A deficiency and believed to protect from cardiovascular diseases and colon and prostate cancers.

95

Yam

Starch rich yams are staple tuber vegetables of West African origin. In addition to their use as food, there is a lot of interesting ritualism and symbolism has been associated around these humble starchy root vegetables all over Africa, Asia, and Latin Americas. Botanically they belong to the family *Dioscoreaceae*. In the genus: *Dioscorea*. There are several hundred species of Dioscorea exist; however, some of commercially important varieties of yam tuber are *Dioscorea rotundata* (white guinea), *D. alata* (yellow), *D. bulbifera* (aerial), *D. opposita* (Chinese), *D. esculenta* (Southeast Asian) and *D. dumenterum* (trifoliate).

The plant is a perennial vine cultivated for its large, edible, underground tuber, which can grow up to 120 pounds in weight and up to 2 meters in length. They are the tropical crops and never grow where the temperature dips below 68 degrees F. Yams are similar in appearance to sweet potato; however, they are not at all related to it. Important differences that distinguish them; yams are monocotyledons, larger in size, features thick, rough, dark brown to pink skin depending up on the cultivar type whereas sweet potatoes (*Ipomoea batatas*) are dicotyledonous, relatively smaller in size and possess very thin peel. Although the tuber is grown throughout Africa, Nigeria is the world's largest producer and exporter of yam, accounting for over 70 percent of the world total output.

Health Benefits of yams

- ☆ Yam is a good source of energy; 100 g provides 118 calories. It mainly composed of complex carbohydrates and soluble dietary fiber. Together, they raise blood sugar levels rather very slowly than simple sugars and therefore, recommended as the low glycemic index healthy food. In addition, dietary fiber helps reduce constipation, decrease bad (LDL) cholesterol levels by binding to it in the intestines and prevent colon cancer risks by preventing toxic compounds in the food from adhering to the colon mucosa.

Yam (*Dioscorea* spp.), Raw, Nutrition Value per 100 g

Principle	*Nutrient Value*	*Percentage of RDA*
Energy	108 Kcal	5 per cent
Carbohydrates	27.88 g	21 per cent
Protein	1.53 g	3 per cent
Total Fat	0.17 g	0.5 per cent
Cholesterol	0 mg	0 per cent
Dietary Fiber	4.1 g	11 per cent
Vitamins		
Folates	23 µg	6 per cent
Niacin	0.552 mg	3.5 per cent
Pantothenic acid	0.314 mg	7 per cent
Pyridoxine	0.293 mg	23 per cent
Riboflavin	0.032 mg	2.5 per cent
Thiamin	0.112 mg	9.5 per cent
Vitamin C	17.1 mg	28.5 per cent
Vitamin A	138 IU	5 per cent
Vitamin E	0.35 mg	2 per cent
Vitamin K	2.3 µg	2 per cent
Electrolytes		
Sodium	9 mg	0.5 per cent
Potassium	816 mg	17 per cent
Minerals		
Calcium	17 mg	2 per cent
Copper	0.178 mg	20 per cent
Iron	0.54 mg	7 per cent
Magnesium	21 mg	5 per cent
Manganese	0.397 mg	17 per cent
Phosphorus	55 mg	8 per cent
Selenium	0.7 µg	0.5 per cent
Zinc	0.24 mg	2 per cent
Phyto-nutrients		
Carotene-β	83 µg	—
Crypto-xanthin	0 µg	—
Lutein-zeaxanthin	0 µg	—

✰ The tuber is an excellent source of B-complex group of vitamins. It provides adequate daily requirements of pyridoxine (vitamin B6), thiamin (vitamin B1), riboflavin, folic acid, pantothenic acid and niacin. These vitamins mediate various metabolic functions in the body.

☆ Fresh root also contains good amounts of anti-oxidant vitamin; vitamin-C. Provides about 29 per cent of recommended levels per 100 g. Vitamin C has important roles in anti-ageing, immune function, wound healing, bone growth.

☆ They contain small amounts of vitamin-A, and beta-carotene levels. Carotenes convert to vitamin A in the body. Both these compounds are strong antioxidants. Vitamin A has many functions like maintaining healthy mucus membranes and skin, night vision, growth and protection from lung and oral cavity cancers.

☆ Further, the tuber is indeed one of the vegetables rich sources of minerals like copper, calcium, potassium, iron, manganese, and phosphorus. 100 g provides about 816 mg of Potassium. Potassium is an important component of cell and body fluids, which helps controlling heart rate and blood pressure by countering hypertensive effects of sodium. Copper is required in the production of red blood cells. Manganese is used by the body as a co-factor for the antioxidant enzyme, *superoxide dismutase*. Iron is required for red blood cell formation.

96

Zucchini

Zucchini squash *or* courgette is the most popular summer squashes in Americas and Europe. Like gourds, it belongs within the Cucurbitaceae (*Cucurbita pepo*) family. Summer squashes are believed to be originating in the Central America and Mexico. Several different cultivars of summer squash are grown throughout the United States during the warm, frost-free season. Almost all the members of the squash family vegetables feature smooth, tender skin, and flesh with small edible seeds and high moisture content. Like other members of the summer squash group, zucchini features bush habit rather than the vine spread of the winter-squashes. Its fruits shall be ready for harvesting about 40-50 days after seed implantation.

Some popular varieties are:

- ☆ **Golden zucchini** features brilliant golden-yellow skin that retains its colour even after cooking.
- ☆ Round varieties are dense, heavy, and nearly seedless with smooth surface.
- ☆ **Tatume**, which is common in Mexico, has similar features of round variety but has the large oval shape.
- ☆ **Costata romanesco** also known as cocozelle is a long, narrow variety with the slight bulge at the bottom end. It features pale raised ribs in mottled green skin. When solid and young, this squash is juicy and sweet.
- ☆ **Middle-Eastern** types are stocky, pale green, tapering ends with a thick dark-green stem. They have smooth shiny skin and solid, crispy and flavourful flesh.
- ☆ **Yellow crooknecks** have thick warty skin with markedly curved neck. They are crunchy in texture with sweet delicate flavour.

Health Benefits of Zucchini

- ☆ Zucchini is one of the very low calorie vegetables; provide only 17 calories per 100 g. It contains no saturated fats or cholesterol. Its peel is good source of dietary fiber that helps reduce constipation and offers some protection against colon cancers.
- ☆ Zucchinis have anti-oxidant value (Oxygen radical absorbance capacity-ORAC) of 180 Trolex Equivalents (TE) per 100g, the value which is far below to some of the berries, and vegetables. Nonetheless, the pods are one of the common vegetables included in weight reduction and cholesterol control programs by the dieticians.
- ☆ Furthermore, zucchinis, especially golden skin varieties, are rich in flavonoid poly-phenolic antioxidants such as *carotenes, lutein and zeaxanthin*. These compounds help scavenge harmful oxygen-derived free radicals and reactive oxygen species (ROS) from the body that play a role in ageing and various disease processes.
- ☆ Courgette is a relatively moderate source of folates, provides of 24 µg or 6 per cent of RDA per 100 g. Folates are important in cell division and DNA synthesis. When taken adequately before pregnancy, it can help prevent neural tube defects in the fetus.
- ☆ It is a very good source of potassium, an important intra-cellular electrolyte. Potassium is a heart friendly electrolyte and helps bring the reduction in blood pressure and heart rates by countering pressure-effects of sodium.
- ☆ Fresh fruits are rich in vitamin A; provide about 200 IU per 100 g.
- ☆ Fresh pods, indeed, are good source of anti-oxidant vitamin-C. Provide about 17.9 µg or 30 per cent of daily-required levels per 100 g.
- ☆ In addition, they contain moderate levels of B-complex group of vitamins like thiamin, pyridoxine, riboflavin and minerals like iron, manganese, phosphorus, and zinc.

Zucchini (*Cucumis pepo*), Raw with Skin, Nutrition Value per 100 g

Principle	*Nutrient Value*	*Percentage of RDA*
Energy	17 Kcal	<1 per cent
Carbohydrates	3.11 g	2.5 per cent
Protein	1.21 g	2 per cent
Total Fat	0.32 g	
		1 per cent
Cholesterol	0 mg	0 per cent
Dietary Fiber	1 g	—

C. Nutritional Value and Health Benefits of Nuts and Seeds

Introduction

Enrich your daily diet with nuts nutrition. Just munch a handful of nuts a day and you will be doing more than good to keep yourself healthy and stay fit. Nuts have it all. Rich in energy, protein, packed with antioxidants, vitamins, minerals and much discussed omega-3 fatty acids. Crunchy yet buttery, wonderfully delicious nuts are wonderful gifts to humankind by Mother Nature.

Nuts are Friendly

- ☆ You can carry them along with you wherever you go.
- ☆ There is no need of special containers or carry boxes.
- ☆ They have long a shelf life. Unlike fruits and vegetables, nuts are not easily perishable.
- ☆ You can enjoy them whenever you like and wherever you want! Nothing to worry about preparation, in fact, nuts hardly needs a kitchen.

Nature's friendly! Almost all the varieties of nuts we eat come in safe envelopes (shells). There is nothing to worry too much about contamination or handling issues. Just grab healthy-looking nuts of your choice and enjoy.

Nuts are Must

Why and How?

- ☆ Nuts are rich in energy and nutrients. Nuts nutrition loaded with excellent source of monounsaturated-fatty acids (MUF) such as oleic and palmitoleic acids, which help to lower LDL or "bad cholesterol" and increase HDL or "good cholesterol." Research studies suggest that Mediterranean diet that is rich in MUF to prevent coronary artery disease, strokes by favouring healthy blood lipid profile.

- They are rich source of all important omega-3 essential fatty acids like *Linoleic acid, α-Linolenic acid*(ALA), *Eicosapentaenoic acid, Docosahexonic acid* etc. Research studies have suggested that n-3 or ù-3 fats by their virtue of anti-inflammatory action help to lower the risk of blood pressure, coronary artery disease, strokes and breast, colon and prostate cancers. Omega-3 fatty acids also offer some benefits in conditions such as rheumatoid arthritis and in cases of Schizophrenia, depression and Alzheimer's disease.
- Nuts are the storehouse of health benefiting *poly-phenolic flavonoid antioxidants* such as carotenes, resveratrol, lutein, cryptoxanthin, etc. These compounds have been found to offer protection against cancers, heart disease, degenerative nerve disease, Alzheimer's disease, and viral/ fungal infections. Studies suggests that resveratrol in peanuts reduces stroke risk by alteration of molecular mechanisms in the blood vessels, reducing their susceptibility to vascular damage through decreased activity of angiotensin (a systemic hormone causing blood vessel constriction that would elevate blood pressure) and increased production of the vasodilator hormone, nitric oxide.
- Nuts nutrition is complete in the sense that in addition to calories, and vitamins, they are rich source of minerals like manganese, potassium, calcium, iron, magnesium, zinc, fluoride and selenium. *Manganese* is a co-factor for the enzyme superoxide dismutase, which is a very powerful free radical scavenger. *Potassium* is an important component of cell and body fluids that helps controlling heart rate and blood pressure. *Copper* is required for the production of red blood cells. *Iron* is required for red blood cell formation. *Fluoride* is a component of bones and teeth and plays important role in prevention of dental caries.
- They contain very good levels of vitamin-E, a powerful lipid soluble antioxidant. Vitamin E is required for maintaining the integrity of cell membrane of mucus membranes and skin; thus, protecting it from harmful oxygen-free radicals.
- Nuts nutrition provides many vital B-complex groups of vitamins such as riboflavin, niacin, thiamin, pantothenic acid, vitamin B-6, and folates. These vitamins are essential for optimum health and well-being.

Culinary Uses of Nuts

- Nuts can be eaten as they are or either salted, or sweetened.
- Sprinkle over desserts, particularly sundaes, ice cream and enjoy!
- They can be employed in confectionery, as an addition to breads, biscuits, sweets, muffins, and cakes.
- Many kinds' nuts are also used to make nutritious butter.

97

Almonds

Wonderfully delicious, almonds have long been revered as the epitome of wellness and health. The nuts are among the richest sources of health-benefiting nutrients essential for optimum health. Its plant is a small deciduous tree, native to mineral-rich West-Asian mountain ranges that provide fertile conditions favourable for their growth. In recent years, it is being cultivated in many regions of the world as an important commercial crop. Botanically, they are the fruits from species of tree belonging to the family of *Rosaceae,* of Genus: *Prunus*. Scientific name: *Prunus dulcis*. During spring season, the tree bears whitish-pink flowers that ultimately become fruits by autumn. In structure, the fruit is a drupe. A single edible seed (kernel), known as 'almond nut', is actually enclosed inside the stony hard shell. Almond kernel features oval to conical in shape, brown in colour, measuring about 2 cms in length and 1 cm in breadth and weigh about 1 to 1.5 g.

Health Benefits of Almonds

- ✰ Almond nuts are rich in dietary fiber, vitamins, and minerals and packed with numerous health promoting phyto-chemicals; the kind of well-balanced food ensuring protection against diseases and cancers.
- ✰ These nuts are the complete source of energy as well as nutrients. They are especially, rich in mono-unsaturated fatty acids like *oleic* and *palmitoleic acids* that help to lower LDL or "bad cholesterol" and increase HDL or "good cholesterol." Research studies suggest that Mediterranean diet, which is excellent in monounsaturated fatty acids help to prevent coronary artery disease and strokes by favouring healthy blood lipid profile.
- ✰ The nuts are an excellent source of vitamin E; contain about 25 g per100 g (about 170 per cent of RDA). Vitamin E is a powerful lipid soluble antioxidant, required for maintaining cell membrane integrity of mucus membranes and skin by protecting from harmful oxygen-free radicals.

Almonds (*Prunus dulcis*), Nutritional Value per 100 g

Principle	*Nutrient Value*	*Percentage of RDA*
Energy	575 Kcal	29 per cent
Carbohydrates	21.67 g	16 per cent
Protein	21.22 g	38 per cent
Total Fat	49.42 g	165 per cent
Cholesterol	0 mg	0 per cent
Dietary Fiber	12.20 g	30 per cent
Vitamins		
Folates	50 µg	12.5 per cent
Niacin	3.385 mg	21 per cent
Pantothenic acid	0.47 mg	9 per cent
Pyridoxine	0.143 mg	11 per cent
Riboflavin	1.014 mg	78 per cent
Thiamin	0.211 mg	16 per cent
Vitamin A	1 IU	0 per cent
Vitamin C	0 mg	0 per cent
Vitamin E	26 mg	173 per cent
Electrolytes		
Sodium	1 mg	0 per cent
Potassium	705 mg	15 per cent
Minerals		
Calcium	264 mg	26 per cent
Copper	0.996 mg	110 per cent
Iron	3.72 mg	46.5 per cent
Magnesium	268 mg	67 per cent
Manganese	2.285 mg	99 per cent
Phosphorus	484 mg	69 per cent
Selenium	2.5 µg	4.5 per cent
Zinc	3.08 mg	28 per cent
Phyto-nutrients		
Carotene-β	1 µg	—
Crypto-xanthin-β	0 µg	—
Lutein-zeaxanthin	1 µg	—

☆ Almonds are free in *gluten* and therefore, are one of the popular ingredients in the preparation of gluten-free food formulas. Such formula preparations are, in fact, healthy alternatives in people with wheat food allergy and *celiac disease*.

- These nuts are packed with many important B-complex groups of vitamins such as riboflavin, niacin, thiamin, pantothenic acid, vitamin B-6, and folates. Together, these vitamins work as co-factors for enzymes during cellular substrate metabolism.
- Further, they are also an incredible source of minerals such as manganese, potassium, calcium, iron, magnesium, zinc, and selenium.
- Almond oil is extracted from the nuts used as an emollient. Applied regularly, it helps keep skin well protected from dryness. Additionally, the oil is used in cooking in Iran, and Turkey. It is also used as "carrier or base oil" in traditional medicines in aromatherapy, in pharmaceutical and cosmetic industries.

Hand full of almonds a day provides much of recommended levels of minerals, vitamins, and protein. Besides, almond oil extracted from the nuts has been used in cooking, and in medicine.

98

Brazil Nuts

Amazon's forest holds some of the unique plant species like Brazil nuts, acai berry, etc., that can be found nowhere else on the planet earth. Native Amazonian cherished these delicious nuts since ages, which provide them much-needed protein, fats and other essential nutrients. Botanically, brazil-nut tree belongs within the family of *Lecythidaceae*, of the genus: *Bertholletia*. Scientific name: *Bertholletia excelsa.* Some of the common names are *castanha-do-pará, castania, para-nut, cream-nut, Castaña-de-Brazil* (*chestnuts of Brazil*) etc.

Brazil nut trees are found conspicuously in the non-flooded forest regions of Brazil, Bolivia, and Peru. They are indeed one of the long living, tallest trees in all the tropical rain forests; grow to 50 meters tall with large erect stem and wide umbrella like foliage at the top 1/3. They thought to have the life span of about 500 to 700 years. Each mature tree bears up to 300 fruit pods in a season. The pod is a large shelled fruit, similar to coconut, takes about 14 months to mature after pollination. Each pod can weigh up to 2.5 kg. The mature fruit falls from the tree usually with a loud crashing sound. The pods remain intact even after fall from such a height. In their natural habitat, they exclusively depend on caviomorph rodents (agoutis, Dasyprocta spp.) that have the ability to gnaw open the woody shell and subsequently disperse leftover seeds for germination. Internally, each fruit features 10-25 seeds arranged in orange-like segments. Each nut is in turn encased within its own thick dark-brown colour individual shell. The edible kernel feature three-sided shape with sweet nutty flavour white meat and weigh about five g.

Health Benefits of Brazil Nuts

- ☆ Brazil nuts are high in calories, contains good quantities of vitamins, anti-oxidants and minerals. The nuts in-fact has been staple diet of Amazonian.
- ☆ 100 g of Brazil nuts provide about 656 calories. Their high caloric content comes from their fats. However, the nuts are an especially excellent source

of mono-unsaturated fatty acids (MUFA) like *palmitoleic acid* (16:1) and *oleic acid* (18:1) that helps to lower LDL or "bad cholesterol" and increases HDL or "good cholesterol" in the blood. Research studies suggest that Mediterranean diet, which is rich in monounsaturated fatty acids help to prevent coronary artery disease and strokes by favouring healthy blood lipid profile.

Brazil Nuts (*Bertholletia excelsa*), Raw, Nutritional Value per 100 g

Principle	*Nutrient Value*	*Percentage of RDA*
Energy	656 Kcal	33 per cent
Carbohydrates	12.27 g	10 per cent
Protein	14.32 g	26 per cent
Total Fat	66.43 g	221 per cent
Cholesterol	0 mg	0 per cent
Dietary Fiber	7.5 g	20 per cent
Vitamins		
Folates	22 mcg	5.5 per cent
Niacin	0.295 mg	2 per cent
Pantothenic acid	0.184 mg	3.5 per cent
Pyridoxine	0.101 mg	8 per cent
Riboflavin	0.035 mg	3 per cent
Thiamin	0.617 mg	51 per cent
Vitamin A	0 IU	0 per cent
Vitamin C	0.7 mcg	1 per cent
Vitamin E-gamma	7.87 mg	52 per cent
Electrolytes		
Sodium	2 mg	0 per cent
Potassium	597 mg	13 per cent
Minerals		
Calcium	160 mg	16 per cent
Copper	1.743 mg	194 per cent
Iron	2.43 mg	30 per cent
Magnesium	376 mg	94 per cent
Manganese	1.223 mg	53 per cent
Phosphorus	725 mg	103 per cent
Selenium	1917 mcg	3485 per cent
Zinc	4.06 mg	36 per cent
Phyto-nutrients		
Carotene-β	0 mcg	—
Crypto-xanthin-β	0 mcg	—
Lutein-zeaxanthin	0 mcg	—

- ☆ The nuts are also a very good source of vitamin-E; contain about 7.87 mg per 100 g (about 52 per cent of RDA). Vitamin-E is a powerful lipid soluble antioxidant. It is required for maintaining the integrity of cell membrane of mucus membranes and skin by protecting it from harmful oxygen-free radicals.
- ☆ Brazil nuts contain exceptionally high levels of selenium. 100 g nuts provide about 1917 µg of selenium and 3485 per cent of recommended daily intake making them as the highest natural source of this mineral. Selenium is an important cofactor for anti-oxidant enzyme *glutathione-peroxidase.* Just 1-2 nuts a day provides enough of this trace element. Adequate selenium foods in the diet help prevent coronary artery disease, liver cirrhosis, and cancers.
- ☆ Furthermore, like almonds and pine nuts, Brazil nuts too are free from *gluten* and therefore, are one of the popular ingredients for the preparation of gluten-free food formulas. These formula preparations are, in fact, healthy alternatives in people with wheat food allergy and *celiac disease.*
- ☆ Additionally, these creamy nuts are an excellent source of B-complex group of vitamins such as thiamin (51 per cent of RDA per 100 g), riboflavin, niacin, pantothenic acid, vitamin B-6 (pyridoxine) and folates. Altogether, they work as co-factors for enzymes during cellular substrate metabolism inside the body.
- ☆ In addition to selenium, they contain very good levels of other minerals such as copper, magnesium, manganese, potassium, calcium, iron, phosphorus, and zinc. Copper helps prevent anemia and bone weakness (osteoporosis). Manganese is an all-important co-factor for antioxidant enzyme, superoxide dismutase.

Brazil nut oil, obtained from these nuts, has many traditional medicinal applications as emollient and massage therapy. It has clear yellow colour with a pleasant, sweet smell and taste. Its emollient property helps to keep skin well protected from dryness. It has also been used in cooking, and as "carrier or base oil" in traditional medicines in aromatherapy, in pharmaceutical and cosmetic industry.

99

Cashew Nut

Delicately sweet yet crunchy and delicious cashew nut is packed with energy, antioxidants, minerals and vitamins that are essential for robust health! Cashew, or "caju" in Portuguese, is one of the popular ingredients in sweet as well savory dishes worldwide. Botanically, cashew is an average size tropical evergreen tree belonging within the *Anacardiaceae* family of the genus: *Anacardium*. Scientific name: *Anacardium occidentale.*

The cashew tree is native to Brazil's Amazon rain forest, which spread all over the world by Portuguese explorers. Today, it is cultivated commercially in Brazil, Vietnam, and India and in many African countries. Cashew tree bears numerous, edible, pear shaped false fruits or "accessory fruits'" called "cashew apples." A small bean shaped, grey colour "true-fruit" is firmly adhering to lower end of these cashew-apples appearing like a clapper in the bell. Botanically, this "true fruit" is a drupe, featuring hard outer shell enclosing a single edible seed or the "cashew nut." The exterior shell composes a phenolic resin, urushiol, which is a potent caustic skin irritant toxin. In the processing units, this outer shell is roasted in order to destroy the urushiol resin, and then; the edible cashew kernel is extracted. Cashew nut measures about an inch in length and 1/2 inches in diameter with kidney or bean shape, and smooth curvy pointed tip. Each nut has two equal halves as in legumes. The nuts are cream white colour with the firm yet delicate texture and smooth surface. Cashews have buttery texture with a pleasant sweet fruity aroma.

Health Benefits of Cashew Nuts

- ✰ Cashews are high in calories. 100 g of nuts provide 553 calories. They are packed with soluble dietary fiber, vitamins, minerals and numerous health-promoting phyto-chemicals that help protect from diseases and cancers.
- ✰ They are rich in "heart-friendly" monounsaturated-fatty acids like *oleic, and palmitoleic acids.* These essential fatty acids help lower harmful LDL-

cholesterol while increasing good HDL cholesterol. Research studies suggest that Mediterranean diet, which is rich in monounsaturated fatty acids help to prevent coronary artery disease and strokes by favouring healthy blood lipid profile.

Cashew Nut (*Anacardium occidentale*), Nutrition Value per 100 g

Principle	*Nutrient Value*	*Percentage of RDA*
Energy	553 Kcal	28 per cent
Carbohydrates	30.19 g	23 per cent
Protein	18.22 g	32.5 per cent
Total Fat	43.85 g	146 per cent
Cholesterol	0 mg	0 per cent
Dietary Fiber	3.3 g	8.5 per cent
Vitamins		
Folates	25 µg	6 per cent
Niacin	1.062 mg	6.5 per cent
Pantothenic acid	0.864 mg	17 per cent
Pyridoxine	0.417 mg	32 per cent
Riboflavin	0.058 mg	4.5 per cent
Thiamin	0.423 mg	35 per cent
Vitamin A	0 IU	0 per cent
Vitamin C	0.5 mg	1 per cent
Vitamin E	5.31 mg	35 per cent
Vitamin K	4.1 µg	3 per cent
Electrolytes		
Sodium	12 mg	1 per cent
Potassium	660 mg	14 per cent
Minerals		
Calcium	37 mg	4 per cent
Copper	2.195 mg	244 per cent
Iron	6.68 mg	83.5 per cent
Magnesium	292 mg	73 per cent
Manganese	1.655 mg	72 per cent
Phosphorus	593 mg	85 per cent
Selenium	19.9 µg	36 per cent
Zinc	5.78 mg	52.5 per cent
Phyto-nutrients		
Carotene-β	0 µg	—
Crypto-xanthin-β	0 µg	—
Lutein-zeaxanthin	22 µg	—

- Cashew nuts are very rich source of essential minerals. Minerals, especially manganese, potassium, copper, iron, magnesium, zinc and selenium are concentrated in these nuts. A handful of cashew nuts a day in the diet would provide enough of these minerals and prevent deficiency diseases. Selenium is an important micronutrient, which functions as a co-factor for antioxidant enzymes such as *Glutathione peroxidases, one of the most powerful antioxidants in the body.* Copper is a cofactor for many vital enzymes, including cytochrome c-oxidase and superoxide dismutase (other minerals function as co-factors for this enzyme are manganese and zinc). Zinc is a co-factor in many enzymes that regulate *growth and development, sperm generation, digestion and nucleic acid synthesis.*
- Cashews are also rich in many essential vitamins such as pantothenic acid (vitamin B5), pyridoxine (vitamin B-6), riboflavin, and thiamin (vitamin B-1). 100 g nuts provide 0.147 mg or 32 per cent of daily-recommended levels of pyridoxine. Pyridoxine reduces the risk of homocystinuria, and sideroblastic anaemia. Niacin helps prevent "pellagra" or dermatitis. Additionally, these vitamins are essential for metabolism of protein, fat, and carbohydrates at cellular levels.
- Further, the nuts are also containing a small amount of zeaxanthin, an important pigment flavonoid antioxidant, which selectively absorbed into the retinal macula lutea in the eyes. It is thought to provide antioxidant and protective UV ray filtering functions and helps prevent age-related macular degeneration (ARMD) in the elderly.

100

Chestnuts

Richly flavoured, starchy, chestnuts are popular cool season edible nuts of the northern hemisphere. The nuts are native to hilly forest of China, Japan, Europe, and North America. Botanically they belong to the beech or*Fagaceae* family of the genus: *Castenea*. Scientific name: *Castanea sativa*. Castaneas are very large deciduous trees. They are monoecious, bearing both male and female flowers ("catkins"), in the same tree. They have a remarkable history to narrate. Early in the 20th century, the once mighty American-chestnut tree was mostly wiped out by pathogenic fungus chestnut blight (*Cryphonectria parasitica*). A renewed interest has been growing since then to revive native chestnut trees throughout of the USA.

Once pollinated, female flowers develop into large spiny burr or involucres, each enclosing about 2-3 edible kernels. The fruit is quite larger compared to its peers like cashews, macadamia, etc. Each nut features smooth, glossy dark-brown colour outer shell, 1-1.5 inch in diameter and weighing 8-12 g depending up on the species. Inside they have a creamy white, sweet and starchy kernel. Four main species of chestnut trees are being cultivated around the world for their nuts; *Castanea sativa* in Europe, *C. dentata* in North America, *C. mollissima* in China and *C. crenata* in Japan. United States is the chief importer of chestnuts from European Union, although China has been the largest exporter of nuts worldwide, especially to Japan.

Health Benefits of Chestnut

- ☆ Chestnuts, unlike other nuts and seeds, are relatively low in calories; contain less fat but are rich in minerals, vitamins and phyto-nutrients that benefit health.
- ☆ Chestnuts are chiefly made of starch when compared to other seeds and nuts, which are high in calorie, protein, and fat. Their nutrition composition is almost similar to that of other staple starch foods such as sweet potato, sweet corn, potatoes, plantain, etc., Nevertheless; they are still good sources of minerals, vitamins and some good-quality protein.

Chestnuts, European, Raw, Unpeeled (*Castanea sativa*), Nutritional Value per 100 g

Principle	*Nutrient Value*	*Percentage of RDA*
Energy	213 Kcal	11 per cent
Carbohydrates	45.54 g	35 per cent
Protein	2.42 g	4 per cent
Total Fat	2.26 g	10 per cent
Cholesterol	0 mg	0 per cent
Dietary Fiber	8.1 g	21 per cent
Vitamins		
Folates	62 µg	15.5 per cent
Niacin	1.179 mg	7 per cent
Pantothenic acid	0.509 mg	11 per cent
Pyridoxine	0.376 mg	29 per cent
Riboflavin	0.168 mg	13 per cent
Thiamin	0.238 mg	20 per cent
Vitamin A	28 IU	1 per cent
Vitamin C	43 mg	72 per cent
Electrolytes		
Sodium	3 mg	0 per cent
Potassium	518 mg	11 per cent
Minerals		
Calcium	27 mg	3 per cent
Copper	0.447 mg	50 per cent
Iron	1.01 mg	13 per cent
Magnesium	32 mg	8 per cent
Manganese	0.952 mg	41 per cent
Phosphorus	93 mg	19 per cent
Zinc	0.52 mg	5 per cent
Phyto-nutrients		
Phyto-sterols	22 µg	—

- ☆ They are a good source of dietary fiber; provide 8.1 g (about 21 per cent of RDI) per 100 g. Fiber diet helps lower blood cholesterol levels by limiting excess cholesterol absorption in the intestines.
- ☆ Chestnuts stand out from other nuts and seeds for their distinct nutrition profile. They are exceptionally rich in vitamin-C. 100 g nuts provide 43 mg of vitamin C (72 per cent of DRI). Vitamin C is essential for matrix formation in teeth, bones and blood vessels. Being a strong anti-oxidant, it offers protection from harmful free radicals.

- Again, as in green-leafy vegetables, chestnuts are rich in folates, which is quite unique feature for nuts and seeds. 100 g nuts provide 62 µg of folates (or 15.5 per cent). Folic acid is required for the formation of red blood cells, DNA synthesis. Adequate consumption of food rich in folates during the peri-conception period helps prevent neural tube defects in the fetus.
- They are rich source of mono-unsaturated fatty like oleic acid (18:1) and palmitoleic acids (16:1). Studies suggest that monounsaturated fats (MUFs) in the diet help lower total as well as LDL (bad cholesterol) and increase HDL (good cholesterol) levels within the blood. Mediterranean diet which is rich in dietary-fiber, MUFs, omega fatty acids and antioxidants help prevent coronary artery disease and strokes by favouring healthy blood lipid profile.
- The nuts are an excellent source of minerals such as iron, calcium, magnesium, manganese, phosphorus and zinc, besides providing a very good amount of potassium (518 mg / 100 g). Potassium helps counter hypertensive action of sodium, lowers heart rate and blood pressure. Iron helps prevent *microcytic-anemia.* Magnesium and phosphorus are important components of bone metabolism.
- Further, they are also rich in many important B-complex groups of vitamins. 100 g of nuts provide 11 per cent of niacin, 29 per cent of pyridoxine (vitamin B-6), 100 per cent of thiamin, and 12 per cent of riboflavin.
- Chestnuts, like hazelnuts and almonds, etc., are free in gluten. And for the same reason, they are one of the popular ingredients in the preparation of gluten-free food formulas for gluten-sensitive, wheat allergy, and celiac disease patients.
- Chinese chestnuts (*C. mollissima*) are good in vitamin A; provide 202 IU per 100 g.

101

Chia Seeds

Have chia seeds packed with all the nutrients to label them as number one, ultimate food grain? Or they just another novel entrant to the spectrum of so called super foods? Chia are tiny, oblong shaped, oil-seeds consumed as one of the staple crops by ancient Aztecs. Indeed, the crop was cultivated in the same measure to corn and amaranth by native Mexicans. Many nutrition experts believe chia (*Salvia hispanica* L.) seeds top the list of functional foods because of their wholesome nutritional properties. In fact, the seeds are incredible sources of phyto-nutrients such as omega-3 fatty acids (α-Linolenic acid), anti-oxidants, minearls, vitamins and dietary fiber in right proportions that a human body need for. Botanically, chia is a low-growing herb in the mint family of plants. Scientific name: *Salvia hispanica.*

Chia (Salvia hispanica) is a small, annual, drought-tolerant, flowering herb and flourishes under well-drained, fertile soil. The plant grows up to 1 meter in height and bears broad, green colour leaves with serrated margins. Purple, pink-violet colour flowers in spikes appear after about 4 months after seedling. The seeds come in variegated colours depending upon cultivar type and vary in colour from black, brown to off white. The seeds measure about 1 mm in diameter and resemble like miniature pinto beans. Its smooth outer coat made of hygroscopic mucilage which absorbs water and swells up several times the original size.

Health Benefits of Chia Seeds

- ☆ Chia seeds compose almost of all the essential nutrients such as protein, omega-3 fatty acids, antioxidants, dietary fiber, vitamins and minerals that are essential for optimum growth and development.
- ☆ Being an oil seed, chia is rich in calories. 100 g seeds provide 388 calories. However, much of their calories come from poly-unsaturated fatty acids (PUF). Chia seeds have excellent proportion of omega-3 to 6 poly-unsaturated fats; in the recommended ratio of 1:4.

Chia Seeds (*Salvia hispanica*), Whole, Dried, Nutritional Value per 100 g

Principle	*Nutrient Value*	*Percentage of RDA*
Energy	486 Kcal	24 per cent
Carbohydrates	42.12 g	32 per cent
Protein	16.54 g	30 per cent
Total Fat	30.74 g	102 per cent
Cholesterol	0 mg	0 per cent
Dietary Fiber	34.4 g	91 per cent
Vitamins		
Folates	49 µg	12 per cent
Niacin	8.83 mg	55 per cent
Riboflavin	0.17 mg	13 per cent
Thiamin	0.62 mg	52 per cent
Vitamin A	210 IU	7 per cent
Vitamin C	1.6	2 per cent
Vitamin E	0.5 mg	3 per cent
Electrolytes		
Sodium	16 mg	1 per cent
Potassium	407 mg	9 per cent
Minerals		
Calcium	631 mg	63 per cent
Iron	7.72 mg	97 per cent
Magnesium	335 mg	83.5 per cent
Manganese	2.723 mg	118 per cent
Phosphorus	860 mg	223 per cent

☆ An important omega 3-fatty acid in chia is alpha-linoleic acid (ALA). Studies have found that ALA and other omega 3 fatty acids by their virtue of anti-inflammatory action help lower risk of blood pressure, coronary artery disease, strokes and breast, colon, and prostate cancers. Adequate quantities of omega-3 in the diet are essential for normal development and maturation of nervous system in infants and young children.

☆ Chia seeds compose several health benefiting anti-oxidants namely *ferulic acid, caffeic acid, quercetin*, etc.

☆ 100 g seeds provide about 91 per cent of daily recommended intake levels of dietary fiber. The outer coat of chia is made of water-soluble mucilage, a non starch polysaccharide (NSP), which swells up several times and acquires gel-like consistency. Mucilage helps in digestion and smooth bowel movements.

- ☆ Chia is gluten free grain. People with known sensitivity to gluten celiac's disease can safely add it in their diet.
- ☆ Chias are one of the low glycemic index foods and according to Nutritiondata.com has a fullness facor of 2.8. Foodstuffs low in glycemic index help in better regulation of daily blood glucose levels. Substitution of chia for rice and other cereal grains may benefit in individuals with diabetes.
- ☆ It is an excellent source of vitamins like niacin, riboflavin, thiamin, folic acid. Niacin is an important B-complex vitamin found abundantly in chia more than doubles the amount in sesame seeds. 100 g of chia provide about 8.83 mg or 55 per cent of daily-required levels of niacin. Niacin helps reduce LDL-cholesterol levels in the blood. In addition, it enhances GABA activity inside the brain, which in turn helps reduce anxiety and neurosis.
- ☆ The seeds are incredibly rich sources of many essential minerals. Calcium, phosporous, iron, manganese, and magnesium are particularly concentrated in chia. Many of these minerals have a vital role in bone mineralization, red blood cell production, enzyme synthesis, as well as regulation of cardiac and skeletal muscle activities.

Just a few table spoons full of chia a day provides enough recommended levels of phenolic anti-oxidants, minerals, vitamins and protein.

102

Coconut

Coconut is a mature fruit of the *Cocos nucifera* palm. Since the ancient times, the nut holds unique place among the millions of inhabitants in South-East Asia, and Pacific islands. It is one of the most sought-after ingredients in the kitchen as it is found in almost every delicacy prepared in these parts of the world. *Cocos nucifera* belongs to the large *Palmaceae* family of palm trees. Coco palm grows well under tropical climates. The palm requires moist, sandy, well-drained soil and flourishes well all along the saline-rich coastal regions.

The coconut palm is an un-branched, erect, tall-growing tree that may reach more than 100 feet and has a lifetime of about 75 to 100 years. Once planted, it may take about 4-5 years to begin their first produce, and often, quite longer time in some species. Several hundred species of the coconut palms grown all over the tropics, and their taste and flavour of water thus may vary according to saline content in the soil, distance from sea-shore, rain-fall, etc. In a season, a single coconut palm may produce 20-150 mature nuts. The fruit is almost spherical to oval in shape and measure between 5-10 inches in width. Its rough outer husk is light green, and turns gray as the nut matures. The husk is about 1-2 inches in thickness and made of tough fibers. Underneath the husk, there is a woody shell enclosing inner edible kernel (endosperm). Just harvested mature fruits contain some amount of sweet water inside its central hollow cavity surrounded by the white meat (endosperm). The fruit with its shell, kernel, and water together constitutes a "coconut" at the commercial level.

Health Benefits of Coconut

- ☆ Coconut is a very versatile and indispensable fruit for most people under the tropical belt. It is a complete food is rich in calories, vitamins, and minerals. An average-size nut weighing 400 g edible meat and water provide almost all the daily-required essential minerals, vitamins, and energy for a medium-sized person.

Coconut (*Cocus nucifera*), Fresh, Nutrition Value per 100 g

Principle	*Nutrient Value*	*Percentage of RDA*
Energy	354 Kcal	18 per cent
Carbohydrates	15.23 g	12 per cent
Protein	3.3 g	6 per cent
Total Fat	33.49 g	167 per cent
Cholesterol	0 mg	0 per cent
Dietary Fiber	9 g	24 per cent
Vitamins		
Folates	26 µg	6.5 per cent
Niacin	0.540 mg	3 per cent
Pantothenic acid	0.300 mg	6 per cent
Pyridoxine	0.054 mg	4 per cent
Riboflavin	0.020 mg	1.5 per cent
Thiamin	0.066 mg	5.5 per cent
Vitamin C	3.3 mg	5.5 per cent
Vitamin A	0 IU	0 per cent
Vitamin E	0.24 mg	2 per cent
Vitamin K	0.2 µg	<1 per cent
Electrolytes		
Sodium	20 mg	1 per cent
Potassium	356 mg	7.5 per cent
Minerals		
Calcium	14 mg	1.4 per cent
Copper	0.435 mg	48 per cent
Iron	2.43 mg	30 per cent
Magnesium	32 mg	8 per cent
Manganese	1.500 mg	65 per cent
Phosphorus	113 mg	16 per cent
Selenium	10.1 µg	18 per cent
Zinc	1.10 mg	10 per cent
Phyto-nutrients		
Carotene-β	0 µg	—
Phytosterols	47 mg	—

☆ 100 g kernel consists of 354 calories. Much of this comes from the fats and protein. Although, its meat is disproportionately high in saturated fats on comparison to other common edible nuts, coconut has many bioactive compounds that are essential for better health.

- The important saturated fatty acid in the coconut is lauric acid (1:12 carbon fatty acid). Lauric acid increases HDL cholesterol levels in the blood. HDL is a high-density lipoprotein, which has beneficial effects on the coronary arteries by preventing vessel blockade (atherosclerosis). Medicine recommends high HDL to total cholesterol levels in the blood for the same reason.
- Coconut water is a very refreshing drink to beat tropical summer thirst. The juice is packed with simple sugar, electrolytes, minerals, and bioactive compounds such as cytokinin, and enzymes such as acid phosphatase, catalase, dehydrogenase, peroxidase, polymerases, etc. Altogether, these enzymes aid in digestion and metabolism.
- Coconut oil extracted from the dry nut is an excellent emollient agent. It is used in cooking, to help scalp hair nourishment, in pharmacy and in medicines.
- Research studies suggest that cytokinins (*e.g.*, kinetin and trans-zeatin) in coconut water showed significant anti-ageing, anti-carcinogenic, and anti-thrombotic effects.
- The kernel is an excellent source of minerals such as copper, calcium, iron, manganese, magnesium, and zinc.
- It is also a very good source of B-complex vitamins such as folates, riboflavin, niacin, thiamin, and pyridoxine. These vitamins are essential in the sense that body requires them from external sources to replenish.
- Coconut meat and water contain a very good amount of potassium. 100 g of fresh meat contains 356 mg per cent or 7.5 per cent of daily required levels of potassium.

103

Flax Seed/Linseed

Flax seed, also known as linseed, is one of the ancient cultivated crops since Mesopotamian times, grown for its oil seeds, and fiber. The chewy seeds are packed with full of nutrients, omega-3 fatty acids, antioxidants, minerals, and essential vitamins. Off late, nutritional and Health Benefits of flax have widely drawn the attention of nutrition researchers as well as health enthusiasts alike across the planet. Flax belongs to the family of *Linaceae,* of the genus of *Linum,* and botanically named as *Linum usitatissimum.* Flax is one of the easily cultivated crops, flourish well both in tropical and subtropical climates, with its higher production as a field crop noted in fertile river valleys. It is an erect annual plant growing about 1 to 1.5 meters tall and bears light-blue coloured attractive flowers. The fruit pod is a round, dry capsule 6–9 mm diameter, containing several brown or golden-yellow seeds (depending on cultivar type). The seeds feature smooth, glossy surface and flat shape that somewhat appear like sesame seeds, but quite larger, measuring about 5–7 mm in length. In general, there exist two common cultivars of flax; one is predominantly grown for its oil seeds and the other variety for fiber. Seed flax generally features brown, and yellow or golden-yellow colour seeds, with most types having similar nutritional values and equal amounts of short-chain omega-3 fatty acids.

Health Benefits of Flax Seed

- ☆ As in other oil seeds, flax too is one of the very high-calorie foods. 100 g of seeds contain 534 calories or 27 per cent of daily-required levels. Further, the seeds are an excellent source of numerous health-benefiting nutrients, dietary fiber, minerals, antioxidants and vitamins that are essential for optimum health.
- ☆ Flax seed is rich in monounsaturated fatty acids like oleic acid. It is also one of the top vegetable sources of omega-3 essential fatty acids such as *linoleic acid, alpha-linolenic acid* (*ALA*) and *arachidonic acids*. Regular intake

of small portions of flax seeds in the diet helps to lower total as well as LDL or "bad cholesterol" and increases HDL or "good cholesterol" levels in the blood. Research studies suggest that Mediterranean diet that is rich in fiber, monounsaturated fatty acids and omega-3 fatty acids help to prevent coronary artery disease and strokes by favouring healthy blood lipid profile.

Flax Seeds (*Linum usitatissimum*), Nutritional Value per 100 g

Principle	*Nutrient Value*	*Percentage of RDA*
Energy	534 Kcal	27 per cent
Carbohydrates	28.8 g	22 per cent
Protein	18.3 g	32.5 per cent
Total Fat	42.16 g	170 per cent
Cholesterol	0 mg	0 per cent
Dietary Fiber	27.3 g	68 per cent
Vitamins		
Folates	87 µg	22 per cent
Niacin	3.08 mg	19 per cent
Pantothenic acid	0.985 mg	20 per cent
Pyridoxine	0.473 mg	36 per cent
Riboflavin	0.161 mg	12 per cent
Thiamin	1.64 mg	137 per cent
Vitamin A	0 IU	0 per cent
Vitamin C	0.6 mg	1 per cent
Vitamin E	19.95 mg	133 per cent
Vitamin K	4.3 µg	3.5 per cent
Electrolytes		
Sodium	30 mg	2 per cent
Potassium	813 mg	17 per cent
Minerals		
Calcium	255 mg	22.5 per cent
Copper	1.12 mg	124 per cent
Iron	5.73 mg	72 per cent
Magnesium	392 mg	98 per cent
Manganese	2.48 mg	108 per cent
Zinc	4.34 mg	39 per cent
Phyto-nutrients		
Carotene-β	0 µg	—
Lutein-zeaxanthin	651 µg	—

☆ Flax seeds are perhaps one of the most widely available botanical sources of n-3 or ω (omega)-3 fatty acids. Flax seed oil consists of approximately 55 per cent ALA (α-linolenic acid). One spoonful of flax seed oil provides about 8 g of omega-3 fatty acids. Research studies have suggested that n-3 fatty acids by their virtue of anti-inflammatory action help lower the risk of blood pressure, coronary artery disease, strokes and breast, colon and prostate cancers. Adequate quantities of n-3 oils are required for normal infant development and maturation of nervous system.

- The seeds contain *lignans,* a class of phytoestrogens considered to have antioxidant and cancer preventing properties.
- Flax are an excellent source of vitamin E, especially rich in *gamma-tocopherol;* containing about 20 g (133 per cent of daily-recommended values) per 100 g. vitamin E is a powerful lipid soluble antioxidant, required for maintaining the integrity of cell membrane of mucus membranes and skin by protecting it from harmful oxygen-free radicals.
- The seeds are packed with many important B-complex groups of vitamins such as riboflavin, niacin, thiamin, pantothenic acid, vitamin B-6, and folates. Thiamin is an essential co-factor for carbohydrate metabolism and helps prevent *beri-beri* disease. Folates help prevent neural tube defects in the fetus when consumed during pre-conception period and pregnancy.
- Furthermore, flax seed is rich source of minerals like manganese, potassium, calcium, iron, magnesium, zinc and selenium.
- Flax or linseed oil has flavourful nutty aroma and has been used in cooking, and as "carrier" *or* "base oil" in traditional medicines and in pharmaceutical uses.

104

Hazelnuts

Hazelnuts are sweet in taste and incredibly nutritious. Botanically, the nuts are fruits from the "birch" or *Betulaceae* family of trees. The "Filbert" (*Corylus maxima*) is similar kind and related to common hazel but only differing in having the nut more fully enclosed by the tubular involucre. In Britain, the nuts are usually enjoyed as "cobnuts." Scientific name: *Corylus avellana.* The hazel is a small deciduous tree, originated in southern Europe and Turkey. It is now being cultivated in many regions of the world, including USA as an important commercial crop.

Hazel tree begin producing fruits about three years after plantation. During each spring season, the tree bears attractive inflorescence (catkins), consisting of the cluster of monoecious flowers arranged closely along a central stem that ultimately becomes fruits by autumn. The hazel fruit is a nut produced in clusters; with each nut held in a short leafy involucre or capsule enclosing about three-quarters of the nut. Each yellow-brown colour nut, roughly spherical to oval in shape, is about 1.5-2 cm long and 1.2 -2 cm broad, featuring a light scar at the base. They fall out of the leafy involucre or capsule when ripe, about 7-8 months after pollination. Hazelnut oil, extracted from the nuts, has been used in as base or carrier oil in medicine, and in aromatherapy.

Health Benefits of Hazelnuts

☆ Hazelnuts are very high in energy and loaded with numerous health-benefiting nutrients that are essential for optimum health. 100 g nuts provide 628 calories. The nuts are rich in mono-unsaturated fatty acids like *oleic* as well as essential fatty acid, *linoleic acid* that helps lower LDL or bad cholesterol and increase HDL or good cholesterol. Research studies suggest that Mediterranean diet that is plentiful in monounsaturated fatty acids help to prevent coronary artery disease and strokes by favouring healthy blood lipid profile.

Hazelnuts (*Corylus avellana*), Nutritional Value per 100 g

Principle	*Nutrient Value*	*Percentage of RDA*
Energy	628 Kcal	31 per cent
Carbohydrates	16.7 g	13 per cent
Protein	14.95g	26.5 per cent
Total Fat	60.75 g	202 per cent
Cholesterol	0 mg	0 per cent
Dietary Fiber	9.7 g	25.5 per cent
Vitamins		
Folates	113 µg	28 per cent
Niacin	1.8 mg	11 per cent
Pantothenic acid	0.918 mg	18 per cent
Pyridoxine	0.563 mg	43 per cent
Riboflavin	0.113 mg	9 per cent
Thiamin	0.643 mg	53.5 per cent
Vitamin A	20 IU	<1 per cent
Vitamin C	6.3 mg	10.5 per cent
Vitamin E	15 mg	100 per cent
Vitamin K	14.2 µg	12 per cent
Electrolytes		
Sodium	0 mg	0 per cent
Potassium	680 mg	14 per cent
Minerals		
Calcium	114 mg	11 per cent
Copper	1.725 mg	192 per cent
Iron	4.7 mg	59 per cent
Magnesium	163 mg	41 per cent
Manganese	6.17 mg	268 per cent
Phosphorus	290 mg	41 per cent
Zinc	2.45 mg	22 per cent
Phyto-nutrients		
Carotene-α	3 µg	—
Carotene-β	11 µg	—
Lutein-zeaxanthin	92 µg	—

☆ These nuts are rich in dietary fiber, vitamins, and minerals and packed with numerous health promoting phyto-chemicals. Altogether, they help protect from diseases and cancers.

- ✰ Hazels are exceptionally rich in folate, which is a unique feature for the nuts. 100 g fresh nuts contain 113 µg; that is about 28 per cent recommended daily intake of this vitamin. Folate is an important vitamin that helps prevent *megaloblastic anemia* and most importantly, neural tube defects in the newborn. Good news for expectant mothers!
- ✰ Hazel nuts are an excellent source of vitamin E; contain about 15 g per 100 g (providing 100 per cent of RDA). Vitamin E is a powerful lipid soluble antioxidant, required for maintaining the integrity of cell membrane of mucus membranes and skin by protecting it from harmful oxygen-free radicals.
- ✰ The nuts, like almonds, are free from gluten and therefore, are safe alternative food sources that can be used in the preparation of gluten-free food formulas for gluten-sensitive, wheat allergy, and celiac disease patients.
- ✰ The nuts are packed with many important B-complex groups of vitamins such as riboflavin, niacin, thiamin, pantothenic acid, pyridoxine (vitamin B-6), and folates.
- ✰ They are rich source of minerals like manganese, potassium, calcium, copper, iron, magnesium, zinc, and selenium. Copper and manganese are essential co-factors for anti-oxidant enzyme, *superoxide dismutase*. Iron helps prevent microcytic-anemia. Magnesium and phosphorus are important components of bone metabolism.
- ✰ Hazelnut oil has nutty aroma and has excellent astringent properties. It helps to keep skin well protected from dryness. The oil has also been used in cooking, and as "carrier or base oil" in traditional medicines in massage therapy, aromatherapy, in pharmaceutical and cosmetic industry.

105

Macadamia

Sweet, delicious and flavourful macadamia nut is one of the popular edible nuts packed with notable health benefiting nutrients. Botanically, macadamia belongs to the family of *Proteaceae*, genus: *Macadamia*. Scientific name: *Macadamia integrifolia.* Common names include Australia nut, Queensland nut, bush nut etc. Macadamia is native to the East Coast rainforests of North Eastern parts of Australia. Several parts of mineral rich, tropical and subtropical areas of Australia, Hawaiian Islands, Middle Americas, Brazil and South African regions are favourite regions where these sweet and crunchy nuts grow in abundance.

Macadamia tree grows to about 15 meters in height. In general, it reaches maturity and begins to produce fruit at around the age of seventh year of plantation. There are about seven species of macadamia grown in their wild natural habitat. However, only two of which are edible and cultivated in the horticulture farms around the world. *Macadamia integrifolia* produces smooth-shelled nuts, whereas *Macadamia tetraphylla* has nuts with a rough shell. During each summer, the tree bears elongated chains of attractive cream-white coloured flower racemes, which subsequently develop into tough shelled, round to spherical shaped fruits with a conical tip. Depending upon the cultivar type its shell may possess smooth or rough outer surface. Each fruit features 0.5 to 1 inch in diameter consisting of husk enveloping around a single kernel. Inside, the edible kernel is white, has smooth buttery surface and sweet taste.

Health Benefits of Macadamia Nuts

- Macadamia nuts have sweet taste and are rich source of energy. 100 g of nuts provide about 718 calorie/100 g, which is one of the highest values among nuts.
- They are packed with numerous health-benefiting nutrients, minerals, antioxidants and vitamins that are essential for optimum health and

wellness

Macadamia Nuts (*Macadamia integrifolia*), Nutritional Value per 100 g

Principle	*Nutrient Value*	*Percentage of RDA*
Energy	718 Kcal	36 per cent
Carbohydrates	13.82 g	10.5 per cent
Protein	7.91 g	14 per cent
Total Fat	75.77 g	253 per cent
Cholesterol	0 mg	0 per cent
Dietary Fiber	8.6 g	23 per cent
Vitamins		
Folates	11 mcg	3 per cent
Niacin	2.473 mg	15 per cent
Pantothenic acid	0.758 mg	15 per cent
Pyridoxine	0.275 mg	21 per cent
Riboflavin	0.162 mg	12 per cent
Thiamin	1.195 mg	100 per cent
Vitamin A	0 IU	0 per cent
Vitamin C	1.2 mg	2 per cent
Vitamin E	0.24 mg	1.5 per cent
Electrolytes		
Sodium	5 mg	0 per cent
Potassium	368 mg	8 per cent
Minerals		
Calcium	85 mg	8.5 per cent
Copper	0.756 mg	84 per cent
Iron	3.69 mg	46 per cent
Magnesium	130 mg	32.5 per cent
Manganese	4.131 mg	180 per cent
Phosphorus	188 mg	27 per cent
Selenium	3.6 mcg	6.5 per cent
Zinc	1.30 mg	11 per cent
Phyto-nutrients		
Phyto-sterols	116 mcg	—
β-sitosterol	108 mcg	—

- ☆ 100 g of macadamia provides 8.6 g or 23 per cent of daily-recommended levels of dietary fiber. Additionally, they are a very good source of phytosterols such as β-sitosterol. However, the nuts contain no cholesterol.
- ☆ Since macadamia is free from *gluten*, it is one of the popular ingredients in the preparation of gluten-free food formulas. Such formula preparations

are a healthy alternative in patients with wheat gluten allergy and *celiac disease*.

- The nuts are rich source of mono-unsaturated fatty (MUF) like oleic acid (18:1) and palmitoleic acids (16:1). Studies suggest that MUF fats in the diet help lower total as well as LDL (bad) cholesterol and increase HDL (good) cholesterol levels in the blood. The Mediterranean diet, which is rich in fiber, MUF fats and antioxidants work favourably in maintaining healthy blood lipid profile and thus offer protection from coronary artery disease and strokes.
- Macadamias are an excellent source of minerals such as calcium, iron, magnesium, manganese and zinc. 100 g nuts provide 3.6 µg of selenium. Selenium is a cardio-protective micro-mineral and an important anti-oxidant cofactor for *glutathione peroxidase* enzyme.
- Furthermore, the nuts are also rich in many important B-complex vitamins that are vital for metabolic functions. 100 g of nuts provide 15 per cent of niacin, 21 per cent of pyridoxine (vitamin B-6), 100 per cent of thiamin, and 12 per cent of riboflavin.
- They contain small amounts of vitamin-A, and vitamin E. Both these fat-soluble vitamins possess potent anti-oxidant activities, which serve to protect cell membranes and DNA damage from harmful oxygen-free radicals.

In concise, sweet, refreshing macadamias are brimming with essential minerals, vitamins and heart-friendly mono-unsaturated fatty acids.

106

Peanuts

Delicious, crunchy, and nutty peanuts are one of the popular oil seeds known to humankind since ancient times. The nuts are enriched with many noteworthy health-benefiting nutrients that are essential for optimum health. They are actually *legumes* but have almost all the qualities that true nuts like almonds have. Botanically, the nuts are small size fruit pods of plant belonging to the *Fabaceae* family of the genus, *Arachis.*Some of the common names are *groundnut, earthnuts,* etc. Scientific name: *Arachis hypogaea*. Peanut is a small annual herb growing up to a foot above the ground. It is thought to have originated in the Central Americas and from where it spread to rest of the world through Spanish explorers. Today, the nuts are widely cultivated oil seeds and has established as prime commercial crop in China, India, African nations, and the United States of America.

After the seedling, the plant takes approximately 120 to 150 days to produce the crop. The process of peanut development is quite interesting! The yellow flowers, after self-pollination, develop in to "ovaries," called *pedicels,* which elongate rapidly to turn downward to bury several inches deep under the ground from where the fruits develop into peanut pods we know. To harvest, the entire plant, including roots, is dug out from the soil. Each plant may bear 10-150 fruit pods. The pods have wrinkled shells that are constricted between pairs of the two to four seeds per pod. Each seed is covered with thin brown colour cover and can be split into two equal halves as in any other legumes. Bambara groundnuts (*Vigna subterranea* L. Verdc) are different from American groundnuts (*Arachis hypogaea*) in that they appear somewhat like round cowpeas. Bambara groundnuts are indeed one of the popular legume crops in many African regions, providing much-needed protein, fat and other essential nutrients of vegetable origin.

Health Benefits of Peanuts

- ✰ Peanuts are rich in energy and contain health benefiting nutrients, minerals, antioxidants and vitamins that are essential for optimum health.

Groundnuts (*Arachis hypogaea*), all Types, Nutritional Value per 100 g

Principle	*Nutrient Value*	*Percentage of RDA*
Energy	567 Kcal	29 per cent
Carbohydrates	16.13 g	12 per cent
Protein	25.80 g	46 per cent
Total Fat	49.24 g	165 per cent
Cholesterol	0 mg	0 per cent
Dietary Fiber	8.5 g	22 per cent
Vitamins		
Folates	240 µg	60 per cent
Niacin	12.066 mg	75 per cent
Pantothenic acid	1.767 mg	35 per cent
Pyridoxine	0.348 mg	27 per cent
Riboflavin	0.135 mg	10 per cent
Thiamin	0.640 mg	53 per cent
Vitamin A	0 IU	0 per cent
Vitamin C	0	0 per cent
Vitamin E	8.33 mg	55.5 per cent
Electrolytes		
Sodium	18 mg	1 per cent
Potassium	705 mg	15 per cent
Minerals		
Calcium	92 mg	9 per cent
Copper	1.144 mg	127 per cent
Iron	4.58 mg	57 per cent
Magnesium	168 mg	42 per cent
Manganese	1.934 mg	84 per cent
Phosphorus	76 mg	54 per cent
Selenium	7.2 µg	13 per cent
Zinc	3.27 mg	30 per cent
Phyto-nutrients		
Carotene-α	0 µg	—
Crypto-xanthin-β	0 µg	—
Lutein-zeaxanthin	0 µg	—

☆ They compose sufficient levels of mono-unsaturated fatty acids especially oleic acid. It helps to lower LDL or "bad cholesterol" and increases HDL or "good cholesterol" level in the blood. Research studies suggest that Mediterranean diet, which is rich in monounsaturated fatty acids help to prevent coronary artery disease and strokes by favouring healthy blood lipid profile.

- ☆ These nuts are a good source of dietary protein composes fine quality amino acids that are essential for growth and development.
- ☆ Research studies have shown that peanuts contain high concentrations of poly-phenolic antioxidants, primarily p-coumaric acid. This compound has been thought to reduce the risk of stomach cancer by limiting the formation of carcinogenic nitrosamines in the stomach.
- ☆ Peanuts are an excellent source of resveratrol, another polyphenolic antioxidant. Resveratrol has been found to have protective function against cancers, heart disease, degenerative nerve disease, Alzheimer's disease, and viral/fungal infections.
- ☆ Furthermore, studies suggest that resveratrol reduce stroke risk by altering molecular mechanisms in the blood vessels (reducing susceptibility to vascular damage through decreased activity of angiotensin, a systemic hormone causing blood vessel constriction that would elevate blood pressure), and by increasing production of the vasodilator hormone, *nitric oxide.*
- ☆ Recent research studies suggest that boiling enhances antioxidant concentration in the peanuts. It has been found that boiled peanuts have two and four-fold increase in isoflavone antioxidants *biochanin-A* and *genistein* content, respectively.
- ☆ The nuts are an excellent source of vitamin E (α-tocopherol); containing about 8 g per100 g. vitamin E is a powerful lipid soluble antioxidant which helps maintain the integrity of cell membrane of mucus membranes and skin by protecting from harmful oxygen free radicals.
- ☆ The nuts are packed with many important B-complex groups of vitamins such as riboflavin, niacin, thiamin, pantothenic acid, vitamin B-6, and folates. 100 g of peanuts provide about 85 per cent of RDI of niacin, which contribute to brain health and blood flow to brain.
- ☆ The nuts are rich source of minerals like copper, manganese, potassium, calcium, iron, magnesium, zinc, and selenium.

Just a hand full of peanuts per day provides enough recommended levels of phenolic anti-oxidants, minerals, vitamins, and protein.

107

Pecans

Delicious, buttery yet rich-flavoured pecans are one of the popular edible tree nuts known to American aborigines since centuries ago. The nuts are enriched with many health-benefiting nutrients, minerals, vitamins that are essential for optimum health. Pecan is a very large sized; deciduous tree belongs to the member of hickory family, *Juglandaceae*. The tree is native to central and southern parts of the United States of America. Today, however, it is being cultivated in many regions of the world as an important commercial tree nuts crop. Scientific name: *Carya illinoinensis*.

Several cultivars exist; some of the highly recommended species include *Cape fear, Elliott* and *Moreland*. During each spring season, the tree bears catkins, consisting of the cluster of monoecious flowers arranged closely along a central stem that ultimately develops into fruits by autumn. Pecan nuts, like the fruit of all other members of the hickory genus, are not true nuts but, botanically, are a *drupe*. Each fruit features oval to oblong shape, 1 to 3 inches long and 0.5–1 inch broad and dark-brown colour with rough 3-4 mm thick husk or involucres. Its husk splits off into four sections at maturity to release a thin-shelled nut. Nuts are generally harvested from October through December. Raw nuts are then subjected to dehydration; the process which is essential to remove moisture and to improve the keeping quality of pecans.

Health Benefits of Pecans

- ☆ Pecan nuts are rich source of energy, provide 690 calories/100 g and contain health benefiting nutrients, minerals, antioxidants and vitamins that are essential for wellness.
- ☆ The nuts are rich in monounsaturated fatty acids like oleic acid and an excellent source of phenolic antioxidants. Regular addition of pecan nuts in the diet helps to decrease *total* as well as *LDL* or "bad cholesterol" and increases *HDL* or "good cholesterol" levels in the blood. Research studies suggest that Mediterranean diet, which is rich in dietary-fiber,

mono-unsaturated fatty acids, and antioxidants would help to prevent coronary artery disease and strokes by favouring healthy blood lipid profile.

Pecans (*Carya illinoinensis*), Nutritional Value per 100 g

Principle	*Nutrient Value*	*Percentage of RDA*
Energy	691 Kcal	34.5 per cent
Carbohydrates	13.86 g	11 per cent
Protein	9.17 g	17 per cent
Total Fat	71.9 g	360 per cent
Cholesterol	0 mg	0 per cent
Dietary Fiber	9.6 g	25 per cent
Vitamins		
Folates	22 μg	5.5 per cent
Niacin	1.167 mg	7 per cent
Pantothenic acid	0.863 mg	17 per cent
Pyridoxine	0.210 mg	16 per cent
Riboflavin	0.130 mg	10 per cent
Thiamin	0.660 mg	55 per cent
Vitamin A	56 IU	2 per cent
Vitamin C	1.1	2 per cent
Vitamin E	24.44 mg	163 per cent
Electrolytes		
Sodium	0 mg	0 per cent
Potassium	410 mg	9 per cent
Minerals		
Calcium	70 mg	7 per cent
Copper	1.2 mg	133 per cent
Iron	2.53 mg	32 per cent
Magnesium	121 mg	30 per cent
Manganese	4.5 mg	196 per cent
Phosphorus	277 mg	40 per cent
Selenium	3.8 μg	7 per cent
Zinc	4.53 mg	41 per cent
Phyto-nutrients		
Carotene-β	29 μg	—
Crypto-xanthin-β	9 μg	—
Lutein-zeaxanthin	17 μg	—

✰ Pecan nuts are rich source of many phyto-chemical substances that may contribute to their overall antioxidant activity, including polyphenolic antioxidant ellagic acid, vitamin E, beta-carotene, lutein and zeaxanthin. Research studies have been suggestive of that these compounds help the body remove toxic oxygen-free radicals and thus, protect the body from diseases, cancers, as well as infections.

- Anti-proliferative properties of *ellagic acid* are because of its ability to inhibit DNA binding of certain carcinogens such as nitrosamines and polycyclic aromatic hydrocarbons; thus protecting the human body from cancers.
- Pecans are an excellent source of vitamin-E, especially rich in gamma-tocopherol; provide about 25 g per100 g. Vitamin E is a powerful lipid soluble antioxidant, required for maintaining the integrity of cell membrane of mucus membranes and skin by protecting it from harmful oxygen-free radicals.
- The nuts are very rich sources of several important B-complex groups of vitamins such as riboflavin, niacin, thiamin, pantothenic acid, vitamin B-6, and folates. Together, these vitamins work as co-factors for the enzyme metabolism inside the human body.
- The nuts are also rich source of minerals like manganese, potassium, calcium, iron, magnesium, zinc, and selenium.

Add hand full of pecan nuts every day in the diet to keep your diet meter complete with sufficient levels of minerals, vitamins and protein.

108

Pine Nuts

Crunchy, sweet and delicious pine nuts are small edible seeds of pine tree. Pine kernels are indeed very good source of plant derived nutrients, essential minerals, vitamins and "heart friendly" mono-unsaturated fatty acids that help benefit in reducing cholesterol levels in the blood. Botanically, pine trees belong to the family of *Pinaceae*, in the genus: *Pinus*. Some of the common names are pinon nuts, pignoli, cedar nuts, chilgoza, pinyon pinenuts etc.

Pine trees grow in the wild cold and taiga forest regions of the northern hemisphere, particularly of Siberia and Canada. They straight erect trees with large stem and reach up to 75 feet in height with pyramidal or umbrella like dense foliage spread. The flowers of pine tree subsequently develop into a cone. The female cones take about two–three years to mature after pollination. At maturity, the female-cones may reach from as small as 3 cm, to a very large cone reaching about 35 cm. Scales at the base and tip of the cone tends to be small and sterile, and therefore, bear no seeds. Once mature and dry, the cones naturally open (split) up to release the seeds.

The two prominent pine species known for their large edible kernels include Pinus sibirica and Pinus koraiensis. Pine nuts feature tough dark-brown colour coat or shell. Stone pines have a long slender kernel in comparison to oriental pines, which are broad, large and have higher fat content. Inside, the kernel has cream white colour, delicate buttery flavour and sweet taste. Chilgoza pine nuts (*Pinus gerardiana*) is another distinct pine variety found in western Himalayan forests of Pakistan, India (Kashmir, Himachal Pradesh), and Afghanistan. Chilgoza pines have similar appearance as stone pines; featuring long slender, pointed kernels.

Health Benefits of Pine Nuts

- ☆ Pine nuts are one of the calorie-rich edible nuts. 100 g of dry-kernels provide 673 calories. Additionally, they comprise of numerous health promoting phyto-chemicals, vitamins, antioxidants, and minerals.

Pine Nuts (*Pinus gerardiana*), Raw, Nutritional Value per 100 g

Principle	*Nutrient Value*	*Percentage of RDA*
Energy	673 Kcal	34 per cent
Carbohydrates	13.08 g	10 per cent
Protein	13.69 g	24 per cent
Total Fat	68.37 g	228 per cent
Cholesterol	0 mg	0 per cent
Dietary Fiber	3.7 g	10 per cent
Vitamins		
Folates	34 µg	9 per cent
Niacin	4.387 mg	27 per cent
Pantothenic acid	0.313 mg	6 per cent
Pyridoxine	0.094 mg	7 per cent
Riboflavin	0.227 mg	17 per cent
Thiamin	0.364 mg	30 per cent
Vitamin A	29 IU	1 per cent
Vitamin C	0.8 mg	1 per cent
Vitamin E	9.33 mg	62 per cent
Electrolytes		
Sodium	2 mg	0 per cent
Potassium	597 mg	13 per cent
Minerals		
Calcium	16 mg	1.5 per cent
Copper	1.324 mg	147 per cent
Iron	5.53 mg	69 per cent
Magnesium	251 mg	63 per cent
Manganese	8.802 mg	383 per cent
Phosphorus	575 mg	82 per cent
Selenium	0.7 µg	1 per cent
Zinc	6.45 mg	58 per cent
Phyto-nutrients		
Carotene-β	17 µg	—
Crypto-xanthin-β	0 µg	—
Lutein-zeaxanthin	9 µg	—

☆ Their high caloric content chiefly comes from fats. Indeed, the nuts, especially rich in mono-unsaturated fatty acids like *oleic acid* (18:1 undifferentiated fat) that helps to lower LDL or "bad cholesterol" and increases HDL or "good-cholesterol" in the blood. Research studies suggest that Mediterranean diet, which contains good amounts of monounsaturated fatty acids, vitamins and antioxidants, helps to prevent coronary artery disease and strokes by favouring healthy blood lipid profile.

- Pine or cedar nuts contain essential fatty acid (ù-6 fat), pinolenic acid. Recent research has shown its potential use in weight loss by curbing the appetite. Pinolenic acid triggers the release of hunger-suppressant enzymes cholecystokinin and glucagon-like peptide-1 (GLP-1) in the gut. In addition, pinolenic acid was thought to have LDL-lowering properties by enhancing hepatic LDL uptake.
- Like almonds, pines are an excellent source of vitamin E; contain about 9.33 mg per 100 g (about 62 per cent of RDA). Vitamin E is a powerful lipid soluble antioxidant, required for maintaining the integrity of cell membrane of mucus membranes and skin by protecting it from harmful oxygen-free radicals.
- Furthermore, like almonds and peanuts, pines too are free from gluten and therefore, are a popular ingredient in the preparation of gluten-free food formulas. Such formula preparations are, in fact, healthy alternatives in people with wheat food allergy and *celiac disease.*
- Pine nuts are an excellent source of B-complex group of vitamins such as thiamin, riboflavin, niacin, pantothenic acid, vitamin B-6 (pyridoxine) and folates. These vitamins work as co-factors for enzymes in cellular substrate metabolism inside the human body.
- Furthermore, pine nuts contain healthy amounts of essential minerals like manganese, potassium, calcium, iron, magnesium, zinc and selenium. At 8.802 mg per 100 g (about 383 per cent of daily recommended intake), pines are one of the richest sources of manganese. Manganese is an all-important co-factor for antioxidant enzyme, superoxide dismutase. It is therefore, consumption of pines helps the body develop resistance against infectious agents and scavenge harmful oxygen-free radicals.

Pine nut oil has a delicate flavour with sweet aroma and is used in many traditional medicinal applications since ancient times. The main chemical components in pine oil are *borneol, bornyl acetate,* α *and* β*-phellandrene,* α*-pinene* and β*-pinene*. Its emollient property helps to keep skin well protected from dryness. It has also been used in cooking, and as "carrier or base oil" in traditional medicines in aromatherapy, in pharmaceutical and cosmetic industry.

109

Pistachio

Wonderfully delicious pistachio nuts have been revered as the symbol of wellness and robust health since ancient times. The nuts are enriched with many health-benefiting nutrients that are essential for optimum health. Pistachio nuts are dry fruits of species of trees belonging in the *Anacardiaceae* family, of the genus: Pistacia. The plant is a medium sized broad, bushy, dioecious, deciduous tree, believed to be originating in the mountain ranges of West-Asian region. Several cultivars exist; however, the most popular variety grown for the commercial purpose is the *Kerman*.

Pistachios grow well under hot, dry climates with cool winters. They are currently being cultivated in large scale in the orchards in the USA, Iran, Syria, Turkey, and China. After plantation, it takes approximately eight to ten years until it produces its first major crop. Once established, it keeps bearing fruits for centuries. The fruit, in fact, is a drupe (a fruit with a large, central located single seed), and the seed kernel is actually the edible portion. Each season, the tree bears heavy clusters of fruits, which appear somewhat like that of a grape bunch. Exteriorly, the mature fruit features off white coloured, hard, shell with splits apart exposing yellow-light green coloured oblong kernel, which measures about 1 inch in length and 1/2 inch in diameter.

Health Benefits of Pistachios

- ✰ Pistachios are delicious tree nuts recognized for their wholesome nutrition principles. Together with walnuts, almonds, and cashew, they form important source of protein, fats and minerals to otherwise dry and arid regions of Central, West and South Asian regions.
- ✰ Pistachios are rich source of energy; 100 g of nuts contain 557 calories. In addition, they are rich in mono-unsaturated fatty acids like oleic acid and an excellent source of antioxidants. Regular consumption of pistachios in the diet helps to lower total as well as bad LDL cholesterol and increases good HDL cholesterol levels within the blood. Research studies suggest

that Mediterranean diet that is rich in dietary-fiber, mono-unsaturated fatty acids, and antioxidants help to prevent coronary artery disease and strokes by favouring healthy blood lipid profile.

Pistachio (*Pistacia vera*), Nutritional Value per 100 g

Principle	*Nutrient Value*	*Percentage of RDA*
Energy	557 Kcal	29 per cent
Carbohydrates	27.97 g	21.5 per cent
Protein	20.60 g	37 per cent
Total Fat	44.44 g	148 per cent
Cholesterol	0 mg	0 per cent
Dietary Fiber	10.3g	27 per cent
Vitamins		
Folates	51 µg	13 per cent
Niacin	1.3 mg	8 per cent
Pantothenic acid	0.520 mg	10 per cent
Pyridoxine	1.7 mg	131 per cent
Riboflavin	0.160 mg	12 per cent
Thiamin	0.870 mg	72.5 per cent
Vitamin A	553 IU	18 per cent
Vitamin C	5 mg	8 per cent
Vitamin E-γ	22.60 mg	150 per cent
Electrolytes		
Sodium	1 mg	0 per cent
Potassium	1025 mg	22 per cent
Minerals		
Calcium	107 mg	11 per cent
Copper	1.3 mg	144 per cent
Iron	4.15 mg	52 per cent
Magnesium	121 mg	30 per cent
Manganese	1.2 mg	52 per cent
Phosphorus	376 mg	54 per cent
Selenium	7 µg	13 per cent
Zinc	2.20 mg	20 per cent
Phyto-nutrients		
Carotene-β	332 µg	—
Crypto-xanthin-β	0 µg	—
Lutein-zeaxanthin	0 µg	—

- They are rich source of many phyto-chemical substances that may contribute to their overall antioxidant activity, including carotenes, vitamin E, and polyphenolic antioxidant compounds. Research studies have been suggestive of that these compounds help the human body remove toxic oxygen-free radicals and thus, protect the body from diseases, cancers, as well as infections.
- Pistachios are an excellent source of vitamin-E, especially rich in *gamma-tocopherol;* contain about 23 g per100 g. vitamin E is a powerful lipid soluble antioxidant, essential for maintaining the integrity of cell membrane of mucus membranes and skin; offer protection from harmful oxygen-free radicals.
- The nuts are packed with many important B-complex groups of vitamins such as riboflavin, niacin, thiamin, pantothenic acid, vitamin B-6, and folates.
- These nuts are the storehouse of minerals like copper, manganese, potassium, calcium, iron, magnesium, zinc, and selenium. 100 g nuts provide 144 per cent of daily-recommended levels of copper. Copper is an essential trace mineral that is required in neuro-transmission, metabolism, as well as red blood cell (RBC) synthesis.
- Pistachio nut oil features flavourful, pleasant nutty aroma and has excellent emollient properties. It helps to keep skin well protected from dryness. It has also been used in cooking, and as "carrier or base oil" in traditional medicines in massage therapy, aromatherapy, in pharmaceutical, and cosmetic industry.

Just a hand full of pistachios a day provides enough recommended levels of phenolic anti-oxidants, minerals, vitamins, and protein.

110

Poppy Seeds

Nutty and pleasant in taste, poppy seeds are nutritious oilseeds used as condiment in cooking. They are the seeds obtained from the dry fruits (pods) of the poppy plant (opium poppy) and entirely free from any sinister side effects of other poppy plant products such as opium poppy. Poppy plant is a biennial herb of East Mediterranean, and Asia Minor origin belonging to the *Papaveraceae* family of the genus: Papaver. Scientific name: *Papaver somniferum.*

Ancient Egyptians were aware of seeds extraction from the poppy head. Through the Arab traders, opium cultivation spread to Persia, ancient Khorasan, and India. Today seeds poppy are a well established commercial crop in many parts of the world Mediterranean countries including Turkey, France, India, and East European region. The poppy plant grows up to 5 feet in height. It requires full sunlight and fertile soil to flourish. Depending upon variety, lilac, blue, red or white flowers appear during spring on long peduncles which subsequently turn into globular or oval shaped fruits (capsules). Individual fruit head measures about 4–6 cm in length and 3–4 cm in diameter, contains numerous tiny, bean (kidney) shaped seeds, which rattle when shaken inside dried capsules. Seeds poppy can be light gray to dark gray, black, or bluish depending on cultivar type. Its seeds, which are used as condiment spice and to press oil, indeed are very safe to use as food and contain negligible quantities of toxic alkaloids of the opium poppy.

Health Benefits of Poppy Seeds

- ✰ Poppy seeds contain many plant derived chemical compounds that are known to have anti-oxidant, disease preventing and health promoting properties.
- ✰ Their unique nutty aromatic flavour is because of many fatty acids and essential volatile oils, which comprise about 50 per cent of net weight. The seeds are especially rich in *oleic* and *linoleic* acids. Oleic acid, a mono-

unsaturated fatty acid, helps lower LDL or "bad cholesterol" and increase HDL or "good cholesterol" levels in the blood. Research studies suggest that Mediterranean diet which is rich in monounsaturated fatty acids helps to prevent coronary artery disease and strokes by favouring healthy blood lipid profile.

Poppy Seeds (*Papaver somniferum*), Nutritional Value per 100 g

Principle	*Nutrient Value*	*Percentage of RDA*
Energy	525 Kcal	26 per cent
Carbohydrates	28.13 g	22 per cent
Protein	17.99 g	32 per cent
Total Fat	41.56 g	139 per cent
Cholesterol	0 mg	0 per cent
Dietary Fiber	19.5 g	51 per cent
Vitamins		
Folates	82 µg	20 per cent
Niacin	0.896 mg	5.5 per cent
Pantothenic acid	0.324 mg	65 per cent
Pyridoxine	0.247 mg	19 per cent
Riboflavin	0.100 mg	8 per cent
Thiamin	0.854 mg	71 per cent
Vitamin A	0 IU	0 per cent
Vitamin C	1 mg	2 per cent
Vitamin E	1.77 mg	12 per cent
Vitamin K	0 mg	0 per cent
Electrolytes		
Sodium	26 mg	2 per cent
Potassium	719 mg	15 per cent
Minerals		
Calcium	1438 mg	144 per cent
Copper	01.627 mg	181 per cent
Iron	9.76 mg	122 per cent
Magnesium	347 mg	87 per cent
Manganese	6.707 mg	292 per cent
Phosphorus	870 mg	124 per cent
Selenium	13.5 µg	24 per cent
Zinc	7.9 mg	72 per cent

☆ Poppy seeds outer coat is a good source of dietary fiber. 100 g raw seeds provide 19.5 g or 51 per cent of recommended daily levels (RDA) of fiber. Much of this fiber is metabolically inert content which helps increase bulk of the food by absorbing water down the digestive tract and thereby easing constipation problem.

- ☆ Additionally, dietary fiber binds to bile salts (a product of cholesterol) and decreases their re-absorption in the colon. It thus helps in a further decrease in blood LDL cholesterol levels.
- ☆ The seeds are excellent source B-complex vitamins such as thiamin, pantothenic acid, pyridoxine, riboflavin, niacin, and folic acid. Many of these vitamins functions as co-factors in substrate metabolism especially fat and carbohydrates.
- ☆ Poppy seeds contain good levels of minerals like iron, copper, calcium, potassium, manganese, zinc and magnesium. *Copper* is required in the production of red blood cells. *Zinc* is a co-factor in many enzymes that regulate growth and development, sperm generation, digestion and nucleic acid synthesis. *Potassium* is an important component of cell and body fluids that helps controlling heart rate and blood pressure. *Manganese* is used by the body as a co-factor for the powerful anti-oxidant enzyme, *superoxide dismutase.*
- ☆ Dried poppy seeds contain very small levels of opium alkaloids such as morphine, thebaine, codiene, papaverine etc. When consumed in food, these compounds produce a minimal effect when consumed on the human nervous system. On the contrary, the chemicals have beneficial effects on the human body; soothe nervous irritability, act as painkillers, and used in pharmacy as well as in many traditional medicines in the preparations of cough mixtures, expectorants etc.

111

Pumpkin Seeds

Pumpkin seeds (pepita) are edible kernels of fruit pumpkin. The seeds, in-fact, are concentrated sources of many health-benefiting vitamins, minerals, anti-oxidants, and all important essential amino acids like *tryptophan and glutamate.* Pumpkin fruit is a squash-like gourd of the *Cucurbitaceae* family native to Mexico. Scientific name: *Cucurbita pepo*. In the Central Americas, hulled and gently roasted pumpkin kernels are popularly known as Pepita.

The pumpkin fruit, in general, is grown as a field vegetable crop; and its seeds, nevertheless, are used as food and to extract useful pumpkin seed oil. In fact, in some parts of central Europe (Styrian province in Austria, Slovenia and Hungary), it is cultivated as a major oil-seed crop at a commercial scale. Generally, the pumpkin fruit is allowed to mature completely in order to obtain good-quality seeds. Each fruit contains up to 500 cream-white husky seeds located at its central hollow cavity interspersed in between net like mucilaginous fibers. The seeds are semi-flat, have a typical oval shape with a conical tip. Inside, the edible kernel has olive-green colour. The kernels have sweet, creamy nutty flavour enjoyed in deserts, as a snack, in savory dishes, etc.

Health Benefits of Pumpkin Seeds

- ☆ Crunchy, delicious pumpkin seeds are high in calories, about 559 calories per 100 g. In addition; they are packed with fiber, vitamins, minerals, and numerous health promoting antioxidants.
- ☆ Their high caloric content mainly comes from protein and fats. On the positive side, the nuts are especially rich in mono-unsaturated fatty acids (MUFA) like oleic acid (18:1) that helps lower bad LDL cholesterol and increases good HDL cholesterol in the blood. Research studies suggest that Mediterranean diet, which is liberal in monounsaturated fatty acids help to prevent coronary artery disease and strokes by favouring healthy blood lipid profile.

Pumpkin Seeds (*Cucurbita pepo*), Dried, Nutritional Value per 100 g

Principle	*Nutrient Value*	*Percentage of RDA*
Energy	559 Kcal	28 per cent
Carbohydrates	10.71 g	8 per cent
Protein	30.23 g	54 per cent
Total Fat	49.05 g	164 per cent
Cholesterol	0 mg	0 per cent
Dietary Fiber	6 g	16 per cent
Vitamins		
Folates	58 µg	15 per cent
Niacin	4.987 mg	31 per cent
Pantothenic acid	0.750 mg	15 per cent
Pyridoxine	0.143 mg	11 per cent
Riboflavin	0.153 mg	12 per cent
Thiamin	0.273 mg	23 per cent
Vitamin A	16 IU	0.5 per cent
Vitamin C	1.9 µg	3 per cent
Vitamin E-γ	35.10 mg	237 per cent
Electrolytes		
Sodium	7 mg	0.5 per cent
Potassium	809 mg	17 per cent
Minerals		
Calcium	46 mg	4.5 per cent
Copper	1.343 mg	149 per cent
Iron	8.82 mg	110 per cent
Magnesium	592 mg	148 per cent
Manganese	4.543 mg	198 per cent
Phosphorus	1233 mg	176 per cent
Selenium	9.4 µg	17 per cent
Zinc	7.81 mg	71 per cent
Phyto-nutrients		
Carotene-β	9 µg	—
Crypto-xanthin-β	1 µg	—
Lutein-zeaxanthin	74 µg	—

☆ The seeds contain good-quality protein. 100 g seeds provide 30 g or 54 per cent of recommended daily allowance. In addition, the seeds are an excellent source of amino acid tryptophan and glutamate. Tryptophan is converted into serotonin and niacin. Serotonin is a beneficial neuro-chemical often labeled as nature's sleeping pill. Further, tryptophan is a precursor of B-complex vitamin, niacin (60 mg of tryptophan = 1mg niacin).

- ✰ Glutamate is required in the synthesis of γ-amino butyric acid (GABA). GABA, an anti-stress neurochemical in the brain, helps reducing anxiety, nervous irritability, and other neurotic conditions.
- ✰ Pumpkin seeds are a very good source of anti-oxidant vitamin E; contain about 35.10 mg of tocopherol-gamma per 100 g (about 237 per cent of RDA). Vitamin E is a powerful lipid soluble antioxidant. It prevents tissue cells from the free radical mediated oxidant injury. Thus, it helps maintain the integrity of mucus membranes and skin by protecting from harmful oxygen-free radicals.
- ✰ Pumpkin kernels are an also excellent source of B-complex group of vitamins such as thiamin, riboflavin, niacin, pantothenic acid, vitamin B-6 (pyridoxine) and folates. These vitamins work as co-factors for various enzymes during cellular substrate metabolism in the human body. In addition, niacin helps to reduce LDL-cholesterol levels in the blood. Along with glutamate, it enhances GABA activity inside the brain, which in turn reduces anxiety and neurosis.
- ✰ Furthermore, its seeds contain very good levels of essential minerals like copper, manganese, potassium, calcium, iron, magnesium, zinc and selenium. Just as in pine nuts, pumpkin seeds too are very rich in manganese (provide 4543 mg per 100 g, about 198 per cent of daily-recommended intake). Manganese is an all-important co-factor for antioxidant enzyme, superoxide dismutase. It is therefore, consumption of pumpkin kernels helps the body develop resistance against infectious agents and scavenge harmful oxygen-free radicals.

112

Quinoa

Quinoa (pronounced as kinwah; quinua in Spanish) is incredibly nutritious grain-like seed obtained from the goosefoot family plant growing in highland plains of South-American Andes origin. The Incas believed quinua "the mother grain (chisaya mama)" as eating it in food would confer them long and healthy life. The crop was once the staple food sources of Incas before the Spanish explorers forced them to abandon its cultivation. The plant is a goosefoot (*Chenopodiaceae*) family of annual herbs. White goosefoot (*Chenopodium album*), and Good King Henry (*C. bonus-henricus*) are some of the close related species. Scientific name: *Chenopodium quinoa.*

Chenopodium quinoa reaches about 3-6 feet in height and prefers well-drained, neutral PH and sandy soil; however, it thrives well under scanty environmental conditions and alkaline (salty) soils. It is cultivated as one of the chief crops in rain-shadow highlands of Bolivia, Peru, Chile and Ecuador. In its natural habitat, quinoa begins flowering (inflorescence) in July, which subsequently develops into edible seeds by August end through September. Its seeds, which appear like amaranth seeds, feature white, gray, pink to light-brown in colour depending upon the cultivar type and measure about two mm in diameter. The seeds feature outer seed coat containing bitter substances known as saponins, which should be removed before eating. Quinoa indeed has found renewed interest, particularly among the food scientists because of its sustainable characteristic of thriving under stress growing conditions. Additionally, it's well-balanced protein, fat and mineral composition may be an answer for malnutrition alleviation in famine-prone regions around the world. Quinoa does not belong to the cereal-class of staple grains. Instead, it is a seed, much similar to other dicotyledons such as amaranth, and legume family plants. Nonetheless, its seeds feature proportionately less fat content than oil seeds and hence treated like any other staple grains.

Health Benefits of Quinoa

☆ 100 g quinoa seeds (grains) provide 368 calories. The grains are moderate sources of energy, and their calorie content may be compared to that of major cereals such as wheat, maize, rice and that of pulses like chickpea, mung bean, cowpea (black-eye pea), etc.

Quinoa (*Chenopodium quinoa*), Nutritional Value per 100 g

Principle	*Nutrient Value*	*Percentage of RDA*
Energy	368 Kcal	18.5 per cent
Carbohydrates	64.16 g	49 per cent
Protein	14.12 g	35 per cent
Total Fat	6.07 g	20 per cent
Cholesterol	0 mg	0 per cent
Dietary Fiber	7 g	18.5 per cent
Vitamins		
Folates (B9)	184 µg	46 per cent
Niacin (B3)	1.520 mg	9.5 per cent
Riboflavin (B2)	0.318 mg	24 per cent
Thiamin (B1)	0.360 mg	30 per cent
Vitamin A	14 IU	0.5 per cent
Vitamin E	2.44 mg	17 per cent
Vitamin K	0	0 per cent
Electrolytes		
Sodium	5 mg	<1 per cent
Potassium	563 mg	12 per cent
Minerals		
Calcium	47 mg	5 per cent
Copper	0.590 mg	65.5 per cent
Iron	4.57 mg	57 per cent
Magnesium	197 mg	42 per cent
Manganese	2.003 mg	87 per cent
Phosphorus	457 mg	65 per cent
Selenium	8.5 µg	15 per cent
Zinc	3.10 mg	28 per cent
Phyto-nutrients		
Carotene-β	8 µg	—
Crypto-xanthin-β	1 µg	—
Lutein-zeaxanthin	163 µg	—

- ✰ Quinoa grains provide 14-18 g of protein per 100 g. They compose all the required essential amino acids for the human body at excellent proportions, especially in lysine, which is otherwise a limiting amino acid in grains like wheat, maize, rice, etc.
- ✰ Quinoa is one of the gluten-free staple foods. Gluten is the protein present in certain *grass family* grains may induce stomach upset and diarrhea in individuals with *celiac disease.*
- ✰ They are rich source of soluble and insoluble dietary fiber. 100 g provide 7 g or 18.5 per cent of daily requirement of fiber. Fiber increase bulkiness of the food and help prevent constipation by speeding up its movement through the gut. In fact, gastro-intestinal transit time of food is greatly decreased. Fiber also binds to toxins and aid in their excretion from the gut and helps protect the colon mucus membrane from cancers. In addition, dietary fibers bind to bile salts (produced from cholesterol) and decrease their re-absorption in colon, thus help lower serum LDL cholesterol levels.
- ✰ Quinoa grains are indeed very good sources of B-complex group of vitamins, vitamin E (α-tocopherol) and essential fatty acids such as *linoleic* and *α-linolenic acid* (18:3). 100 g of grains contain 184 µg or 46 per cent of daily-required levels of folates (B9).
- ✰ Tender quinoa leaves and flower-heads are excellent sources of vitamin A, folate and antioxidants such as *lutein, carotene, crypto-xanthin,* and *zeaxanthin.* Yellow and red cultivar seeds also rich in these pigment antioxidants. Altogether, these compounds indeed functions as powerful anti-oxidants by removing harmful free radicals from the body thus protect from cancers, infection, ageing and degenerative neurological diseases.
- ✰ Quinoa is an excellent source of minerals like iron, copper, calcium, potassium, manganese, and magnesium. *Copper* is required for the production of red blood cells. *Iron* is required for red blood cell formation. *Magnesium* relaxes blood vessels leading to brain and found to have curative effects on depression, and headache. *Zinc* is a co-factor in many enzymes that regulate growth and development, sperm generation, digestion and nucleic acid synthesis. *Potassium* is an important component of cell and body fluids that helps controlling heart rate and blood pressure. *Manganese* is used by the body as a co-factor for the powerful anti-oxidant enzyme, superoxide dismutase.

Quinoa is not a cereal grain, but a dicotyledon seed (minute fruit) of goosefoot plant. Nonetheless, it has been treated like grain and consumed as one of the chief food sources along with potato and corn since ancient times by native Andean inhabitants. It features all the characteristics of a seed and composes good proportions of all the essential amino acids, fats, starch, fiber, minerals, and vitamins. Quinoa oil, extracted from its seeds indeed is an excellent source of essential fatty acids, and vitamin E.

113

Sesame Seeds

One of the first oil seeds known to humankind, sesame seeds are used in culinary as well as in traditional medicines for their nutritive, preventive, and curative properties. Its oil seeds are sources for some phyto-nutrients such as omega-6 fatty acids, flavonoid phenolic anti-oxidants, vitamins and dietary fiber with potent anti-cancer as well as health promoting properties. Sesame plant is a tall annual herb of the *Pedaliaceae* family, which grows extensively in Asia, particularly in Burma, China, and India. It is also one of the chief commercial crops in Nigeria, Sudan and Ethiopia. Its scientific name: *Sesamum indicum.*

The plant requires well-drained sandy soil and tropical environment to flourish. It grows to about 5 feet tall and bears plenty of pink-white foxglove type flowers. The pods appear soon containing white, brown, or black seeds depending up on the cultivar type, arranged in rows inside. Each pod (2-5 cm in length) is a long rectangular box like capsule with deep grooves on its sides. A single pod (1 to 3 inches in length) may contain up to 100 or more seeds. Sesame seeds are small, almost oblate in shape featuring pleasant nutty flavour and high oil content.

Health Benefits of Sesame Seeds

- ☆ Delicious, crunchy sesame seeds are widely considered healthful foods. 100 g of seeds provide 573 calories. Although, much of its calorie comes from fats, sesame contains several notable health-benefiting nutrients, minerals, antioxidants and vitamins that are essential for wellness.
- ☆ The seeds are especially rich in mono-unsaturated fatty acid *oleic acid*, which comprise up to 50 per cent fatty acids in them. Oleic acid helps to lower LDL or "bad cholesterol" and increases HDL or "good cholesterol" in the blood. Research studies suggest that Mediterranean diet which is rich in mono-unsaturated fats help to prevent coronary artery disease and stroke by favouring healthy lipid profile.

Sesame Seeds (*Sesamum indicum*), Whole, Dried, Nutritional Value per 100 g

Principle	*Nutrient Value*	*Percentage of RDA*
Energy	573 Kcal	29 per cent
Carbohydrates	23.45 g	18 per cent
Protein	17.73 g	32 per cent
Total Fat	49.67 g	166 per cent
Cholesterol	0 mg	0 per cent
Dietary Fiber	11.8 g	31 per cent
Vitamins		
Folates	97 µg	25 per cent
Niacin	4.515 mg	28 per cent
Pantothenic acid	0.050 mg	1 per cent
Pyridoxine	0.790 mg	61 per cent
Riboflavin	0.247 mg	19 per cent
Thiamin	0.791 mg	66 per cent
Vitamin A	9 IU	<1 per cent
Vitamin C	0	0 per cent
Vitamin E	0.25 mg	2 per cent
Electrolytes		
Sodium	11 mg	1 per cent
Potassium	468 mg	10 per cent
Minerals		
Calcium	975 mg	98 per cent
Copper	4.082 mg	453 per cent
Iron	14.55 mg	182 per cent
Magnesium	351 mg	88 per cent
Manganese	2.460 mg	107 per cent
Phosphorus	629 mg	90 per cent
Selenium	34.4 µg	62.5 per cent
Zinc	7.75 mg	70 per cent
Phyto-nutrients		
Carotene-β	5 µg	—
Crypto-xanthin-β	0 µg	—
Lutein-zeaxanthin	0 µg	—

☆ The seeds are an also very valuable source of dietary proteins with fine quality amino acids that are essential for growth, especially in children. Just 100 g of seeds provide about 18 g of protein (32 per cent of daily-recommended values).

- In addition, sesame seeds contain health benefiting compounds such as sesamol (3, 4-methylene-dioxyphenol), sesaminol, furyl-methanthiol, guajacol (2-methoxyphenol), phenylethanthiol and furaneol, vinylguacol and decadienal. *Sesamol* and *sesaminol* are phenolic anti-oxidants. Together, these compounds help stave off harmful free radicals from the human body.
- Sesame is among the seeds rich in quality vitamins and minerals. They are very good sources of B-complex vitamins such as niacin, folic acid, thiamin (vitamin B1), pyridoxine (vitamin B6), and riboflavin.
- 100 g of sesame contains 97 µg of folic acid, about 25 per cent of recommended daily intake. Folic acid is essential for DNA synthesis. When given to expectant mothers during the peri-conception period, it may prevent neural tube defects in the baby.
- Niacin is another B-complex vitamin found abundantly in sesame. About 4.5 mg or 28 per cent of daily-required levels of niacin is provided by just 100 g of seeds. Niacin helps reduce LDL-cholesterol levels in the blood. In addition, it enhances GABA activity inside the brain, which in turn helps reduce anxiety and neurosis.
- The seeds are incredibly rich sources of many essential minerals. Calcium, iron, manganese, zinc, magnesium, selenium, and copper are especially concentrated in sesame seeds. Many of these minerals have a vital role in bone mineralization, red blood cell production, enzyme synthesis, hormone production, as well as regulation of cardiac and skeletal muscle activities.

Just a hand full of sesame a day provides enough recommended levels of phenolic anti-oxidants, minerals, vitamins and protein.

114

Sunflower

Sweet, nutty sunflower seeds are an excellent source of calories, essential fatty acids, vitamins, and minerals. The seeds are mainly employed in the extraction of edible oil at a commercial scale all over the world. Its seeds are eaten as delicious snacks by humans and birds alike. Sunflower plant is a tall, erect, herbaceous annual plant belonging to the family of *Asteraceae* of the genus, *Helianthus*. Its botanical name is *Helianthus annuus*. It is native to Middle American region from where it spread as an important commercial crop all over the world by the European explorers. At present, Russian Union, China, The USA, and Argentina are the leading producers of sunflower crop.

Sunflower flourishes well in well-drained moist limey soil. The plant prefers good sunlight. Domesticated varieties bear single large flower head at the top. Unlike its domestic cultivar type, wild sunflower exhibits several branches with each branch ending with their own individual flower head. The flower head in sunflower plant consists of two types of flowers. While its perimeter consists of sterile yellow colour large petals (ray flowers), the central disk is made of numerous small fertile flowers arranged in concentric whorls, which subsequently convert to achenes (seeds). Sunflower seeds are about 6 mm to 10 mm in length and feature conical shape with smooth surface. Their outer coat (hull) is black with gray-white edible-kernel inside. Each sunflower head may contain several hundreds of edible oil seeds.

Health Benefits of Sunflower Seeds

- ☆ Delicious, nutty, and crunchy sunflower seeds are widely considered healthful foods. They are high in energy, 100 g seeds consist of 584 calories. Nonetheless, they are incredible sources of health benefiting nutrients, minerals, antioxidants and vitamins.
- ☆ Much of their calories come from fatty acids. The seeds are especially rich in poly-unsaturated fatty acid linoleic acid, which comprise more 50 per

cent fatty acids in them. They are also good in mono-unsaturated oleic acid that helps lower LDL or "bad cholesterol" and increases HDL or "good-cholesterol" in the blood. Research studies suggest that Mediterranean diet which is rich in monounsaturated fats help to prevent coronary artery disease and stroke by favouring healthy lipid profile.

Sunflower Seeds (*Helianthus annuus*), Kernels, Dried, Nutritional Value per 100 g

Principle	*Nutrient Value*	*Percentage of RDA*
Energy	584 Kcal	29 per cent
Carbohydrates	20 g	15 per cent
Protein	20.78 g	37 per cent
Total Fat	51.46 g	172 per cent
Cholesterol	0 mg	0 per cent
Dietary Fiber	8.6 g	23 per cent
Vitamins		
Folates	227 µg	57 per cent
Niacin	8.335 mg	52 per cent
Pantothenic acid	1.130 mg	22 per cent
Pyridoxine	1.345 mg	103 per cent
Riboflavin	0.355 mg	27 per cent
Thiamin	1.480 mg	123 per cent
Vitamin A	50 IU	1.6 per cent
Vitamin C	1.4	2 per cent
Vitamin E	35.17 mg	234 per cent
Electrolytes		
Sodium	9 mg	1 per cent
Potassium	645 mg	14 per cent
Minerals		
Calcium	78 mg	8 per cent
Copper	1.800 mg	200 per cent
Iron	5.25 mg	63 per cent
Magnesium	325 mg	81 per cent
Manganese	1.950 mg	85 per cent
Phosphorus	660 mg	94 per cent
Selenium	53 µg	96 per cent
Zinc	5.00 mg	45 per cent
Phyto-nutrients		
Carotene-β	30 µg	—
Crypto-xanthin-β	0 µg	—

- Like other nuts, they are also a very good source of proteins with fine quality amino acids such as tryptophan that are essential for growth, especially in children. Just 100 g of seeds provide about 21 g of protein (37 per cent of daily-recommended values).
- In addition, the sunflower seeds contain health benefiting poly-phenol compounds such as chlorogenic acid, quinic acid, and caffeic acids. These are natural anti-oxidants, which help remove harmful oxidant molecules from the body. Further, chlorogenic acid helps reduce blood sugar levels by limiting glycogen breakdown in the liver.
- Further, the seeds are indeed a very rich source of vitamin E; contain about 35.17 g per100 g (about 234 per cent of RDA). Vitamin E is a powerful lipid soluble antioxidant, required for maintaining the integrity of cell membrane of mucus membranes and skin by protecting it from harmful oxygen-free radicals.
- Sunflower kernels among is one of the finest sources of B-complex group of vitamins. They are very good sources of B-complex vitamins such as niacin, folic acid, thiamin (vitamin B1), pyridoxine (vitamin B6), pantothenic acid, and riboflavin.
- Sunflowers are incredible sources of folic acid. 100 g of kernels contains 227 µg of folic acid, which is about 37 per cent of recommended daily intake. Folic acid is essential for DNA synthesis. When given to an expectant mother during the peri-conception period, it may prevent neural tube defects in the baby.
- Niacin and pyridoxine are other B-complex vitamins found abundantly in sunflower seeds. About 8.35 mg or 52 per cent of daily-required levels of niacin is provided by just 100 g of seeds. Niacin helps reduce LDL-cholesterol levels in the blood. In addition, it enhances GABA activity inside the brain, which in turn helps reduce anxiety and neurosis.
- The seeds are incredibly rich sources of many essential minerals. Calcium, iron, manganese, zinc, magnesium, selenium, and copper are especially concentrated in sunflower. Many of these minerals have a vital role in bone mineralization, red blood cell production, enzyme synthesis, hormone production, as well as regulation of cardiac and skeletal muscle activities.

Just a hand full of sunflower kernels a day provides much of the recommended level of phenolic anti-oxidants, minerals, vitamins, and protein.

115

Walnuts

Walnuts are revered since ancient times as a symbol of intellectuality, since their kernels have convoluted surface inside the shell resembling as that of brain! The nuts are enriched with many health-benefiting nutrients, especially Ù-3 fatty acids that are essential for optimum health. The nuts are edible kernels of the fruits from tree belonging to *Juglandaceae* family, in the genus: *Juglan.* Juglan species plants are medium sized, semi-tropical, deciduous trees believed to be originating in the mountain ranges of Central Asian or southern Europe region.

There exist at least 30 varieties of walnut cultivars. The three popular verities grown for the commercial purpose are the *English or Persian walnut* (*Juglans regia*), the *Black walnut* (*Juglans nigra*), and the *White or butternut walnut* (*Juglans cinerea*). Commercially, the nuts are being cultivated in the United States of America, Romania, France, Turkey, and China. After plantation, the tree takes approximately four years until it produces its first major crop. During each season, the fruits are ready for harvesting by August when the thick green hull begins to crack open to expose light brown coloured hard-shelled "walnut." Each nut features roughly spherical in shape, about the size of medium-sized lemon weighing about 10-15 g and enclosing single edible kernel inside. Structurally, the walnut kernel consists of two uneven, corrugated lobes, off white in colour and covered by a paper thin, light brown skin. The lobes are partially attached to each other. Oil extracted from the nuts, apart from cooking, has also been used as base or carrier oil in medicine, and aromatherapy.

Health Benefits of Walnuts

- The nuts are rich source of energy and contain health benefiting nutrients, minerals, antioxidants and vitamins that are essential for optimum health.
- They are rich in monounsaturated fatty acids (about 72 per cent) like oleic acid and an excellent source of all important omega-3 essential fatty acids like *linoleic acid, alpha-linolenic acid* (*ALA*) *and arachidonic acids.* Regular intake

of walnuts in the diet helps to lower total as well as LDL or "bad cholesterol" and increases HDL or "good cholesterol" levels in the blood. Research studies suggest that Mediterranean diet that is rich in mono-unsaturated fatty acids, and omega-3 fatty acids help to prevent coronary artery disease and strokes by favouring healthy blood lipid profile.

Walnuts (*Juglans regia*), Nutritional Value per 100 g

Principle	*Nutrient Value*	*Percentage of RDA*
Energy	654 Kcal	33 per cent
Carbohydrates	13.71 g	11 per cent
Protein	15.23 g	27 per cent
Total Fat	65.21 g	217 per cent
Cholesterol	0 mg	0 per cent
Dietary Fiber	6.7 g	18 per cent
Vitamins		
Folates	98 µg	24 per cent
Niacin	1.125 mg	7 per cent
Pantothenic acid	0.570 mg	11 per cent
Pyridoxine	0.537 mg	41 per cent
Riboflavin	0.150 mg	11.5 per cent
Thiamin	0.341 mg	28 per cent
Vitamin A	20 IU	0.5 per cent
Vitamin C	1.3 mg	2 per cent
Vitamin E-γ	20.83 mg	139 per cent
Vitamin K	2.7 µg	2 per cent
Electrolytes		
Sodium	2 mg	0 per cent
Potassium	441 mg	9 per cent
Minerals		
Calcium	98 mg	10 per cent
Copper	1.5 mg	167 per cent
Iron	2.9 mg	36 per cent
Magnesium	158 mg	39.5 per cent
Manganese	3.4 mg	148 per cent
Phosphorus	346 mg	49 per cent
Selenium	4.9 µg	9 per cent
Zinc	3.09 mg	28 per cent
Phyto-nutrients		
Carotene-β	12 µg	—
Crypto-xanthin-β	0 µg	—
Lutein-zeaxanthin	9 µg	—

- ☆ Eating just as much as 25 g each day provides about 90 per cent of RDI (recommended daily intake) of omega-3 fatty acids. Research studies have suggested that n-3 fatty acids by their virtue of anti-inflammatory action help to lower the risk of blood pressure, coronary artery disease, strokes and breast, colon and prostate cancers.
- ☆ They are rich source of many phyto-chemical substances that may contribute to their overall anti-oxidant activity, including *melatonin, ellagic acid, vitamin E, carotenoids, and poly-phenolic compounds*. These compounds have potential health effects against cancer, ageing, inflammation, and neurological diseases.
- ☆ Scientists at University of Scranton, Pennsylvania had recently discovered that walnuts have highest levels of popyphenolic antioxidants than any other common edible nuts. 100 g of walnuts contain 13541 µmol TE (Trolex equivalents) of oxidant radical absorbance capacity (ORAC). Eating as few as six to seven average size nuts a day could help scavenge disease causing free radicals from the body.
- ☆ In addition, they are an excellent source of vitamin E, especially rich in gamma-tocopherol; contain about 21 g per 100 g (about 140 per cent of daily-required levels). Vitamin E is a powerful lipid soluble antioxidant, required for maintaining the integrity of cell membrane of mucus membranes and skin by protecting it from harmful oxygen-free radicals.
- ☆ These nuts are packed with many important B-complex groups of vitamins such as riboflavin, niacin, thiamin, pantothenic acid, vitamin B-6, and folates.
- ☆ They also very are rich source of minerals like manganese, copper, potassium, calcium, iron, magnesium, zinc, and selenium. Copper is a cofactor for many vital enzymes, including cytochrome c-oxidase and superoxide dismutase (other minerals function as co-factors for this enzyme are manganese and zinc). Zinc is a co-factor in many enzymes that regulate growth and development, sperm generation, digestion and nucleic acid synthesis. Selenium is an important micronutrient, which functions as a co-factor for anti-oxidant enzymes such as *glutathione peroxidases.*
- ☆ Walnut oil has flavourful nutty aroma and exhibits excellent astringent properties. Applied locally, it helps to keep skin well protected from dryness. It has also been used in cooking, and as "carrier or base oil" in traditional medicines in massage therapy, aromatherapy, in pharmaceutical and cosmetic industry.

Munch a handful of walnuts a day and you will have enough recommended levels of minerals, vitamins, and protein.

D. Nutritional Value and Health Benefits of Spices

Introduction

Spices not only just excite your taste buds but are composed of an impressive list of phyto-nutrients, essential oils, antioxidants, minerals and vitamins that are essential for our wellness. Spices have been in part of our food since centuries, and now become integral part of our life. Thanks to the European explorers, whose contributions in spreading them from their place of origin greatly help spices popular all over the world?

Spices can be categorized botanically as follows:

- ☆ Leaves of aromatic plants: Examples include bay leaf, rosemary, thyme, etc.
- ☆ Fruits or seeds: Examples include fennel, coriander, fenugreek, mustard and black pepper etc.
- ☆ Roots or bulbs: Examples include garlic, galangal, turmeric, ginger, etc.

Why Spices in our Diet?

- ☆ Spices contain an impressive list of plant-derived chemical compounds that are known to have disease preventing and health promoting properties. They have been in use since ancient times for its anti-inflammatory, carminative, anti-flatulent properties.
- ☆ The components in the spices have been found to have anti-clotting action, prevent clogging of platelets in the blood vessels, thus helps easing blood flow, preventing stroke and coronary artery disease.
- ☆ The active principles in the spices may augment the motility of the intestinal tract as well as increases the digestion power by stimulating copious gastro-intestinal enzyme secretions.
- ☆ Throat gargling with tepid thyme water can help relieve sore throat and bronchitis symptoms. Thyme is also being used as an anti-septic mouthwash in the treatment of caries and gingivitis.

- ☆ Decoction of certain healthy spices is taken by mouth for the treatment of colds, influenza, mild fevers, indigestion, stomach upset, and painful menstruation.
- ☆ Spices are also known to have natural anti-helminthes (control worm infestation) function in traditional medicines.
- ☆ The essential volatile oils in certain spices work as a rubefacient (soothes skin around the site of application and improves the local blood circulation), increasing the flow of blood to make the skin feel warmer. They are being applied as a popular home remedy for arthritis and sore muscles, used either as a poultice or in hot baths.
- ☆ Spices essential oils are being used in aromatherapy as well as de-odorants in the perfume industry.
- ☆ Spices contain a good amount of minerals like potassium, manganese, iron, and magnesium. Potassium is an important component of cell and body fluids that helps controlling heart rate and blood pressure. Manganese is used by the body as a co-factor for the antioxidant enzyme, superoxide dismutase.

Culinary Uses of Spices

Spices can be aromatic or pungent in flavours and peppery or slightly bitter in taste. In order to keep their fragrance and flavour intact, they are generally added in the cooking recipes at the last moments since, prolonged cooking results in evaporation of essential oils.

- ☆ Spices are being used in the preparation of season soups, barbecue sauces, pickling and as a main ingredient in a variety of curry powders.
- ☆ Spices along with some seasonal herbs are being used to enhance the flavour and taste of vegetable, chicken, fish and meat dishes.
- ☆ Some healthy spices like cloves, coriander etc., are also been used in flavour drinks.

116

Allspice

Allspice, also known popularly as Jamaican pepper or pimento, is one of the widely used spice ingredients in the Mexican and other Central American cuisines. The spice corn is a dried "unripe" fruit obtained from the evergreen tropical shrub belonging to the family of *Myrtaceae* of the genus of *pimento*. Scientific name: *Pimenta dioica.* The pimento tree is native to tropical evergreen rain forest of Central American region and Caribbean islands. Generally, the plant starts bearing fruits after about five years of implantation. Unripe green berries, generally, picked up when they reach full size and then subjected to sun light drying thoroughly. Thus, shriveled berries appear similar to brown peppercorns, measuring about 6 mm in diameter but contain two seeds unlike peppercorns, which have only one centrally placed seed. Ground allspice has strong spicy taste and aroma that closely resemble a mixture of black-pepper, nutmeg, cloves, and cinnamon.

Health Benefits of Allspice

- ☆ The active principles in the allspice found to have been anti- nflammatory, *rubefacient* (warming and soothing), carminative and anti-flatulent properties.
- ☆ Allspice contains health benefiting essential oils such as eugenol, a phenylpropanoids class of chemical compound, which gives pleasant, sweet aromatic fragrances to this spice. It also contains *caryophyllene, methyleugenol, glycosides, tannins, quercetin, resin,* and *sesquiterpenes.* At the processing units, these volatile essential oils are obtained through distillation process using this spice corn. The outer coat of the allspice-berries is believed to have the greatest concentration of some of the compounds of medicinal activities.
- ☆ As in black peppercorns, the active principles in the allspice may increase the motility of the gastro-intestinal tract as well as augment the digestion power by increasing enzyme secretions inside the stomach and intestines.

Allspice (*Pimenta dioica*), Nutrition Value per 100 g

Principle	*Nutrient Value*	*Percentage of RDA*
Energy	236 cal	13 per cent
Carbohydrates	72.12 g	55 per cent
Protein	6.09 g	11 per cent
Total Fat	8.69 g	29 per cent
Cholesterol	0 mg	0 per cent
Dietary Fiber	21.6 g	54 per cent
Vitamins		
Folates	36 µg	9 per cent
Niacin	2.860 mg	18 per cent
Pantothenic acid	0.210 mg	16 per cent
Pyridoxine	0.210 mg	16 per cent
Riboflavin	0.063 mg	8.5 per cent
Thiamin	0.101 mg	2.5 per cent
Vitamin A	540 IU	18 per cent
Vitamin C	39.2 mg	65 per cent
Electrolytes		
Sodium	77 mg	5 per cent
Potassium	1044 mg	22 per cent
Minerals		
Calcium	661 mg	66 per cent
Copper	0.553 mg	61 per cent
Iron	7.06 mg	88 per cent
Magnesium	135 mg	34 per cent
Manganese	2.943 mg	128 per cent
Phosphorus	113 mg	16 per cent
Zinc	1.01 mg	9 per cent

☆ *Eugenol,* has local anesthetic and antiseptic properties, hence; useful in gum and dental treatment procedures. Recent research studies have shown that preparation made from allspice oil mixed with extractions from garlic, and oregano can work against *E. coli, Salmonella* and *L. monocytogenes* infections.

☆ The spice is enriched with the good amount of minerals like potassium, manganese, iron, copper, selenium, and magnesium. Iron is an important co-factor for cytochrome-oxidase enzymes during cellular metabolism. It is also required for red blood cell production in the bone marrow. Being an important component of cell and body fluids, potassium helps control heart rate and blood pressure. Manganese is used by the body as a co-factor for the powerful antioxidant enzyme, superoxide dismutase.

- ☆ Further, the spice also contains very good amounts of vitamin A, vitamin B-6 (pyridoxine), riboflavin, niacin and vitamin-C. Vitamin C is a powerful natural antioxidant; regular consumption of foods rich in vitamin C helps the body develop resistance against infectious agents and scavenge harmful, pro-inflammatory free radicals.

117

Anise Seed

Are you suffering from gripping stomach pain? Or with cold and cough? Take a few sips of anise seed mixed in hot-water and you feel better instantly. This humble herbal plant known for its aromatic spicy seeds is quite popular as well for its distinctive culinary and medicinal values. Anise or aniseed botanically belongs to the Apiaceae family of the Genus Pimpinella and known scientifically as *Pimpinella anisum*. This spicy plant is native to Middle-East and Mediterranean region, probably originated in Egypt.

Anise is a perennial herbal plant; generally, grows up to a height of about 2 feet. It bears white coloured umbelli form flowers by July, and is harvested by cutting the whole plant once the seeds ripened. Then the seeds are separated from the flower heads by threshing method. The seeds feature oblong or curved like comma shape, about 3-4 mm long, light brown colour and fine stripes over its outer surface. The seeds feature delicately sweet and aromatic bouquet with a distinctive liquorice flavour. The special fragrance is due to essential oil, *anethole* in them.

Star anise (*Illicium verum*) is the spicy fruit, obtained from evergreen tree native of South-West China. It has similar flavour and taste like that of anise seed. The tree bears star-shaped fruits, which turn rust-red colour when ripen and envelope inside amber-coloured seeds. Both the seed as well as the husks are used as spice in cooking.

Health Benefits of Anise

- ✰ Exotic anise spice contains some of the important plants derived chemical compounds that are known to have been anti-oxidant, disease preventing, and health promoting properties.
- ✰ The chief component that gives the characteristic sweet, aromatic flavour to anise is anethole, the essential volatile oil present in the seeds. Other important compounds found in these seeds include *estragol, p-anisaldehyde, anise alcohol, acetophenone, pinene* and *limonene*.

Anise Seeds (*Pimpinella anisum*), Nutritional Value per 100 g

Principle	*Nutrient Value*	*Percentage of RDA*
Energy	337 Kcal	17 per cent
Carbohydrates	50.02 g	38 per cent
Protein	17.60 g	31 per cent
Total Fat	79 per cent	15.90 g
Cholesterol	0 mg	0 per cent
Dietary Fiber	14.6 g	38 per cent
Vitamins		
Folates	10 µg	2.5 per cent
Niacin	3.060 mg	19 per cent
Pantothenic acid	0.797 mg	16 per cent
Pyridoxine	0.650 mg	50 per cent
Riboflavin	0.290 mg	22 per cent
Thiamin	0.340 mg	28 per cent
Vitamin A	311 IU	10.5 per cent
Vitamin C	21 mg	35 per cent
Electrolytes		
Sodium	16 mg	1 per cent
Potassium	1441 mg	31 per cent
Minerals		
Calcium	646 mg	65 per cent
Copper	0.910 mg	101 per cent
Iron	36.96 mg	462 per cent
Magnesium	170 mg	42.5 per cent
Manganese	2.300 mg	100 per cent
Phosphorus	440 mg	63 per cent
Selenium	5.0 mg	9 per cent
Zinc	5.30 mg	48 per cent

- ☆ Anise seed oil obtained from extraction of the seeds is found application in many traditional medicines as*stomachic, antiseptic, antispasmodic, carminative, digestive, expectorant, stimulant* and *tonic*.
- ☆ The seeds are an excellent source of many essential B-complex vitamins such as pyridoxine, niacin, riboflavin, and thiamin. Pyridoxine (vitamin B-6) helps increase GABA neuro-chemical levels in the brain.
- ☆ The spicy seeds are the great source of minerals like calcium, iron, copper, potassium, manganese, zinc and magnesium. 100 g dry seeds contain 36.96 mg or 462 per cent daily required levels of iron. Potassium is an important component of cell and body fluids that helps control heart rate and blood pressure.

- ☆ Copper is a cofactor for many vital enzymes, including *cytochrome C-oxidase* and *superoxide dismutase* (other minerals function as cofactors for this enzyme are manganese and zinc). Copper is also required in the production of red blood cells.
- ☆ The spice also contains good amounts of anti-oxidant vitamins such as vitamin-C and vitamin-A.

118

Bay Leaf

Pleasantly aromatic bay leaf or bay laurel is one of the well-recognized leafy-spices since ancient times, revered for its medicinal as well as the culinary values. In legends, bay laurel is highly regarded in astrology science as the tree of the *Sun* god, under the celestial sign of *Leo*. Botanically, bay tree belongs to the family of *Lauraceae* of the genus; *Laurus*. Scientific name: *Laurus nobilis*. Bay laurel is thought to have originated in Asia Minor region and from where it spread to all over the Mediterranean and other parts of Asia during earlier times. The bay plant is a tall, conical, evergreen tree growing more than 30 feet in height. Yellow or greenish white, star shaped flowers appear in clusters during early spring, which subsequently produce dark green-purplish single seed berry. The leaves are elliptic and shiny dark green about 3-4 inches long, rather thick, and leathery. The leaves give off a sweet aroma when broken and added to dish. When slightly wilted and dried, they are strongly aromatic. The dried fruit is also being used as a flavouring agent in cuisines.

Health Benefits of Bay Leaf

- ☆ Bay leaf was highly praised by the Greeks and the Romans, who deeply believed that the herb symbolizes wisdom, peace, and protection.
- ☆ The spice contains many notable plants derived compounds, minerals and vitamins that are essential for optimum health.
- ☆ This spice has many volatile active components such as *α-pinene, β-pinene, myrcene, limonene, linalool, methyl chavicol, neral, α-terpineol, geranyl acetate, eugenol,* and *chavicol.* These compounds are known to have been antiseptic, anti-oxidant, digestive, and thought to have anti-cancer properties.
- ☆ Fresh leaves are very rich source of vitamin-C; provide 46.5 mg or 77.5 per cent of RDA per 100 g. Vitamin-C (ascorbic acid) is one of the powerful natural anti-oxidant that help remove harmful free radicals from the body. Ascorbic acid also has immune booster, wound healing and anti-viral effects.

Bay Leaf (*Laurus nobilis*), Nutritional Value per 100 g

Principle	*Nutrient Value*	*Percentage of RDA*
Energy	313 Kcal	15.5 per cent
Carbohydrates	74.97 g	57 per cent
Protein	7.61 g	13 per cent
Total Fat	8.36 g	29 per cent
Cholesterol	0 mg	0 per cent
Dietary Fiber	26.3 g	69 per cent
Vitamins		
Folates	180 mcg	45 per cent
Niacin	2.005 mg	12.5 per cent
Pyridoxine	1.740 mg	133 per cent
Riboflavin	0.421 mg	32 per cent
Vitamin A	6185 IU	206 per cent
Vitamin C	46.5 mg	77.5 per cent
Electrolytes		
Sodium	23 mg	1.5 per cent
Potassium	529 mg	11 per cent
Minerals		
Calcium	834 mg	83 per cent
Copper	0.416 mg	46 per cent
Iron	43 mg	537 per cent
Magnesium	120 mg	30 per cent
Manganese	8.167 mg	355 per cent
Phosphorus	113 mg	16 per cent
Selenium	2.8 mcg	5 per cent
Zinc	3.70 mg	33 per cent

- ☆ Furthermore, its fresh leaves and herb parts are very good in folic acid; contain about 180 mg or 45 per cent of daily-recommended values per 100 g. Folates are important in DNA synthesis and when given during the peri-conception period, they can help prevent neural tube defects in the baby.
- ☆ Bay leaves are an excellent source of vitamin A; contain 6185 IU or 206 per cent of recommended daily levels per 100 g. Vitamin A is a natural antioxidant and is essential for healthy visual sight. It is also required for maintaining mucus membranes and skin health. Consumption of natural foods rich in vitamin A has been found to help to protect from lung and oral cavity cancers.
- ☆ The spice is indeed a very good source of many vitamins such as niacin, pyridoxine, pantothenic acid and riboflavin. These B-complex groups of

vitamins help in enzyme synthesis, nervous system function, and regulating body metabolism.

- This noble spice is a good source of minerals like copper, potassium, calcium, manganese, iron, selenium, zinc and magnesium. Potassium is an important component of cell and body fluids that helps control heart rate and blood pressure. Manganese and copper are used by the body as co-factors for the antioxidant enzyme, superoxide dismutase. Iron is essential for red blood cell production and as a co-factor for cytochrome-oxidase enzymes.

Medicinal Uses of Bay Leaf

- Medicinally, the benefits of the bay leaf and its berries are plentiful. It has astringent, diuretic, and appetite stimulant properties.
- Essential oil from the bay leaves contains mostly cineol (50 per cent); furthermore, *eugenol, chavicol, acetyl eugenol, methyl eugenol, α- and β-pinene, phellandrene, linalool, geraniol* and *terpineol* are also found.
- Infusions of herb parts are reputed to soothe the stomach and relieve flatulence and colic pain.
- The lauric acid in the bay laurel leaves has insect repellent properties.
- Bay laurel infusions are used to soothe the stomach ulcers and relieve flatulence.
- The components in the essential oil can also be used in many traditional medicines in the treatment of arthritis, muscle pain, bronchitis and flu symptoms.

119

Black Pepper

Incredibly popular black pepper, often referred as "king of spice" is a well-known spice since ancient times. The peppercorn plant is native to tropical evergreen rain forest of South Indian state, Kerala, from where it spread to rest of the world. The Pepper fruit, also known as the peppercorn, is actually a berry obtained from this plant. Botanically, peppercorn belongs to the family of *Piperaceae* of the genus of piper; and known scientifically as *Piper nigrum*. This perennial vine and climber requires supporting tree or pole to grow in height; thus has similar growth characteristics that of *beetle leaf* plant. It begins to bear small round berries after about three to four years of implantation. Technically, the pepper fruit is a drupe, measuring about 5 mm in diameter, containing a single large seed at its center.

Several colour peppercorns found in the markets are nothing but the same fruit, which picked up from the plant at different stages of maturity and subjected to different methods of processing. Generally, the peppercorns are harvested when they are half-matured and just about to turn red. They are then left to dry under the sun light that causes them to shrivel and turns black. Alternatively, green pepper corns are picked while the berries still unripe and green. The white peppercorns are derived when fully ripe berries soaked in brine to remove their dark outer coat, leaving the inner white-colour pepper seed. Black peppers have a strong spicy taste that comes to them from volatile-oils such as piperine. In ground peppers, these volatile oils may evaporate if kept open in the air for longer periods.

Cubeb or tailed pepper berries are dried unripe fruits of the *Piper cubeba* vine that is grown mainly Indonesian rain forest. They appear similar to black peppercorns but have a characteristic stalk which is often interpreted as a "tail." Cubeb berries have a distinctive flavour rich in monoterpene essential oil, *cubebene.*

Health Benefits of Black Pepper

☆ Peppercorns contain an impressive list of plant derived chemical compounds that are known to have disease preventing and health promoting properties. Peppers have been in use since ancient times for its anti-inflammatory, carminative, anti-flatulent properties.

Black Peppers (*Piper nigrum*), Nutritional Value per 100 g

Principle	*Nutrient Value*	*Percentage of RDA*
Energy	255 Kcal	13 per cent
Carbohydrates	64.81 g	49 per cent
Protein	10.95 g	19.5 per cent
Total Fat	3.26 g	11 per cent
Cholesterol	0 mg	0 per cent
Dietary Fiber	26.5 g	69 per cent
Vitamins		
Choline	11.3 mg	2 per cent
Folic acid	10 mcg	2.5 per cent
Niacin	1.142 mg	7 per cent
Pyridoxine	0.340 mg	26 per cent
Riboflavin	0.240 mg	18 per cent
Thiamin	0.109 mg	9 per cent
Vitamin A	299 IU	10 per cent
Vitamin C	21 mg	35 per cent
Vitamin E-γ	4.56 mg	30 per cent
Vitamin K	163.7 mcg	136 per cent
Electrolytes		
Sodium	44 mg	3 per cent
Potassium	1259 mg	27 per cent
Minerals		
Calcium	437 mg	44 per cent
Copper	1.127 mg	122 per cent
Iron	28.86 mg	360 per cent
Magnesium	194 mg	48.5 per cent
Manganese	5.625 mg	244.5 per cent
Phosphorus	173 mg	25 per cent
Zinc	1.42 mg	13 per cent
Phyto-nutrients		
Carotene-β	156 mcg	—
Carotene-α	0 mcg	—
Crypto-xanthin-β	48 mcg	—
Lutein-zeaxanthin	205 mcg	—
Lycopene	6 mcg	—

- ☆ Peppercorns are composed of health benefiting essential oils such as piperine, an amine alkaloid, which gives strong spicy pungent character to the pepper. It also contains numerous monoterpenes hydrocarbons such as *sabinene, pinene, terpenene, limonene, mercene,* etc., which gives aromatic property to the pepper.
- ☆ The above-mentioned active principles in the pepper may increase the gut motility as well as the digestion power by increasing gastro-intestinal enzyme secretions. It has also been found that piperine can increase absorption of selenium, B-complex vitamins, beta-carotene, as well as other nutrients from the food.
- ☆ Black peppercorns contain a good amount of minerals like potassium, calcium, zinc, manganese, iron, and magnesium. Potassium is an important component of cell and body fluids that helps controlling heart rate and blood pressure. *Manganese* is used by the body as a co-factor for the antioxidant enzyme, *superoxide dismutase. Iron* is essential for cellular respiration and blood cell production.
- ☆ They are also an excellent source of many vital B-complex groups of vitamins such as Pyridoxine, riboflavin, thiamin and niacin.
- ☆ Peppercorns are a good source of many anti-oxidant vitamins such as vitamin-C and vitamin-A. They are also rich in flavonoid polyphenolic anti-oxidants like carotenes, cryptoxanthin, zeaxanthin and lycopene. These compounds help the body removes harmful free radicals and helps protect from cancers and diseases.

120

Capers

Unique, piquant flavoured capers are flowering buds of a low-growing caper shrub. The buds are indeed one of the desired ingredients in the kitchens all over the Mediterranean basin. The spice buds botanically belong to the family of *Capparaceae*, in the genus: *Capparis*. Scientific name: *Capparis spinosa.* The plant is a spiny, trailing, deciduous shrub native to the Mediterranean. It prefers warm humid climate and grows in abundance all around the Cyprus, Italy, Greece, North African and some Asia Minor regions. The shrub begins producing flower (caper) buds from third year of plantation.

In general, their small cream coloured buds are handpicked in the early hours of the day which are otherwise unfold to a beautiful whitish-pink four sepal flower with long tassels of purple stamens. Soon after harvesting, the buds are washed and allowed to wilt for few hours in the sun before putting into jars and covered with salt, vinegar, brine, or olive oil. In commercial practice, capers are categorized and sold by their size in the markets. Smaller sized buds are more valuable than large ones. *Non-pareil* and *surfines* are some small buds while *capucines, capotes* and *grusas* are sold as big sized ones.

Health Benefits of Capers

- ✰ Being flower buds, capers are in fact very low in calories, 23 calories per 100 g. However, the spice contains many phytonutrients, anti-oxidants and vitamins essential for optimum health.
- ✰ Capers are one of the plant sources high in flavonoid compounds rutin (or rutoside) and quercetin. Capers are in-fact very rich source of quarcetin (180 mg/100 g) second only to tea leaf. Both these compounds are powerful anti-oxidants. Research studies suggest that quercetin has anti-bacterial, anti-carcinogenic, analgesic and anti-inflammatory properties.

Capers *Capparis spinosa*, Canned, Nutrient Value per 100 g

Principle	*Nutrient Value*	*Percentage of RDA*
Energy	23 Kcal	1 per cent
Carbohydrates	4.89 g	4 per cent
Protein	2.36 g	4 per cent
Total Fat	0.86 g	3 per cent
Cholesterol	0 mg	0 per cent
Dietary Fiber	3.2 g	8 per cent
Vitamins		
Folates	23 mcg	6 per cent
Niacin	0.652 mg	4.5 per cent
Pantothenic acid	0.027 mg	0.5 per cent
Pyridoxine	0.023 mg	2 per cent
Riboflavin	0.139 mg	11 per cent
Thiamin	0.018 mg	1.5 per cent
Vitamin A	138 IU	4 per cent
Vitamin C	4.3 mg	7 per cent
Vitamin E	0.88 mg	6 per cent
Vitamin K	24.6 mcg	20.5 per cent
Electrolytes		
Sodium	2954 mg	197 per cent
Potassium	40 mg	1 per cent
Minerals		
Calcium	40 mg	4 per cent
Copper	0.374 mg	42 per cent
Iron	1.67 mg	21 per cent
Magnesium	33 mg	8 per cent
Manganese	0.078 mg	3 per cent
Phosphorus	10 mg	1 per cent
Selenium	1.2 mcg	2 per cent
Zinc	0.32 mg	3 per cent
Phyto-nutrients		
Carotene-β	83 mcg	—
Carotene-α	0 mcg	—
Cryptoxanthin-β	0 mcg	—
Lutein-zeaxanthin	0 mcg	—

☆ Furthermore, rutin strengthen capillaries and inhibits platelet clump formation in the blood vessels. Both these actions of rutin help in smooth circulation of blood in very small vessels. Rutin has found application in some in trial treatments for hemorrhoids, varicose veins and in bleeding conditions such as hemophilia. It also found to reduce LDL cholesterol levels in obese individuals

- The spicy buds contain healthy levels of vitamins such as vitamin A, vitamin K, niacin, and riboflavin. Niacin helps lower LDL cholesterol.
- Furthermore, minerals like calcium, iron, and copper are present in them. High sodium levels are because of added granular sea salt (sodium chloride).

Medicinal Uses

- Caper parts have been used to relieve rheumatic pain in traditional medicines.
- The spicy caper pickles traditionally added to recipes as appetite stimulant. In addition, they help relieve stomachache and flatulence conditions.

121

Caraway Seed

In traditional Europe, caraway seed is the chief spice features in savory dishes. Strongly aromatic caraway is a member of parsley or Umbelliferae family, the family that also includes commonly known herbs and spices such as dill, anise, fennel, cumin, etc. Scientific name of caraway is *Carum carvi.*

Caraway is grown extensively all over Europe, North Africa, and Asia Minor. This biennial, herbaceous plant blooms once in every two years to creamy flowers in umbels. It grows to about 2 feet in height and bears small feathery leaves. The seeds which appear like cumin, are crescent in shape and dark brown with up to five vertical ribs. The seeds are harvested during early hours of the day to avoid spilling on the ground. The cut plants are then staked until they dry, and the seeds are threshed.

Health Benefits of Caraway Seeds

- ✰ In addition to its use as medicinal values, caraway indeed has many health benefiting nutrients, minerals, vitamins and anti-oxidants.
- ✰ Caraway seeds are rich source of dietary fiber. 100 g seeds provide 38 g of fiber. They increase bulk of the food and help prevent constipation by speeding up movement of food through the gut. In fact, gastro-intestinal transit time of food is greatly decreased.
- ✰ Fiber also binds to toxins in the food and helps protect the colon mucus membrane from cancers. In addition, dietary fibers bind to bile salts (produced from cholesterol) and decrease their re-absorption in colon, thus help lower serum LDL cholesterol levels.
- ✰ Caraway contains several health benefiting essential oils. Principle volatile compounds are *carvone, limonene, carveol, pinen, cumuninic aldehyde, furfurol,* and *thujone*. These active principles in the caraway seeds are known to have antioxidant, digestive, carminative, and anti-flatulent properties.

Caraway Seed (*Carum carvi*), Nutritional Value per 100 g

Principle	*Nutrient Value*	*Percentage of RDA*
Energy	333 Kcal	17 per cent
Carbohydrates	49.90 g	38 per cent
Protein	19.77 g	35 per cent
Total Fat	14.59 g	48 per cent
Cholesterol	0 mg	0 per cent
Dietary Fiber	38 g	100 per cent
Vitamins		
Folates	10 μg	2.5 per cent
Niacin	3.606 mg	23 per cent
Pyridoxine	0.360 mg	28 per cent
Riboflavin	0.379 mg	29 per cent
Thiamin	0.383 mg	32 per cent
Vitamin A	363 IU	12 per cent
Vitamin C	21 mg	35 per cent
Vitamin E	2.5 mg	17 per cent
Electrolytes		
Sodium	17 mg	1 per cent
Potassium	1351 mg	29 per cent
Minerals		
Calcium	689 mg	69 per cent
Copper	0.910 mg	101 per cent
Iron	16.23 mg	203 per cent
Magnesium	258 mg	64.5 per cent
Manganese	1.300 mg	56.5 per cent
Phosphorus	568 mg	81 per cent
Zinc	5.5 mg	50 per cent
Phyto-nutrients		
Carotene-β	189 μg	—
Crypto-xanthin-β	58 μg	—
Lutein-zeaxanthin	205 μg	—

- ✰ Caraway has several healths benefiting flavonoid antioxidnats such as *lutein, carotene, and crypto-xanthin* and *zeaxanthin*. These compounds are indeed functions as powerful anti-oxidants by removing harmful free radicals from the body thus protect from cancers, infection, ageing and degenerative neurological diseases.
- ✰ Caraway spice is an excellent source of minerals like iron, copper, calcium, potassium, manganese, selenium, zinc and magnesium. *Copper* is required

in the production of red blood cells. Iron is required for red blood cell formation. *Zinc* is a co-factor in many enzymes that regulate growth and development, sperm generation, digestion and nucleic acid synthesis. *Potassium* is an important component of cell and body fluids that helps controlling heart rate and blood pressure. *Manganese* is used by the body as a co-factor for the powerful anti-oxidant enzyme, *superoxide dismutase.*

✰ The seeds indeed are the storehouse for many vital vitamins. Vitamin A, vitamin E, vitamin C as well as many B-complex vitamins like thiamin, pyridoxine, riboflavin, and niacin particularly is concentrated in the caraway seeds.

122

Cardamom

Cardamom is a seed pod, known since antiquity for its culinary and medicinal properties. The spice is native to evergreen rain forest of southern India and now grown in only few tropical countries. Botanically, it belongs to the family of *"Zingiberaceae"* and consists of two genera; *Elettaria* and *Amomum*. Generally, the plant grows up to 4 meters in length in thick clumps and starts bearing prized *seed pods* after about two years of plantation. Each pod measures about 1-2 cm in length. Both the above-named cardamom varieties feature three-sided pods with a thin, yet tough papery outer cover and tiny deep-brown to black seeds that are arranged in vertical rows. *Elettaria* pods are small and light green, while *amomum* pods are larger and dark brown. The pods are being used as flavouring base in both food and drink, in cooking recipes and as well as in medicine. Black cardamom (*Amomum subulatum*) also known as *Nepal cardamom* (*Bari or kali elaichi*) is relatively big sized pod of same Zingiberaceae family. The pod has dark brown rough outer coat, measure about 2-4 cm in length and 1-2 cm in diameter. The pods have camphor-like intense flavour commonly used in spicy stews in sub Himalayan plains of India, Pakistan, and Nepal as well as in China.

Health Benefits of Cardamom

- ✰ This exotic spice contains many plants derived chemical compounds that are known to have been anti-oxidant, disease preventing and health promoting properties.
- ✰ The spicy pods contain many essential volatile oils that include *pinene, sabinene, myrcene, phellandrene, limonene, 1, 8-cineole, terpinene, p-cymene, terpinolene, linalool, linalyl acetate, terpinen-4-oil, a-terpineol, a-terpineol acetate, citronellol, nerol, geraniol, methyl eugenol,* and *trans-nerolidol.*
- ✰ The therapeutic properties of cardamom-oil have found application in many traditional medicines as antiseptic, antispasmodic, carminative, digestive, diuretic, expectorant, stimulant, stomachic and tonic.

Cardamom (*Amomum subulatum*), Nutritional Value per 100 g

Principle	*Nutrient Value*	*Percentage of RDA*
Energy	311 Kcal	15.5 per cent
Carbohydrates	68.47 g	52.5 per cent
Protein	10.76 g	19 per cent
Total Fat	6.7 g	23 per cent
Cholesterol	0 mg	0 per cent
Dietary Fiber	28 g	70 per cent
Vitamins		
Niacin	1.102 mg	7 per cent
Pyridoxine	0.230 mg	18 per cent
Riboflavin	0.182 mg	14 per cent
Thiamin	0.198 mg	16.5 per cent
Vitamin A	0 IU	0 per cent
Vitamin C	21 mg	35 per cent
Electrolytes		
Sodium	18 mg	1 per cent
Potassium	1119 mg	24 per cent
Minerals		
Calcium	383 mg	38 per cent
Copper	0.383 mg	42.5 per cent
Iron	13.97 mg	175 per cent
Magnesium	229 mg	57 per cent
Manganese	28 mg	1217 per cent
Phosphorus	178 mg	25 per cent
Zinc	7.47 mg	68 per cent

- ☆ Cardamom is a good source of minerals like potassium, calcium, and magnesium. A 100 g pod contains 1119 mg of this electrolyte. Potassium is an important component of cell and body fluids that helps control heart rate and blood pressure. Copper is required in the production of red blood cells.
- ☆ Additionally, it is also an excellent source of iron and manganese. 100 g pods contain 13.97 mg or 175 per cent of daily-required levels of iron. Iron is required for red blood cell formation and cellular metabolism. Manganese is a co-factor for the enzyme, superoxide dismutase, which is a very powerful free radical scavenger.
- ☆ Further, these aromatic pods are rich in many vital vitamins, including riboflavin, niacin, vitamin-C that is essential for optimum health.

123

Ajwain Seeds

Strongly pungent and aromatic, ajwain seeds are one of the popular spices commonly feature in Indian and Middle-Eastern cuisines. Botanically, the spicy seeds belong to the family of Apiaceae (*Umbelliferae*), of the genus; Trachyspermum. Scientific name: *Trachyspermum copticum.* The *Umbellifers* are the members of carrot or parsley family, which includes many herbs and spices such as dill,fennel, anise seed, and caraway. Some of the common names for the seeds are ajowan seeds, carom seeds, etc. The Trachyspermum copticum (ajwain) is thought to have originated in the Asia Minor or Persia from where it spread to Indian subcontinent. The plant is a small, cool season annual herb, which grows up to two to three feet in height. It features tiny white-petal flowers that develop into small, oval-shaped seeds that are ready for harvest by the end of winter or early spring. Ajwain seeds are olive green to brown in colour, have similar appearance to cumin or caraway seeds with vertical stripes on their outer surface. Since they comprise an essential oil, *thymol,* their flavour closely resembles that of thyme.

Health Benefits of Ajwain

- ☆ Ajwain seeds contain health benefiting essential oils such as thymol, a monopterone derivative class of chemical compound, which gives aromatic fragrances to seeds. In addition, they also comprise in small amounts other phyto-chemicals such as *pinene, cymene, limonene* and *terpinene.*
- ☆ The active principles in the ajwain may help increase the digestive function of the intestinal tract by increasing gut juices (gastro-intestinal secretions).
- ☆ Thymol, the essential oil obtained from ajwain has local anaesthetic, anti-bacterial and antifungal properties.
- ☆ Like caraway, ajowan seeds are rich in fiber, minerals, vitamins, and anti-oxidants.

Medicinal Uses

- ✰ Ajwain seeds have long been used in traditional ayurvedic and unani medicines for various ailments. Extraction obtained from this spice is sometimes used as carminative in treating flatulence and indigestion.
- ✰ Thymol's germicide and antiseptic properties utilized in many cough remedies. In India, the seeds are used to ease asthma.
- ✰ Ajwain seed oil has the highest percentage of thymol.

Culinary Uses

Ajwain seeds mainly feature in savory Indian, Pakistani, and Middle-Eastern cooking. In order to keep the fragrance and flavour intact, ajowan seeds generally crushed just before preparing dishes and added to the cooking recipes at final stages. This is because prolonged cooking results in evaporation of essential oils.

- ✰ In Punjab province of India and Pakistan, the spice seeds particularly added to make bread known as *ajwain paratha.*
- ✰ Some Indian vegetarian bean/lentil and chicken/fish curries contain this spice and in the Middle East, it is used to flavour meat and rice dishes.
- ✰ The seeds used generously as condiment in snacks, spicy biscuits, to flavour drinks, soups, sauces in India.
- ✰ They also used in pickling along with fenugreek, mustard seeds, turmeric, etc.

124

Cayenne Pepper

Fiercely hot and pungent cayenne pepper is one of the widely used spicy ingredients in many cuisines. Cayenne fruits are long thin pods obtained from the plant belonging to the nightshade family of *Solanaceae*, of the genus; *capsicum*. Scientific name: *Capsicum annuum* var. annuum. Cayenne variety of chili pepper plant is native to Central American region where it was used as a spicy ingredient in Mexican cuisines for several thousand years. The spice was introduced to the rest of the world, especially in the Indian subcontinent, by Spanish and Portuguese explorers during 16th and 17th centuries. They are now one of the important commercial crops in India, Pakistan, China, Argentina and USA.

Cayenne pepper is a perennial small shrub reaching about 90-100 cm in height. The plant prefers well drained sandy soil and warm climate. Its woody stem with numerous branches is covered with thick dark-green foliage. Small creamy- white flowers appear all over the bush which subsequently develops into long, slender, glossy bright green colour fruits (pods). The fruits finally attain mature status when they turn to deep-red colour. The hotness of cayenne is measured in "Scoville heat units" (SHU). On the Scoville scale, cayenne pepper has 30,000 to 50,000 SHU. Interiorly, each cayenne fruit features numerous tiny, flat, disk-shaped, off-white or cream coloured seeds. The seeds are actually clinging to central situated white placenta. In general, the cayenne chili peppers are allowed to ripen completely in the plant and picked up by hand when they are matured and turned red. They are then left to dry, which causes them to shrivel. Cayenne chilies have a strong spicy taste that comes to them from the active alkaloid compounds' *capsaicin, capsanthin* and *capsorubin*.

Health Benefits of Cayenne Peppers

- ☆ Although inherently hot and intolerable even in small amounts, cayennes are one of the health-benefiting spice items concentrated with minerals, vitamins and certain phyto-nutrients. It is no wonder this wonderful spice has been found a place in modern as well as in traditional medicines for its disease preventing and health promoting properties.

Cayenne Peppers (*Capsicum annuum* var. *annuum*), Red, Raw, Nutrition Value per 100 g

Principle	*Nutrient Value*	*Percentage of RDA*
Energy	318 Kcal	16 per cent
Carbohydrates	56.63 g	43 per cent
Protein	12.01 g	21 per cent
Total Fat	17.27 g	57.56 per cent
Cholesterol	0 mg	0 per cent
Dietary Fiber	27.2 g	71 per cent
Vitamins		
Folates	106 mcg	26 per cent
Niacin	8.701 mg	54 per cent
Pyridoxine	2.450 mg	39 per cent
Riboflavin	0.919	71 per cent
Thiamin	0.328 mg	27 per cent
Vitamin A	41610 IU	1387 per cent
Vitamin C	76.4 mg	127 per cent
Vitamin E	29.83 mg	199 per cent
Vitamin K	80.3 mg	67 per cent
Electrolytes		
Sodium	30 mg	2 per cent
Potassium	2014 mg	43 per cent
Minerals		
Calcium	148 mg	15 per cent
Copper	0.373 mg	41 per cent
Iron	7.80 mg	97.5 per cent
Magnesium	152 mg	38 per cent
Manganese	2.00 mg	87 per cent
Phosphorus	293 mg	42 per cent
Selenium	8.8 mcg	18 per cent
Zinc	2.48 mg	22.5 per cent
Phyto-nutrients		
Carotene-β	21840 μg	—
Carotene-α	0 μg	—
Cryptoxanthin-β	6252 μg	—
Lutein-zeaxanthin	13157 μg	—

- ☆ Cayenne contains health benefiting an alkaloid compound capsaicin, which gives strong spicy pungent character. Early laboratory studies on experimental mammals suggest that capsaicin has anti-bacterial, anti-carcinogenic, analgesic and anti-diabetic properties. When used judiciously it also found to reduce triglycerides and LDL cholesterol levels in obese individuals.
- ☆ Fresh cayenne peppers, red or green, are rich source of vitamin-C. 100 g fresh chilies provide about 76.4 mg or about 127 per cent of RDA. Vitamin-C is a potent water soluble antioxidant. It is required for the collagen synthesis in the body. Collagen is the main structural protein in the body required for maintaining the integrity of blood vessels, skin, organs, and bones. Regular consumption of foods rich in vitamin C helps the body protect from scurvy; develop resistance against infectious agents (boosts immunity) and scavenge harmful, pro-inflammatory free radicals from the body.
- ☆ Cayenne chili peppers are perhaps the richest source of vitamin A among spice. Just 100 g of cayenne has 41,610 IU or astoundingly 1387 per cent of vitamin A. In addition, this prized spice is also home for anti-oxidant flavonoids such as carotenes, lutein, zea xanthin and cryptoxanthin. Together, these antioxidants in capsicum help to protect the body from injurious effects of free radicals generated during stress and disease conditions.
- ☆ The spice contains very high levels of essential minerals. The spice if even consumed in small quantities regularly would provide sufficient levels of iron, copper, zinc, potassium, manganese, magnesium and selenium. Manganese is used by the body as a co-factor for the antioxidant enzyme, superoxide dismutase. Selenium is an anti-oxidant trace element required by the human body for smooth heart and liver functions.
- ☆ 100 g of cayenne peppers provides 2014 mg or 47 per cent of a daily-required amount of potassium. Potassium is an important electrolyte in the cells and body fluids that helps controlling heart rate and blood pressure. It thus counters the bad effects of sodium.
- ☆ Cayenne peppers are also good in B-complex group of vitamins such as niacin, pyridoxine (vitamin B-6), riboflavin and thiamin (vitamin B-1). These vitamins are essential in the sense that body requires them from external sources to replenish. B-complex vitamins facilitate cellular metabolism through various enzymatic functions.

125

Chili Pepper

Chili peppers, despite their fiery "hotness" are one of very popular spices known for medicinal and health benefiting properties. The chili is actually a fruit pod from the plant belonging to the nightshade family of *Solanaceae*, within the genus; *capsicum*. Scientific name: *Capsicum annum*. Some common members of nightshade family are tomato, abuergine, potato etc.tc.

Several cultivars of peppers are grown all around the world. The chili plant is native to Central American region where it was used as the chief spice ingredients in Mexican cuisine for centuries. It was introduced to the rest of the world by Spanish and Portuguese explorers during 16th and 17th centuries and now grown widely in many parts of the world as an important commercial crop. *Chili plant* is a perennial small shrub with woody stem growing up to a meter height and bears white coloured flowers. The pods are very variable in size, shape, colour, and pungency. Depending on the cultivar type, they range from the mild, fleshy, Mexican bell peppers to the tiny, fiery, finger-like chili peppers, commonly grown in Indian subcontinent. The hotness of chili is measured in "Scoville heat units" (SHU). On the Scoville scale, a sweet bell pepper scores 0, a jalapeño pepper around 2,500-4,000 units, and a Mexican habañeros have 200,000 to 500,000 units.

Interiorly, each fruit features numerous tiny, white, or cream coloured, circular and flat seeds. The seeds are actually clinging around the central white-placenta. To harvest, chilies can be picked up while they are green or when they reach complete maturity and dried in the plant. Usually, the fruits are picked up by hand when they are matured and turned red. They are then left to dry, which causes them to shrivel. Chilies have a strong spicy taste that comes to them from the active alkaloid compounds: *capsaicin, capsanthin* and *capsorubin*.

Health Benefits of Chili Peppers

☆ Chili pepper contains an impressive list of plant derived chemical compounds that are known to have disease preventing and health promoting properties.

Chili Peppers (*Capsicum annum*), Raw, Nutrient Value per 100 g

Principle	*Nutrient Value*	*Percentage of RDA*
Energy	40 Kcal	2 per cent
Carbohydrates	8.81 g	7 per cent
Protein	1.87 g	3 per cent
Total Fat	0.44 g	2 per cent
Cholesterol	0 mg	0 per cent
Dietary Fiber	1.5 g	3 per cent
Vitamins		
Folates	23 µg	6 per cent
Niacin	1.244 mg	8 per cent
Pantothenic acid	0.201 mg	4 per cent
Pyridoxine	0.506 mg	39 per cent
Riboflavin	0.086 mg	6.5 per cent
Thiamin	0.72 mg	6 per cent
Vitamin A	952 IU	32 per cent
Vitamin C	143.7 mg	240 per cent
Vitamin E	0.69 mg	4.5 per cent
Vitamin K	14 µg	11.5 per cent
Electrolytes		
Sodium	9 mg	0.5 per cent
Potassium	322 mg	7 per cent
Minerals		
Calcium	14 mg	1.5 per cent
Copper	0.129 mg	14 per cent
Iron	1.03 mg	13 per cent
Magnesium	23 mg	6 per cent
Manganese	0.187 mg	8 per cent
Phosphorus	43 mg	6 per cent
Selenium	0.5 µg	1 per cent
Zinc	0.26 mg	2 per cent
Phyto-nutrients		
Carotene-β	534 µg	—
Carotene-α	36 µg	—
Cryptoxanthin-β	40 µg	—
Lutein-zeaxanthin	709 µg	—

- Chilies contain health benefiting an alkaloid compound in them, capsaicin, which gives strong spicy pungent character. Early laboratory studies on experimental mammals suggest that capsaicin has anti-bacterial, anti-carcinogenic, analgesic and anti-diabetic properties. It also found to reduce LDL cholesterol levels in obese individuals.
- Fresh chili peppers, red and green, are rich source of vitamin-C. 100 g fresh chilies provide about 143.7 µg or about 240 per cent of RDA. Vitamin C is a potent water-soluble antioxidant. It is required for the collagen synthesis in the body. Collagen is the main structural protein in the body required for maintaining the integrity of blood vessels, skin, organs, and bones. Regular consumption of foods rich in vitamin-C helps the body protect from scurvy; develop resistance against infectious agents (boosts immunity) and scavenge harmful, pro-inflammatory free radicals from the body.
- They are also good in other antioxidants like vitamin A, and flavonoids like β-*carotene, α-carotene, lutein, zeaxanthin, and cryptoxanthin*. These antioxidant substances in capsicum help to protect the body from injurious effects of free radicals generated during stress, diseases conditions.
- Chilies contain a good amount of minerals like potassium, manganese, iron, and magnesium. Potassium is an important component of cell and body fluids that helps controlling heart rate and blood pressure. Manganese is used by the body as a co-factor for the antioxidant enzyme, superoxide dismutase.
- Chilies are also good in B-complex group of vitamins such as niacin, pyridoxine (vitamin B-6), riboflavin and thiamin (vitamin B-1). These vitamins are essential in the sense that body requires them from external sources to replenish.

126

Cinnamon

Fragrant cinnamon spice is one of the highly prized spices that has been in use since biblical times for its medicinal and culinary properties. This delightfully exotic, sweet-flavoured spice stick is traditionally obtained from the outer brown bark of Cinnamomum trees, which when dried, rolls into a tubular form known commercially as "quill." The cinnamon plant is a small, evergreen bushy tree belonging to the family of *Lauraceae* or laurel within the genus; *Cinnamomum.* This novel spice is native to Sri Lankan island but also found in many other countries such as Myanmar, Bangladesh, India, China, and Indonesia.

Varieties of the cinnamon-tree exist; however, Sri Lankan variety is regarded as "true cinnamon" and scientifically named as *Cinnamonum verum.* Traditionally, the inner bark is bruised with a brass rod, peeled and long incisions are made in the bark. It is then rolled by hand and allowed to dry in the sunlight. It is the bark of the tree from where aromatic essential oil (makes up 0.5 per cent to 1 per cent of its composition) is extracted. Usually, the oil is processed by roughly pounding the bark, macerating it in seawater, and then quickly distilling the whole. The oil features golden-yellow colour, with the characteristic odor of cinnamon and a very hot aromatic taste. The pungent taste and scent in cinnamon spice are due to compound *cinnamic aldehyde* and *cinnamaldehyde* in the oil. Cassia, also known as *chinese cinnamon,* is a different member of *Lauraceae* family and named as *Cinnamomum cassia.* Cassia is coarser, more spicy, and pungent but less fragrant than cinnamon. It is usually substituted for the cinnamon in savory dishes.

Health Benefits of Cinnamon

☆ The active principles in the cinnamon spice are known to have anti-oxidant, anti-diabetic, anti-septic, local anesthetic, anti-inflammatory, rubefacient (warming and soothing), carminative and anti-flatulent properties.

Cinnamon Spice (*Cinnamonum verum*), ORAC Value-267536, Nutritional Value per 100 g

Principle	*Nutrient Value*	*Percentage of RDA*
Energy	247 Kcal	12 per cent
Carbohydrates	50.59 g	39 per cent
Protein	3.99 g	7 per cent
Total Fat	1.24 g	4.5 per cent
Cholesterol	0 mg	0 per cent
Dietary Fiber	53.1 g	133 per cent
Vitamins		
Folates	6 μg	1.5 per cent
Niacin	1.332 mg	8 per cent
Pantothenic acid	0.358 mg	7 per cent
Pyridoxine	0.158 mg	12 per cent
Riboflavin	0.041 mg	3 per cent
Thiamin	0.022 mg	2 per cent
Vitamin A	295 IU	10 per cent
Vitamin C	3.8 mg	6 per cent
Vitamin E	10.44 mg	70 per cent
Vitamin K	31.2 μg	26 per cent
Electrolytes		
Sodium	10 mg	<1 per cent
Potassium	431 mg	9 per cent
Minerals		
Calcium	1002 mg	100 per cent
Copper	0.339 mg	38 per cent
Iron	8.32 mg	104 per cent
Magnesium	60 mg	15 per cent
Manganese	17.466 mg	759 per cent
Phosphorus	64 mg	9 per cent
Zinc	1.83 mg	17 per cent
Phyto-nutrients		
Carotene-β	112 μg	—
Crypto-xanthin-β	129 μg	
Lutein-zeaxanthin	222 μg	—
Lycopene	15 μg	—

- Cinnamon has the highest anti-oxidant strength of all the food sources in nature. The total measured ORAC (Oxygen radical absorbance capacity) value for this novel spice is 267536 trolex equivalents (TE), which is many hundred times more than in chokeberry, apples, etc.
- The spice contains health benefiting essential oils such as eugenol, a phenylpropanoids class of chemical compound, which gives pleasant, sweet aromatic fragrances. Eugenol has got local anesthetic and antiseptic properties, hence; useful in dental and gum treatment procedures.
- Other important essential oils in cinnamon include *ethyl cinnamate, linalool, cinnamaldehyde, beta-caryophyllene*, and *methyl chavicol.*
- *Cinnamaldehyde* in cinnamon-sticks has been found to have anti-clotting action, prevents platelet clogging inside the blood vessels, and thereby helps prevent stroke, peripheral arterial and coronary artery diseases.
- The active principles in this spice may increase the motility of the intestinal tract as well as help aid in the digestion power by increasing gastro-intestinal enzyme secretions.
- This spicy bark is an excellent source of minerals like potassium, calcium, manganese, iron, zinc, and magnesium. Iron is required for cellular metabolism as a co-factor and in RBC's production. Potassium is an important component of cell and body fluids that helps control heart rate and blood pressure. Manganese and copper are chiefly used by the body as co-factors for the antioxidant enzyme, *superoxide dismutase.*
- It also contains very good amounts of vitamin A, niacin, pantothenic acid, and pyridoxine.
- Further, it is also a very good source of flavonoid phenolic anti-oxidants such as *carotenes, zeaxanthin, lutein* and *cryptoxanthin*.

127

Cloves

Cloves are one of the highly prized spices, widely recognized all over the world for their medicinal and culinary qualities. They are the "flower buds" from evergreen rain-forest tree native to Indonesia. Botanically, the spice belongs to the family of *Myrtaceae* of the genus; *Sygyzium*, and scientifically named as *Sygizium aromaticum*. The flower buds are initially pale in colour, gradually turn to green, and, finally develop into bright-red clove buds by the time of harvesting. Buds are generally picked up when they reach 1.5-2 cm in length. Structurally, each bud consists of long calyx; terminating in four spreading sepals, and four unopened petals, which form a small ball (dome) at the center. The sweet aroma of cloves is due to *eugenol*, an essential oil in them.

Medicinal Properties and Health Benefits of Cloves

- ✰ The active principles in the clove are known to have antioxidant, anti-septic, local anesthetic, anti-inflammatory, rubefacient (warming and soothing), carminative and anti-flatulent properties.
- ✰ The spice contains health benefiting essential oils such as eugenol. It is a phenyl-propanoids class of chemical compound, which gives pleasant, sweet aromatic fragrances to the clove-bud. Eugenol has local anesthetic and antiseptic properties, hence; useful in dental care essentials as well as in treatment procedures.
- ✰ The other important constituents in this spice include:

  Essential oils: acetyl eugenol, beta-caryophyllene and vanillin, crategolic acid;

  Tannins: gallotannic acid, methyl salicylate (painkiller);

  The flavonoids: eugenin, kaempferol, rhamnetin, and eugenitin;

  Triterpenoids: such as oleanolic acid, stigmasterol and campesterol and several *sesquiterpenes*.

Cloves (*Sygizium aromaticum*), Nutritive Value per 100 g

Principle	*Nutrient Value*	*Percentage of RDA*
Energy	47 Kcal	2 per cent
Carbohydrates	10.51 g	8 per cent
Protein	3.27 g	6 per cent
Total Fat	0.15 g	0.5 per cent
Cholesterol	0 mg	0 per cent
Dietary Fiber	5.4 g	14 per cent
Vitamins		
Folates	68 µg	17 per cent
Niacin	1.046 mg	6.5 per cent
Pantothenic acid	0.338 mg	7 per cent
Pyridoxine	0.116 mg	9 per cent
Riboflavin	0.066 mg	5 per cent
Thiamin	0.072 mg	6 per cent
Vitamin A	13 IU	0.5 per cent
Vitamin C	11.7 mg	20 per cent
Vitamin E	0.19 mg	1 per cent
Vitamin K	14.8 µg	12 per cent
Electrolytes		
Sodium	94 mg	6 per cent
Potassium	370 mg	8 per cent
Minerals		
Calcium	44 mg	4 per cent
Copper	0.231 mg	27 per cent
Iron	1.28 mg	16 per cent
Magnesium	60 mg	15 per cent
Manganese	0.256 mg	11 per cent
Phosphorus	90 mg	13 per cent
Selenium	7.2 µg	13 per cent
Zinc	2.32 mg	21 per cent
Phyto-nutrients		
Carotene-β	8 µg	—
Crypto-xanthin-β	0 µg	—
Lutein-zeaxanthin	464 µg	—

☆ The active principles in the clove may increase gut motility as well as improve the digestion power by increasing gastro-intestinal enzyme secretions. Thus, helps relieve indigestion and constipation problems.

- ✰ The spice also contains a good amount of minerals like potassium, manganese, iron, selenium and magnesium. Potassium is an important electrolyte of cell and body fluids that helps control heart rate and blood pressure. Manganese is used by the body as a co-factor for the antioxidant enzyme, *superoxide dismutase.*
- ✰ Further, the spice buds contain very good amounts of vitamin A and beta-carotene levels. These compounds are known to have antioxidant properties. Vitamin A is also required by the body for maintaining healthy mucus membranes and skin in addition to essential for vision. Consumption of natural foods rich in flavonoids helps to protect the body from lung and oral cavity cancers.
- ✰ Additionally, this spice is a good source of vitamin-K, vitamin-B6 (pyridoxine), thiamin (vitamin B-1), vitamin-C and riboflavin. Consumption of foods rich in vitamin C helps the body develop resistance against infectious agents and scavenge harmful oxygen-free radicals.

128

Coriander Seeds

Have you ever caught up with gripping stomach pain? Drinking a few sips of extraction obtained from coriander seeds, dill, caraway, fennel, and aniseed from your granny's kitchen spice box is perhaps the most effective carminative remedy for this ailment. Coriander is tender hollow stemmed plant in the *Apiaceae* family, of the genus: *Coriandum*. Its scientific name is *Coriandum sativum*. Pleasant, aromatic and spicy, the seeds have been in use since ancient times in cooking as well as in various traditional medicines.

Coriander is native to Southeastern Europe and grown extensively all over Europe, Middle East, China, India, and Turkey. It is recognized as cilantro in the west. This herbaceous plant grows up to 2 feet in height with branching stems, featuring deep green soft, hairless, bi or tri-lobed leaves. The mature plant bears small light pink colour flowers that subsequently turn into globular or oval-shaped fruits (seeds). The seeds measure about 4-6 mm in diameter with central hollow cavity containing two vertical vittae containing some important essential oils. The seeds are ready for harvest when the plants turn brown, leaves begin to dry and fall. Immature seeds are light green and taste bitter. To harvest, the crop is cut, tied in small bundles, and dried in the sunlight for several days. Traditionally, to separate the seeds, either the sheaves are beaten with stick or a lightweight roller used to wear off the pods.

Health Benefits of Coriander Seeds

- ☆ Coriander seeds contain many plant-derived chemical compounds that are known to have been anti-oxidant, disease preventing, and health promoting properties.
- ☆ The characteristic aromatic flavour of coriander seeds comes from the many fatty acids and essential volatile oils. Some important fatty acids in the dried seeds include *petroselinic acid, linoleic acid* (*omega* 6), *oleic acid,* and *palmitic acid.* In addition, the seeds contain essential oils such as *linalool*

(68 per cent), a-pinene (10 per cent), geraniol, camphene, terpine etc. Together; these active principles are responsible for digestive, carminative, and anti-flatulent properties of the seeds.

Coriander Seeds (*Coriander sativum*), Nutritional Value per 100 g

Principle	*Nutrient Value*	*Percentage of RDA*
Energy	298 Kcal	15 per cent
Carbohydrates	54.99 g	42 per cent
Protein	12.37 g	22 per cent
Total Fat	17.77 g	60 per cent
Cholesterol	0 mg	0 per cent
Dietary Fiber	41.9 g	110 per cent
Vitamins		
Folates	1 µg	<1 per cent
Niacin	2.130 mg	13 per cent
Riboflavin	0.290 mg	22 per cent
Thiamin	0.239 mg	20 per cent
Vitamin A	0 IU	0 per cent
Vitamin C	21 mg	35 per cent
Electrolytes		
Sodium	35 mg	2 per cent
Potassium	1267 mg	27 per cent
Minerals		
Calcium	709 mg	71 per cent
Copper	0.975 mg	108 per cent
Iron	16.32 mg	204 per cent
Magnesium	330 mg	83 per cent
Manganese	1.900 mg	82 per cent
Phosphorus	409 mg	39 per cent
Zinc	4.70 mg	43 per cent

- ☆ As in other spices, coriander is also rich in of dietary fiber. 100 g seeds provide 41.9 g of fiber. Much of this fiber is metabolically inert insoluble fiber, which helps increase bulk of the food by absorbing water throughout the digestive system and help easing constipation condition.
- ☆ In addition, dietary fibers bind to bile salts (produced from cholesterol) and decrease their re-absorption in colon, thus help lower serum LDL cholesterol levels. Together with flavonoid anti-oxidants, fiber composition of coriander helps protect the colon mucus membrane from cancers.
- ☆ The seeds are an excellent source of minerals like iron, copper, calcium, potassium, manganese, zinc and magnesium. *Copper* is required in the

production of red blood cells. Iron is essential for cell metabolism and red blood cell formation. *Zinc* is a co-factor in many enzymes that regulate growth and development, sperm generation, digestion and nucleic acid synthesis. *Potassium* is an important component of cell and body fluids that helps controlling heart rate and blood pressure. *Manganese* is used by the body as a co-factor for the powerful anti-oxidant enzyme, *superoxide dismutase.*

- ☆ Unlike other dry spice seeds that lack in vitamin C, coriander seeds contain an ample amount of this anti-oxidant vitamin. 100 g of dry seeds provide 21 mg or 35 per cent of RDI of this vitamin.
- ☆ Furthermore, the seeds indeed are the storehouse of many vital B-complex vitamins like thiamin, riboflavin, and niacin.

129

Cumin

One of the special spices known for their distinctive aroma, cumin seeds are popular in North African, Middle Eastern, Western Chinese, Indian, Cuban and Northern Mexican cuisines. The spice is native to middle-east Asian region, and now grown all over the world for its flavourful seeds. The plant is the small flowering herbaceous plant belonging to the family of *Apiaceae* of the genus of; *Cuminum,* and scientifically known as *Cuminum cyminum.* The cumin plant flourishes well in sandy, fertile soil along with hot summer weather conditions. It bears small, gray-yellow colour, oblong shaped seeds with vertical ridges on their outer surface; that closely resembles caraway seeds in appearance. The distinctive flavour, strong and warm sense of cumin seeds on taste buds comes from a certain group of essential oils in them. The chief constituent and important aromatic compound are cuminaldehyde (4-isopropylbenzaldehyde). Black cumin (*Bunium persicum*), also popular as wild or shahi jeera in the Perian and Indian sub-continent, is related species to regular cumin. The seeds feature long, slender, curved, dark-brown pods with distinctive earthy flavour.

Health Benefits of Cumin Seeds

- ☆ Cumin seeds contain numerous phyto-chemicals that are known to have antioxidant, carminative and anti-flatulent properties. The seeds are an excellent source of dietary fiber.
- ☆ Its seeds contain certain health-benefiting essential oils such as *cuminaldehyde* (4-isopropylbenzaldehyde), *pyrazines, 2-methoxy-3-sec-butylpyrazine, 2-ethoxy-3-isopropylpyrazine,* and *2-methoxy-3-methylpyrazine.*
- ☆ The active principles in the cumin may augment the motility of the gastro-intestinal tract as well as aids in the digestion power by increasing gut enzyme secretions.

Cumin Seeds (*Cuminum cyminum*), Nutritional Value per 100 g

Principle	*Nutrient Value*	*Percentage of RDA*
Energy	375 Kcal	19 per cent
Carbohydrates	44.24 g	34 per cent
Protein	17.8 g	32 per cent
Total Fat	22.27g	74 per cent
Cholesterol	0 mg	0 per cent
Dietary Fiber	10.5 g	26 per cent
Vitamins		
Folates	10 μg	2.5 per cent
Niacin	4.58 mg	28.5 per cent
Pyridoxine	0.435 mg	33 per cent
Riboflavin	0.32 mg	24.5 per cent
Thiamin	0.628 mg	52 per cent
Vitamin A	1270 IU	42 per cent
Vitamin C	7.7 mg	13 per cent
Vitamin E	3.3 mg	22 per cent
Vitamin K	5.4 μg	4.5 per cent
Electrolytes		
Sodium	1788 mg	38 per cent
Potassium	68 mg	11 per cent
Minerals		
Calcium	931 mg	93 per cent
Copper	0.867 mg	96 per cent
Iron	66.36 mg	829 per cent
Magnesium	366 mg	91 per cent
Manganese	3.3 mg	145 per cent
Phosphorus	499 mg	71 per cent
Zinc	4.8 mg	43.5 per cent
Phyto-nutrients		
Carotene-β	762 μg	—
Crypto-xanthin-β	0 μg	—
Lutein-zeaxanthin	448 μg	—

☆ This spice is an excellent source of minerals like iron, copper, calcium, potassium, manganese, selenium, zinc and magnesium. *Copper* is required in the production of red blood cells. Iron is required for red blood cell formation. *Zinc* is a co-factor in many enzymes that regulate growth and development, sperm generation, digestion and nucleic acid synthesis. *Potassium* is an important component of cell and body fluids that helps controlling heart rate and blood pressure. *Manganese* is used by the body as a co-factor for the powerful anti-oxidant enzyme, *superoxide dismutase.*

- ☆ The spice also contains very good amounts of B-complex vitamins such as thiamin, vitamin B-6, niacin, riboflavin, and other vital anti-oxidant vitamins like vitamin E, vitamin A and vitamin C.
- ☆ The seeds are also rich source of many flavonoid phenolic anti-oxidants such as carotenes, zeaxanthin, and lutein.

130

Fennel Seeds

Sweet, anise-flavoured fennel seed together with thyme, nettle, mugwort, etc., has been revered as one of nine Anglo-Saxon sacred herbs for its health benefits. The spice is one of the most sought-after ingredients in many popular cuisines all over the Mediterranean regions. Fennel is a perennial herb plant belongs to *parsley* or *Umbelliferae* family, a broad family of herbs and spices, which includes some common members such as caraway, dill, anise, cumin etc. Scientific name of fennel is *Foeniculum vulgare* var. dulce.

Fennel is native to Southern Europe and grown extensively all over Europe, Middle-Eastern, China, India, and Turkey. This herbaceous plant reaches up to 2 meters (about 6 feet) in height with deep green feathery (lacy) leaves and bears golden-yellow flowers in umbels. In general, fennel seeds are harvested when the seed heads turn light-brown. The seeds, which resemble to anise seeds in appearance, feature oblong or curved (comma) shape, about 3-4 mm long, light brown-colour with fine vertical stripes over their surface. In general, seeds are harvested during early hours of the day to avoid spilling of seeds on the ground. As in caraway, the cut plants staked until they were dry; then threshed, processed and packed to be sold. Fennel bulb (*Foeniculum vulgare* var. azoricum), used as a vegetable, is closely related to seeding fennel. It has grown for its anise flavoured sweet taste fronds in many parts of Mediterranean region.

Health Benefits of Fennel Seeds

- ✰ Fennel symbolizes longevity, courage, and strength. In addition to its use as medicinal values, fennel has much health benefiting nutrients, essential compounds, anti-oxidants, dietary fiber, minerals, and vitamins.
- ✰ Fennel seeds indeed contain numerous flavonoid anti-oxidants like kaempferol and quercetin. These compounds function as powerful anti-oxidants by removing harmful free radicals from the body thus protect from cancers, infection, ageing and degenerative neurological diseases.

Fennel Seed (*Foeniculum vulgare*), Nutritional Value per 100 g

Principle	*Nutrient Value*	*Percentage of RDA*
Energy	345 Kcal	17 per cent
Carbohydrates	52.29 g	40 per cent
Protein	15.80 g	28 per cent
Total Fat	14.87 g	48 per cent
Cholesterol	0 mg	0 per cent
Dietary Fiber	39.8 g	104 per cent
Vitamins		
Niacin	6.050 mg	37 per cent
Pyridoxine	0.470 mg	36 per cent
Riboflavin	0.353 mg	28 per cent
Thiamin	0.408 mg	34 per cent
Vitamin A	135 IU	4.5 per cent
Vitamin C	21 mg	35 per cent
Electrolytes		
Sodium	88 mg	6 per cent
Potassium	1694 mg	36 per cent
Minerals		
Calcium	1196 mg	120 per cent
Copper	1.067 mg	118 per cent
Iron	18.54 mg	232 per cent
Magnesium	385 mg	96 per cent
Manganese	6.533 mg	284 per cent
Phosphorus	487 mg	70 per cent
Zinc	3.70 mg	33.5 per cent

- ☆ Like in caraway, fennel seeds too are rich source of dietary fiber. 100 g seeds provide 39.8 g of fiber. Much of this roughage is metabolically inert insoluble fiber, which helps increase bulk of the food by absorbing water throughout the digestive system and easing constipation condition.
- ☆ In addition, dietary fibers bind to bile salts (produced from cholesterol) and decrease their re-absorption in colon, thus help lower serum LDL cholesterol levels. Together with flavonoid anti-oxidants, fiber composition of fennel helps protect the colon mucus membrane from cancers.
- ☆ Fennel seeds compose of health benefiting volatile essential oil compounds such as anethole, limonene, anisic aldehyde, pinene, myrcene, fenchone, chavicol, and cineole. These active principles in the fennel are known to have antioxidant, digestive, carminative, and anti-flatulent properties.

- ☆ Fennel seeds are concentrated source of minerals like copper, iron, calcium, potassium, manganese, selenium, zinc, and magnesium. *Copper* is required in the production of red blood cells. Iron is required for red blood cell formation. *Zinc* is a co-factor in many enzymes that regulate growth and development, sperm generation, digestion and nucleic acid synthesis. *Potassium* is an important component of cell and body fluids that helps controlling heart rate and blood pressure. *Manganese* is used by the body as a co-factor for the powerful anti-oxidant enzyme, *superoxide dismutase.*
- ☆ Furthermore, the seeds indeed are the storehouse for many vital vitamins. Vitamin A, vitamin E, vitamin C as well as many B-complex vitamins like thiamin, pyridoxine, riboflavin and niacin particularly is concentrated in these seeds.

131

Fenugreek Seeds

Strongly aromatic and flavourful fenugreek seeds are popular spices widely used for their well recognized for culinary as well as for medicinal properties. Traditionally, fenugreeks are being used to cure digestive problems and to improve breast milk secretion in the nursing mothers. Botanically, the fenugreek is a small annual leguminous herb belonging to the family of *fabaceae*, of the genus: *Trigonella*. Scientific name: *Trigonella foenum-graecum*. Some of the common names include *greek-hay, mehti, bird's foot, greek-clover*, etc.

Fenugreek is native to sub-Himalayan plains of Indian subcontinent, and today; it is widely grown all-over southern and Mediterranean Europe, Middle-East Asia and northern African regions. The plant grows up to about 1-2 feet in height with light green colour tri-foliate leaves and white flowers. It bears long, slender, yellow-brown pods containing about 10-20 golden-yellow colour seeds. Its seeds are small, hard, and resemble tiny, multi-faceted stones. Raw seeds have maple flavour and bitter taste; however, their taste becomes more acceptable once they were gently dry-roasted under light heat.

Health Benefits of Fenugreek Seeds

- ✰ Fenugreek seeds are rich source of minerals, vitamins, and phytonutrients. 100 g seeds provide 323 calories.
- ✰ The seeds are a very good source of soluble dietary fiber. Soaking the seeds in water makes their outer coat soft and mucilaginous. 100 g of seeds provide 24.6 g or over 65 per cent of dietary fiber.
- ✰ Non-starch polysaccharides (NSP) which constitute major fiber content in the fenugreeks include *saponins, hemicellulose, mucilage, tannin,* and *pectin*. These compounds help lower blood LDL-cholesterol levels by inhibiting bile salts re-absorption in the colon. They also bind to toxins in the food and help to protect the colon mucus membrane from cancers.

Fenugreeks (*Trigonella foenum-graecum*), Nutritional Value per 100 g

Principle	*Nutrient Value*	*Percentage of RDA*
Energy	323 Kcal	16 per cent
Carbohydrates	58.35 g	45 per cent
Protein	23 g	41 per cent
Total Fat	6.41 g	21 per cent
Cholesterol	0 mg	0 per cent
Dietary Fiber	24.6 g	65 per cent
Vitamins		
Folates	57 µg	14 per cent
Niacin	1.640 mg	7 per cent
Pyridoxine	0.600 mg	46 per cent
Riboflavin	0.366 mg	28 per cent
Thiamin	0.322 mg	27 per cent
Vitamin A	60 IU	2 per cent
Vitamin C	3 mg	5 per cent
Electrolytes		
Sodium	67 mg	4.5 per cent
Potassium	770 mg	16 per cent
Minerals		
Calcium	176 mg	18 per cent
Copper	1.110 mg	123 per cent
Iron	33.53 mg	419 per cent
Magnesium	191 mg	48 per cent
Manganese	1.228 mg	53 per cent
Phosphorus	296 mg	42 per cent
Selenium	6.3 µg	11 per cent
Zinc	2.50 mg	23 per cent

- ✰ NSPs (non-starch polysaccharides) increase the bulk of the food and augment bowel movements. Altogether, NSPs assist in smooth digestion and help relieve constipation ailments.
- ✰ It has been established that amino-acid 4-hydroxy isoleucine present in the fenugreek seeds has facilitator action on insulin secretion. In addition, fiber in the seeds helps lower rate of glucose absorption in the intestines thus controls blood sugar levels. The seeds are therefore recommended in the diabetic diet.
- ✰ The seeds contain many phytochemical compounds such as *choline, trigonelline diosgenin, yamogenin, gitogenin, tigogenin* and *neotigogens.* Together, these compounds account for the medicinal properties of fenugreeks.

- ☆ This prized spice is an excellent source of minerals like copper, potassium, calcium, iron, selenium, zinc, manganese, and magnesium. Potassium is an important component of cell and body fluids that helps control heart rate and blood pressure by countering action on sodium. Iron is essential for red blood cell production and as a co-factor for cytochrome-oxidases enzymes.
- ☆ It is also rich in many vital vitamins that are essential nutrients for optimum health, including thiamin, pyridoxine (vitamin B6), folic acid, riboflavin, niacin, vitamin A, and vitamin-C.

Medicinal Uses of Fenugreek Seed

- ☆ Its seeds have been used in many traditional medicines as a laxative, digestive, and as a remedy for cough and bronchitis.
- ☆ If used regularly, fenugreeks may help control cholesterol, triglyceride as well as high blood sugar (glycemic) levels in diabetics.
- ☆ Fenugreek seeds added to cereals and wheat flour (bread) or made into gruel, given to the nursing mothers to increase milk synthesis.

132

Horseradish

Horseradish is a long, tapering root used as a condiment in the kitchens. The root has a strong, hot, and sharp flavour, which can be only be described after experiencing its unique taste. Botanically, this spicy, pungent root belongs to the genus: *Armoracia*, of Brassicaceae (mustards) family, sharing with members like mustard, kale, radish, cabbage, etc. Scientific name: *Armoracia rusticana*. Horseradish is native to Eastern and Mediterranean regions of Europe from where it spread to Americas during colonial times. The plant is a small perennial herb but can be grown as annual field crop for its thick, rough, fleshy roots in many parts of Europe, America, and Asia, including Germany, USA, England, Hungary, Japan, and China. The plant features broad, crinkled leaves. It grows best in cool climates with good sun light conditions. In general, the rootlets (root sections) planted in the spring and harvested by autumn. The roots are usually about 6-12 inches long with few rounded knots at the root end. Wasabi (*Wasabia japonica*) is a Japanese variant of horseradish but has stronger flavour and delicate apple green colour flesh.

Health Benefits of Horseradish

- ☆ Horseradish is low in calories and fat; however, it contains a good amount of dietary fiber, vitamins, minerals, and anti-oxidants. Certain active principles in it found to have been anti-inflammatory, *diuretic* (increase urine output), and nerve soothing effects.
- ☆ The root contains many volatile phyto-chemical compounds, which give its much-famed pungent character. Some of the major constituents in the root are *allyl isothiocyanate, 3-butenyl isothiocyanate, 2-propenylglucosinlate* (*sinigrin*), *2-pentyl isothiocyanate, and phenylethyl isothiocyanate.* It has been found that these compounds have been anti-oxidant as well as detoxification functions.

Horseradish Prepared (*Armoracia rusticana*), Nutrition Value per 100 g

Principle	*Nutrient Value*	*Percentage of RDA*
Energy	48 cal	2 per cent
Carbohydrates	11.29 g	9 per cent
Protein	1.18 g	2 per cent
Total Fat	0.69 g	3 per cent
Cholesterol	0 mg	0 per cent
Dietary Fiber	3.3 g	9 per cent
Vitamins		
Folates	57 µg	14 per cent
Niacin	0.386 mg	2.5 per cent
Pantothenic acid	0.093 mg	2 per cent
Pyridoxine	0.073 mg	6 per cent
Riboflavin	0.024 mg	2 per cent
Thiamin	0.008 mg	<1 per cent
Vitamin A	2 IU	<1 per cent
Vitamin C	24.9 mg	41 per cent
Electrolytes		
Sodium	314 mg	21 per cent
Potassium	246 mg	5 per cent
Minerals		
Calcium	56 mg	6 per cent
Copper	0.058 mg	6 per cent
Iron	0.42 mg	5 per cent
Magnesium	27 mg	7 per cent
Manganese	0.126 mg	5.5 per cent
Phosphorus	31 mg	4.5 per cent
Zinc	0.83 mg	8 per cent
Phyto-nutrients		
Carotene-β	1 µg	—
Crypto-xanthin-β	0 µg	—
Lutein-zeaxanthin	10 µg	—

- ☆ It is a potent gastric stimulant, increases appetite, and aids in digestion. The volatile phyto-chemical compounds in the root stimulate salivary, gastric, and intestinal glands to secrete digestive enzymes, thereby facilitate digestion.
- ☆ Horseradish has good amounts of vitamin-C, which is a powerful water soluble anti-oxidant. 100 g fresh root provides 29 mg or 41 per cent of daily-recommended values. Vitamin C helps alleviate viral infections by boosting

immunity. In addition, it helps remove harmful free radicals from the body and protect it from cancers, inflammation, infections, etc.

- ☆ The root spice has some of vital minerals in moderation like sodium, potassium, manganese, iron, copper, zinc, and magnesium. Iron is an important co-factor for cytochrome-oxidase enzymes during cellular metabolism. It is also required for red blood cell production in the bone marrow. Being an important component of cell and body fluids, potassium helps control heart rate and blood pressure. Manganese is used by the body as a co-factor for the powerful antioxidant enzyme, *superoxide dismutase.*
- ☆ In addition, the root has small amounts of essential vitamins such as folate, vitamin B-6 (pyridoxine), riboflavin, niacin, and pantothenic acid.

133

Jalapeno Pepper

Jalapeno peppers are one of the favoured varieties of chili peppers used in the North American cuisine, especially in the Southern states where Tex-Mex cuisine is popular. Just as other chili peppers, jalapenos too originate in the Mexico. In fact, the name "jalapeno" is derived from the provincial capital, *Xalapa* in the Eastern Mexican Veracruz state, where they cultivated by early settlers since ancient times. Later, the crop was introduced to the rest of the world by Spanish explorers. Botanically, jalapenos are the fruit pods from the nightshade family (Solanaceae), of the genus, *capsicum*. Scientific name: *Capsicum annum.*

Jalapeno is a perennial small herbaceous plant with woody stem growing up to a meter in height. It grows well under fertile, well-drained soil conditions. Adequate sunlight and moisture requirements are the other main factors influencing foliage, flower, and fruiting conditions. The plant bears tiny, white colour flowers after about 45 days after transplantation and green pepper pods after about 60 days. Mature jalapenos generally exhibit small cracks around the shoulders of the fruit. Harvesting ripe red fruits may last over several months. The pods measure about 4-10 cm in length and feature smooth shiny skin with blunt, slightly tapering tip. Inside, each fruit features numerous tiny, white, circular, and flat seeds. The seeds cling on around the central, white-placenta. As in other chili peppers, jalapenos also have a strong spicy taste that comes to them from the active alkaloid compounds; capsaicin, capsanthin and capsorubin. On the Scoville scale hotness scale, jalapeños fall in medium-hot range 2,500-4,000 "Scoville heat units" (SHU). On comparison, sweet bell peppers have zero units, and Mexican habañeros have 200,000 to 500,000 units.

Health Benefits of Jalapeno Peppers

☆ Chilies contain health benefiting alkaloid compound, capsaicin, which gives them strong spicy pungent character. Tolerance level of peppers in human beings, including jalapeno peppers may have wide individual variations. Wherever feasible, they should be consumed in moderation to avoid any untoward experiences. Nonetheless, they packed with an impressive list of phto-chemical compounds that are known to have disease preventing and health promoting properties.

Jalapeno Peppers (*Capsicum annum*), Raw, Nutrient Value per 100 g

Principle	*Nutrient Value*	*Percentage of RDA*
Energy	29 Kcal	1.5 per cent
Carbohydrates	6.50 g	5 per cent
Protein	0.91 g	2 per cent
Total Fat	0.37 g	2 per cent
Cholesterol	0 mg	0 per cent
Dietary Fiber	2.8 g	7 per cent
Vitamins		
Folates	27 µg	7 per cent
Niacin	1.280 mg	8 per cent
Pantothenic acid	0.315 mg	6 per cent
Pyridoxine	0.419 mg	32 per cent
Riboflavin	0.070 mg	5 per cent
Thiamin	0.040 mg	3 per cent
Vitamin A	1078 IU	36 per cent
Vitamin C	118.6 mg	198 per cent
Vitamin E	3.58 mg	24 per cent
Vitamin K	18.5 µg	15 per cent
Electrolytes		
Sodium	3 mg	<0.5 per cent
Potassium	248 mg	5 per cent
Minerals		
Calcium	12 mg	1.2 per cent
Copper	0.046 mg	5 per cent
Iron	0.25 mg	3 per cent
Magnesium	15 mg	4 per cent
Manganese	0.097 mg	4 per cent
Phosphorus	26 mg	5 per cent
Selenium	0.4 µg	<1 per cent
Zinc	0.14 mg	1 per cent
Phyto-nutrients		
Carotene-β	561 µg	—
Carotene-α	67 µg	—
Cryptoxanthin-β	105 µg	—
Lutein-zeaxanthin	861 µg	—

- ✰ Capsaicin has been found to have anti-bacterial, anti-carcinogenic, analgesic, and anti-diabetic properties, at least in some early laboratory studies on experimental mammals. It also found to reduce LDL-cholesterol levels in obese individuals.
- ✰ Fresh jalapeno peppers are rich source of vitamin-C. Ripe fruits have more of this vitamin than raw greens. 100 g provide about 118.6 µg or about 198 per cent of RDA. Vitamin C is a potent water-soluble antioxidant. It is required for the collagen synthesis within the body. Collagen is the main structural protein required for maintaining the integrity of blood vessels, skin, tissues, organs, and bones. Regular consumption of foods rich in vitamin C helps the body protect from scurvy; develop resistance against infectious agents (boosts immunity) and scavenge harmful, pro-inflammatory free-radicals from the body.
- ✰ Furthermore, they contain other valuable antioxidants such as vitamin A, and flavonoids like *beta-carotene, alpha-carotene, lutein, zeaxanthin*, and *cryptoxanthin.* These antioxidant substances in capsicum help to protect the body from injurious effects of free-radicals generated from stress and disease conditions.
- ✰ Jalapeno chillies characteristically contain more pyridoxine, vitamin E, vitamin K than other varieties of peppers. Vitamin K increases bone mass by promoting osteotrophic activity in the bones. It also has the beneficial effect in Alzheimer's disease patients by limiting neuronal damage in their brain.

134

Mace Spice

Mace spice is dried outer aril enveloping firmly around the nutmeg kernel. Nutmeg and mace indeed are two separate spice products of same nutmeg fruit. However, mace characteristically has higher concentration of certain essential oils and features refined yet intense aroma than nutmeg and thereby commands special place in the spice box. Mace as well as the nutmeg kernel is obtained from the tropical evergreen tree native to the rain forest Indonesian *Maluku Islands,* also known as the Spice Islands. Binomially the plant belongs to *Myristicaceae* family and known as *Myristica fragrans.* There are several species grown all over the world other than Myristica species, such as *M. argentea, M. malabarica* (*Indian*), *and M. fatua,* are rather similar to M. fragrans, however, have inferior flavour and aroma. Botanically, the nutmeg fruit, in fact, is a drupe like apricot. Once completely ripen, it splits through its bottom (basal) end to reveal single centrally placed oval shaped hard seed (kernel) known as the commercial "nutmeg." Crimson-red coloured lacy or thread like arils known as mace closely envelops the kernel. Mace spice is then carefully peeled off the kernel surface by either hand or using a knife and allowed to dry under shade for 3-4 days. Dried mace arils, which now appear amber in colour are processed and graded before dispatched for sale.

Health Benefits of Mace Spice

- ✰ Essentially employed as an aromatic agent, mace spice greatly enhances colour, taste and flavour of foods. Nonetheless, it contains some of the anti-oxidant compounds essential oils, minerals, and vitamins.
- ✰ Mace features quite different nutritional profile than nutmeg has. It is less in calories, however, has more concentrations of essential oils, vitamin A, vitamin C, carotenes, iron, calcium,
- ✰ The spice contains fixed oil *trimyristine,* and many essential volatile oils, which gives a sweet aromatic flavour such as *myristicin, elemicin, eugenol*

and *safrole*. These oils occur in higher concentration in mace than in nutmeg. The other less important volatile-oils are *pinene, camphene, dipentene, cineole, linalool, sabinene, safrole, terpeniol.*

Mace Spice (*Myristica fragrans*), Ground, Nutritional Value per 100 g

Principle	*Nutrient Value*	*Percentage of RDA*
Energy	475 Kcal	24 per cent
Carbohydrates	50.50 g	39 per cent
Protein	6.71 g	12 per cent
Total Fat	32.38 g	162 per cent
Cholesterol	0 mg	0 per cent
Dietary Fiber	20.2 g	54 per cent
Vitamins		
Folates	76 µg	19 per cent
Niacin	1.350 mg	8 per cent
Pyridoxine	0.160 mg	12 per cent
Riboflavin	0.448 mg	34 per cent
Thiamin	0.312 mg	26 per cent
Vitamin-A	800 IU	27 per cent
Vitamin C	21 mg	35 per cent
Electrolytes		
Sodium	80 mg	5 per cent
Potassium	463 mg	10 per cent
Minerals		
Calcium	252 mg	25 per cent
Copper	2.467 mg	274 per cent
Iron	13.90 mg	174 per cent
Magnesium	163 mg	41 per cent
Manganese	1.500 mg	65 per cent
Phosphorus	110 mg	30 per cent
Zinc	2.15 mg	20 per cent

- ☆ The active principles in ace spice have many therapeutic applications in many traditional medicines as anti-fungal, anti-depressant, aphrodisiac, digestive, and carminative functions.
- ☆ Mace has more vitamin-C content than nutmeg. 100 g mace spice has 21 mg against just 3 mg of nutmegs. Likewise, mace blades contain more riboflavin (vitamin B-2).
- ☆ Mace arils are rather excellent sources of vitamin-A. 100 g of mace provides 800 IU vitamin A, nearly nine times more compared to that in nutmeg.

☆ Mace arils contain more calcium, copper, iron and magnesium than nutmeg. 100 g of mace powder has 13.90 mg of iron when compared to just 3.04 mg of nutmeg. Manganese and copper are used by the body as co-factors for the antioxidant enzyme, *superoxide dismutase.* Iron is essential for red blood cell production and as a co-factor for *cytochrome oxidases* enzymes.

Medicinal Uses

☆ As in nutmeg, mace extraction has also been in Chinese and Indian traditional medicines for illnesses related to the nervous and digestive systems. The compounds in this spice such as *myristicin* and *elemicin* have been soothing as well as stimulant properties on brain.

☆ Nutmeg and mace oil contains *eugenol,* which has been used in dentistry for toothache relief.

☆ The oil is also used as a local massage to reduce muscular pain and rheumatic pain of joints.

☆ Freshly prepared decoction with honey has been used to relief of nausea, gastritis, and indigestion ailments.

135

Mustard Seeds

Mustard seeds have been highly prized medicinal as well as culinary spice being in use since ancient times. The seeds are obtained from mustard plant belonging to Brassica family which also includes cabbage, broccoli, brussels-sprouts, etc. Scientific name: *Brassica juncea*. Mustards are native to Asia Minor, but now cultivated as a main commercial crop in Canada, India, China, and temperate climates of European region. Mustards are winter crops. The plant reaches about 4-5 feet in height and bears golden yellow coloured flowers. Its tiny, round seeds measuring about one mm in diameter is encased inside a fruit pod in a similar fashion like green pea pod.

In general, three main varieties of mustard are grown worldwide for use.

- ☆ White mustard seeds (*Brassica alba*): The seeds are light straw yellow coloured and are slightly larger than the other two varieties. White seeds exhibit mild pungency.
- ☆ Black mustards (*Brassica nigra*): The seeds are commonly seen in South Asia. The seeds are sharp and more pungent than other two varieties.
- ☆ Brown mustards (*Brassica juncea*): The seeds are native to sub-Himalayan plains of Northern India.

Health Benefits of Mustard Seeds

- ☆ Generally perceived as health benefiting spice, mustard seeds are indeed very rich in phyto-nutrients, minerals, vitamins and anti-oxidants.
- ☆ Being one of the chief oil seeds, mustards are indeed very high in calories; 100 g of seeds provide 508 calories. Nonetheless, the seeds are made of quality proteins, essential oils, vitamins, minerals, and dietary fiber.
- ☆ The seeds are high in essential oils as well as plant sterols. Some of important sterols include such as *brassicasterol, campesterol, sitosterol, avenasterol* and

stigmasterol. Some of glucosinolate and fatty acids in the seeds include *sinigrin, myrosin, erucic, eicosenoic, oleic, and palmitic acids.*

Mustard Seeds (*Brassica juncea*), Nutrition Value per 100 g

Principle	*Nutrient Value*	*Percentage of RDA*
Energy	508 Kcal	25 per cent
Carbohydrates	28.09 g	21 per cent
Protein	26.08 g	46 per cent
Total Fat	36.24 g	121 per cent
Cholesterol	0 mg	0 per cent
Dietary Fiber	12.2 g	32 per cent
Vitamins		
Folates	162 µg	40 per cent
Niacin	4.733 mg	30 per cent
Pantothenic acid	0.810 mg	16 per cent
Pyridoxine	0.397 mg	31 per cent
Riboflavin	0.261 mg	20 per cent
Thiamin	0.805 mg	67 per cent
Vitamin A	31 IU	1 per cent
Vitamin C	7.1 mg	12 per cent
Vitamin E	19.82 mg	132 per cent
Vitamin K	5.4 µg	4 per cent
Electrolytes		
Sodium	13 mg	1 per cent
Potassium	738 mg	16 per cent
Minerals		
Calcium	266 mg	27 per cent
Copper	0.645 mg	71 per cent
Iron	9.21 mg	115 per cent
Magnesium	370 mg	92 per cent
Manganese	2.448 mg	106 per cent
Selenium	208.1 µg	378 per cent
Zinc	6.08 mg	55 per cent
Phyto-nutrients		
Carotene-β	18 µg	—
Crypto-xanthin-β	0 µg	—
Lutein-zeaxanthin	508 µg	—

☆ Mustard seeds are an excellent source of essential B-complex vitamins such as folates, niacin, thiamin, riboflavin, and pyridoxine (vitaminB-6), pantothenic acid. These vitamins are essential in the sense that body requires them from external sources to replenish. These B-complex groups of vitamins help in enzyme synthesis, nervous system function and regulating body metabolism.

- 100 g of mustards provide 4.733 mg of niacin (vitamin B-3). Niacin is a part of nicotinamide co-enzymes, helps lower blood cholesterol and triglyceride levels.
- Mustard seeds contain flavonoid and carotenoid antioxidants such as carotenes, zeaxanthin, and lutein. In addition, the seeds compose a small amount of vitamin anti-oxidants such as vitamin A, C, and vitamin K.
- The seeds are an excellent source of vitamin E, gamma tocopherol; contain about 19.82 mg per 100 g (about 132 per cent of RDA). Vitamin E is a powerful lipid soluble antioxidant, required for maintaining the integrity of cell membrane of mucus membranes and skin by protecting it from harmful oxygen-free radicals.
- Mustards are rich source of health benefiting minerals. Calcium, manganese, copper, iron, selenium and zinc are some of the minerals especially concentrated in these seeds. Calcium helps build bone and teeth. Manganese is used by the body as a co-factor for the antioxidant enzyme *superoxide dismutase*. Copper is required in the production of red blood cells. Iron is required for the red blood cell formation and cellular metabolism.

136

Nutmeg

Fragrant rich nutmeg is one of the highly prized spices known since antiquity for its aromatic, aphrodisiac, and curative properties. Nutmegs are evergreen trees, native to the rain forest Indonesian Moluccas Island, also known as the *Spice Islands*. Botanically, the plant belongs to *Myristicaceae* family and known as *Myristica fragrans.* Several species grown all over the world other than Myristica species, such as *M. argentea, M. malabarica* (Indian), and *M. fatua,* are rather similar to *M. fragrans* in appearance; however, they have less intense flavour and aroma. The spice tree is a large evergreen plant that thrives well under tropical climates. A fully-grown tree reaches about 50-60 feet in height and is the source of nutmeg and mace, two valuable spices. The nutmeg fruit, in fact, is a drupe, about the size of an apricot, which when ripen splits up to reveal single centrally situated oval shaped hard kernel known as "nutmeg spice." The seed is closely enveloped by crimson-red coloured lacy or thread like arils known as "mace." Both spices have a similar warm, sweet aromatic flavour.

Health Benefits of Nutmeg

- ☆ Nutmeg and mace spice contains many plant-derived chemical compounds that are known to have been anti-oxidant, disease preventing, and health promoting properties.
- ☆ The spicy nut contains fixed oil *trimyristin* and many essential volatile oils such as which gives a sweet aromatic flavour to nutmeg like *myristicin, elemicin, eugenol* and *safrole*. The other volatile-oils are *pinene, camphene, dipentene, cineole, linalool, sabinene, safrole, terpeniol*.
- ☆ The active principles in nutmeg have many therapeutic applications in many traditional medicines as anti-fungal, anti-depressant, aphrodisiac, digestive, and carminative functions.
- ☆ This prized spice is a good source of minerals like copper, potassium, calcium, manganese, iron, zinc and magnesium. Potassium is an important

component of cell and body fluids that helps control heart rate and blood pressure. Manganese and copper are used by the body as co-factors for the antioxidant enzyme, superoxide dismutase. Iron is essential for red blood cell production and as a co-factor for cytochrome oxidases enzymes.

Nutmeg (*Myristica fragrans*), Ground Form, Nutritional Value per 100 g

Principle	*Nutrient Value*	*Percentage of RDA*
Energy	525 Kcal	26 per cent
Carbohydrates	49.29 g	38 per cent
Protein	5.84 g	10 per cent
Total Fat	36.31 g	180 per cent
Cholesterol	0 mg	0 per cent
Dietary Fiber	20.8 g	55 per cent
Vitamins		
Folates	76 μg	19 per cent
Niacin	1.299 mg	8 per cent
Pyridoxine	0.160 mg	12 per cent
Riboflavin	0.057 mg	4 per cent
Thiamin	0.346 mg	29 per cent
Vitamin-A	102 IU	3.5 per cent
Vitamin C	3 mg	5 per cent
Electrolytes		
Sodium	16 mg	1 per cent
Potassium	350 mg	7.5 per cent
Minerals		
Calcium	184 mg	18 per cent
Copper	1.027 mg	114 per cent
Iron	3.04 mg	38 per cent
Magnesium	183 mg	46 per cent
Manganese	2.900 mg	126 per cent
Phosphorus	213 mg	30 per cent
Zinc	2.15 mg	20 per cent
Phyto-nutrients		
Carotene-β	16 μg	—
Crypto-xanthin-β	90 μg	—
Lutein-zeaxanthin	0 μg	—

☆ It is also rich in many vital B-complex vitamins, including vitamin C, folic acid, riboflavin, niacin, vitamin A and many flavonoid anti-oxidants like beta-carotene and cryptoxanthin that are essential for optimum health.

Medicinal Uses

- Since ancient times, nutmeg and its oil were being used in Chinese and Indian traditional medicines for illnesses related to the nervous and digestive systems. The compounds in this spice such as *myristicin* and *elemicin* have been soothing as well as stimulant properties on brain.
- Nutmeg oil contains *eugenol*, which has been used in dentistry for toothache relief.
- The oil is also used as a local massage to reduce muscular pain and rheumatic pain of joints.
- Freshly prepared decoction with honey has been used to relief of nausea, gastritis, and indigestion ailments.

137

Saffron

Saffron is one of the highly prized spices known since antiquity for its colour, flavour and medicinal properties. It is the dried "stigma" or threads of the flower of the *S. crocus* plant. The plant is a bulbous perennial plant that belongs to the family of *Iridaceae* of the genus, *Crocus* and known botanically as *Crocus sativus*. This exotic spice is a native of Southern Europe and now cultivated worldwide in many countries, particularly in Spain, Italy, France, Greece, Turkey, and Iran and in Jammu and Kashmir state of India.

The *Crocus sativus* plant grows to about 15-20cm in height and bears lavender coloured flowers during each season, which lasts from October until November. Each flower features perianth, which consists of a stalk, known as "style," connecting to the three "stigmas" or threads to the rest of the plant. These orange-yellow coloured stigmas along with the style constitute "saffron" which is used as spice. Good crop production demands cool dry climate with well-drained rich fertile soil and irrigation facilities or sufficient amount of rain fall. The flowers are generally harvested during the early-morning hours and soon the stigma separated, allowed to dry and packed for marketing. Saffron has the distinct flavour because of the chemical components in it picrocrocin, and safranal. It also contains a natural carotenoid chemical compound, crocin, which gives saffron its golden-yellow hue. These traits along with its medicinal properties make it a valuable ingredient in many foods worldwide.

Health Benefits of Saffron

- ☆ Saffron contains many plants derived chemical compounds that are known to have been anti-oxidant, disease preventing and health promoting properties.
- ☆ The flower stigma are composed of many essential volatile oils but the most important being safranal, which gives saffron its distinct hay-like flavour. Other volatile oils in saffron are *cineole, phenethenol, pinene, borneol, geraniol, limonene, p-cymene, linalool, terpinen-4-oil, etc*.

Saffron (*Crocus sativus*), Nutritional Value per 100 g

Principle	*Nutrient Value*	*Percentage of RDA*
Energy	310 Kcal	15.5 per cent
Carbohydrates	65.37 g	50 per cent
Protein	11.43 g	21 per cent
Total Fat	5.85 g	29 per cent
Cholesterol	0 mg	0 per cent
Dietary Fiber	3.9 g	10 per cent
Vitamins		
Folates	93 µg	23 per cent
Niacin	1.46 mg	9 per cent
Pyridoxine	1.010 mg	77 per cent
Riboflavin	0.267 mg	20 per cent
Vitamin A	530 IU	18 per cent
Vitamin C	80.8 mg	135 per cent
Electrolytes		
Sodium	148 mg	10 per cent
Potassium	1724 mg	37 per cent
Minerals		
Calcium	111 mg	11 per cent
Copper	0.328 mg	37 per cent
Iron	11.10 mg	139 per cent
Magnesium	264 mg	66 per cent
Manganese	28.408 mg	1235 per cent
Phosphorus	252 mg	36 per cent
Selenium	5.6 µg	10 per cent
Zinc	1.09 mg	10 per cent

- ☆ This colourful spice has many non-volatile active components; the most important of them is α-crocin, a carotenoid compound, which gives the stigmas their characteristic golden-yellow colour. It also contains other carotenoids, including *zeaxanthin, lycopene, α- and β-carotenes.* These are important antioxidants that help protect the human body from oxidant-induced stress, cancers, infections and acts as immune modulators.
- ☆ The active components in saffron have many therapeutic applications in many traditional medicines as antiseptic, antidepressant, anti-oxidant, digestive, anti-convulsant.
- ☆ This novel spice is a good source of minerals like copper, potassium, calcium, manganese, iron, selenium, zinc and magnesium. Potassium is an important component of cell and body fluids that helps control heart rate and blood

pressure. Manganese and copper are used by the body as co-factors for the antioxidant enzyme, *superoxide dismutase*. Iron is essential for red blood cell production and as co-factor forcytochrome oxidases enzymes.

☆ Additionally, it is also rich in many vital vitamins, including vitamin A, folic acid, riboflavin, niacin, vitamin-C that is essential for optimum health.

Medicinal Uses

☆ The active components present in saffron have many therapeutic applications in many traditional medicines since long time ago as anti-spasmodic, carminative, diaphoretic.

☆ Research studies have shown that, safranal, a volatile oil found in the spice, has antioxidant, cytotoxicity towards cancer cells, anticonvulsant and antidepressant properties.

☆ Alfa-crocin, a carotenoid compound, which gives the spice its characteristic golden-yellow colour, has been anti-oxidant, anti-depressant, and anti-cancer properties.

138

Sichuan Peppercorns

Aromatic, pungent Sichuan peppercorns are the spicy outer coat of the berries belonging to prickly ash family. The rusty-red peppercorns popular as *huajiao*, are one of the five spices of Chinese five spice powder. Scientific name: *Zanthoxylum simulans.*

Prickly-ash tree is a citrus family flowering plant native to mountainous Szechuan region in China. It is a small tree with numerous short spines on both the stems and the leaf petioles, and large knobs on the branches as in lemon. During the winter season, the flowers appear in slender cymes, which subsequently produce tiny reddish-brown berries by the end of spring. The ripe fruits open out to release the round, black seeds from inside. It is this outer coat (husk) which is used as spice whereas seeds are discarded. Sichuan peppercorns are different from black peppercorns, which are native to India in plant characters, aroma and hotness. Sichuan peppers feature anise like flavour with slight tangy taste and produce less pronounced pungency than other type peppers.

Health Benefits of Sichuan Peppercorns

- ☆ One of the popular ingredients of East Asian cooking, the peppers are indeed one of the richest sources of essential oils, minerals, vitamins, and antioxidants.
- ☆ Unlike other pepper verities, Sichuan pepper contains unique essential oils, which give them special citrus-flavour with biting pungent sense. Their aromatic flavour comes from terpenes such as β-*myrcene, limonene, geraniol, linalool, cineol, citronellal, and dipentene;* whereas, their hotness property is due to certain alkamides embedded in their outer shell.
- ☆ Like black peppercorns, Szechuan peppers also aid in the digestion power by increasing gastro-intestinal juice in the gut.

- ✰ Szechuan peppers are an also good source of vitamins such as vitamin A, carotenes, pyridoxine, and thiamin and minerals like copper, potassium, iron, manganese, phosphorus, selenium and zinc.

139

Tamarind

Sweet and tangy tamarind is one of the widely used condiment spices found in every South Asian kitchen. The tree is very large with long heavy drooping branches and dense foliage. Full grown-up tree might reach up to 80 feet in height. During each season, the tree bears irregularly curved pods in abundance all along its branches. Each pod has thick outer shell encasing deep brown colour sticky pulp enveloping 2-10 hard dark-brown colour seeds. Botanically, the tree is among the large tropical trees belonging to the family of *Fabaceae*, of the genus: *Tamarindus*. Scientific name: ***Tamarindus indica.*** Tamarinds are evergreen tropical trees native to Africa. They grow throughout tropical and subtropical regions of Africa, South Asia, South America and Caribbean islands.

Health Benefits of Tamarind

- ☆ Tamarind fruit contains certain health benefiting essential volatile chemical compounds, minerals, vitamins and dietary fiber.
- ☆ Its sticky pulp is rich source of non-starch polysaccharides (NSP) or dietary-fiber such as *gums, hemicelluloses, mucilage, pectin* and *tannins*. 100 g of fruit pulp provides 5.1 or over 13 per cent of dietary fiber. NSP or dietary fiber in the food increases its bulk and augments bowel movements thereby help prevent constipation. The fiber also binds to toxins in the food thereby help protect the colon mucus membrane from cancer-causing chemicals.
- ☆ In addition, dietary fibers in the pulp bind to bile salts (produced from cholesterol) and decrease their re-absorption in the colon; thereby help excretion of "bad" or LDL cholesterol levels from the body.
- ☆ While lemon composes of *citric acid*, tamarind is rich in tartaric acid. Tartaric acid gives a sour taste to food but is also a very powerful antioxidant. (Anti-oxidant E-number is E334). It helps the body protect from harmful free radicals.

Tamarind (*Tamarindus indica*), Nutritional Value per 100 g

Principle	*Nutrient Value*	*Percentage of RDA*
Energy	239 Kcal	12 per cent
Carbohydrates	62.50 g	40 per cent
Protein	2.80 g	5 per cent
Total Fat	0.60 g	3 per cent
Cholesterol	0 mg	0 per cent
Dietary Fiber	5.1 g	13 per cent
Vitamins		
Folates	14 µg	3.5 per cent
Niacin	1.938 mg	12 per cent
Pantothenic acid	0.143 mg	3 per cent
Pyridoxine	0.066 mg	5 per cent
Thiamin	0.428 mg	36 per cent
Vitamin A	30 IU	1 per cent
Vitamin C	3.5 mg	6 per cent
Vitamin E	0.10 mg	<1 per cent
Vitamin K	2.8 µg	2 per cent
Electrolytes		
Sodium	28 mg	2 per cent
Potassium	628 mg	13 per cent
Minerals		
Calcium	74 mg	7 per cent
Copper	0.86 mg	9.5 per cent
Iron	2.80 mg	35 per cent
Magnesium	92 mg	23 per cent
Phosphorus	113 mg	16 per cent
Selenium	1.3 µg	2 per cent
Zinc	0.10 mg	1 per cent
Phyto-nutrients		
Carotene-β	18 µg	—
Crypto-xanthin-β	0 µg	—
Lutein-zeaxanthin	0 µg	—

- ☆ Tamarind fruit contains many volatile phytochemicals such as *limonene, geraniol, safrole, cinnamic acid, methyl salicylate, pyrazine* and *alkyl-thiazoles*. Together these compounds account for the medicinal properties of tamarind.
- ☆ This prized spice is a good source of minerals like copper, potassium, calcium, iron, selenium, zinc and magnesium. Potassium is an important component of cell and body fluids that helps control heart rate and blood

pressure. Iron is essential for red blood cell production and as a co-factor for *cytochrome oxidases* enzymes.

- In addition, it is also rich in many vital vitamins, including thiamin (36 per cent of daily required levels), vitamin A, folic acid, riboflavin, niacin, and vitamin-C. Much of these vitamins play antioxidant, and co-factor functions for enzyme metabolism inside the body.

Medicinal Uses of Tamarind

- Its pulp has been used in many traditional medicines as a laxative, digestive, and as a remedy for biliousness and bile disorders.
- This spice condiment is also used as emulsifying agent in syrups, decoctions, etc., in different pharmaceutical products.

140

Vanilla Beans

Pleasantly fragrant rich vanilla beans are the pods or fruits obtained from a tropical climbing orchid. Mayans used them to flavour chocolate drink centuries before Spanish first set their foot in Mexico in 1520. This highly prized bean is native to tropical rain forest of Central America and only recently spread to other tropical-regions by Spanish explorers. Botanically the plant is a perennial herbaceous climbing vine belonging to the family of *Orchidaceae*, in the genus: *vanilla*. Scientific name: *Vanilla planifolia*.

The vanilla plant has unique growth characteristics. It requires supporting tree or pole to grow in height. A matured vine bears deep trumpet-shaped flowers. The flowers open only for one day and are naturally pollinated either by the Melipona bee or long beaked humming birds. However, in cultivation, they are hand-pollinated and now grown commercially in Madagascar, Indonesia, India, Puerto Rica and West Indies. There exist three main cultivars of vanilla. Among them, *Vanilla planifolia* is the most sought-after bean worldwide. The other species include *Vanilla pompona* and *Vanilla tahitiensis* (grown in Polynesian islands), although the *vanillin* content of these species is inferior to Vanilla planifolia. Unripe vanilla pods are harvested when they are 5-8 inches long and turn light yellow. They are then blanched briefly in boiling water, sweated and dried under sun over a period of weeks until they turn dark-brown and wrinkled. White crystalline efflorescence which is vanillin, the main essence of vanilla, appears inside in some of the superior quality pods.

Health Benefits of Vanilla

- Vanilla beans are one of the expensive non-pungent spices especially used as flavouring agent in a wide array of sweet-drinks and confectionaries.
- Vanilla extract composed of simple and complex sugars, essential oils, vitamins and minerals.

Vanilla Extract (*Vanilla planifolia*), Nutritional Value per 100 g

Principle	*Nutrient Value*	*Percentage of RDA*
Energy	288 Kcal	14 per cent
Carbohydrates	12.65 g	10 per cent
Protein	0.06 g	<1 per cent
Total Fat	0.06 g	<1 per cent
Cholesterol	0 mg	0 per cent
Dietary Fiber	0.0 g	0 per cent
Vitamins		
Folates	0 mcg	0 per cent
Niacin	0.425 mg	3 per cent
Pantothenic caid	0.035 mg	0.5 per cent
Pyridoxine	0.026 mg	2 per cent
Riboflavin	0.095 mg	7 per cent
Vitamin A	0 IU	0 per cent
Vitamin C	0 mg	0 per cent
Electrolytes		
Sodium	9 mg	0.5 per cent
Potassium	148 mg	3 per cent
Minerals		
Calcium	11 mg	1 per cent
Copper	0.072 mg	8 per cent
Iron	0.12 mg	1.5 per cent
Magnesium	12 mg	3 per cent
Manganese	0.230 mg	10 per cent
Phosphorus	6 mg	1 per cent
Selenium	0.0 mcg	0 per cent
Zinc	0.11 mg	1 per cent

- ☆ The chief chemical component in the beans is vanillin. They also include numerous traces of other constituents such as *eugenol, caproic acid, phenoles, phenol ether, alcohols, carbonyl compounds, acids, ester, lactones, aliphatic and aromatic carbohydrates* and *vitispiranes.*
- ☆ Ancient Mayans believed that vanilla drink was supposed to have aphrodisiac qualities. No modern research study, however, establishes its role in the treatment of sexual dysfunctions.
- ☆ Its extract contains small amounts of B-complex groups of vitamins such as niacin, pantothenic acid, thiamin, riboflavin and vitamin B-6. These vitamins help in enzyme synthesis, nervous system function and regulating body metabolism.

- ✰ This spice also contains small traces of minerals such as calcium, magnesium, potassium, manganese, iron and zinc. Potassium is an important component of cell and body fluids that helps control heart rate and blood pressure. Manganese and copper are used by the body as co-factors for the antioxidant enzyme, superoxide dismutase. Iron is essential for red blood cell production and as a co-factor for cytochrome-oxidases enzymes.

E. Nutritional Value and Health Benefits of Herbs

Introduction

Healthy herbs have long held an important place in our wellness. Prized since ancient times, and today we even more depend on them to purify our body, mind, and soul. Of course, we all use herbal parts in our daily lives, one way or the other, whether for their fragrance, for their healing power, or in lovely recipes. Herbal benefits are many; be it for spiritual reasons or to spice up your taste buds, or as a home remedy for ailments like cold, or sore throat herbs are handy for each need. Although, the herbs been in use in our diet since antiquity, only recently have taken the center-stage of nutrition scientific world for their potential health benefiting and detoxification properties.

141

Basil Herb

The king of herbs, basil herb is one of the oldest and popular herbal plants rich in much notable health benefiting phyto-nutrients. This highly prized plant is revered as "holy herb" in many traditions all around the world. Basil belongs to the family of *Lamiaceae*, of the genus: *Ocimum*. Its scientific name is "*Ocimum basilicum.*" Basil is originally native to Iran, India and other tropical regions of Asia. This bushy annual herbal plant is grown for its medicinally useful leaves and seeds. Basil grows best in warm, tropical climates. Fully-grown plant reaches on an average about 100 cm in height. The leaves are light green, silky about 2.5 inches long and 1 inch broad with opposite arrangement. The flowers are quite big, white and arranged in a terminal spike. Varieties of basil herb exist. "Mediterranean" cultivar is typically called sweet basil, has light green leaves as opposite to "Asian basil" (*Ocinum sanctum*) that feature large, hairy stems and stalks with pink flowers, purple or red leaves and has stronger 'clove' like flavour. There is also *lemon basil,* which has "lemon" flavour. Thai basil (*Ocinum basilicum* 'Horapha') is similar in characteristics to Asian basil but features narrow, pointed, light green colour leaves with a sweet licorice aroma.

Health Benefits of Basil Herb

- ☆ Basil leaves contain many notable plants derived chemical compounds that are known to have disease preventing and health promoting properties.
- ☆ Basil herb contains many *polyphenolic flavonoids* like orientin and vicenin. These compounds were tested *in-vitro* laboratory for their possible anti-oxidant protection against radiation-induced lipid per-oxidation in mouse liver.
- ☆ Basil leaves contain much health benefiting essential oils such as *eugenol, citronellol, linalool, citral, limonene* and *terpineol.* These compounds are known to have anti-inflammatory and anti-bacterial properties.

Basil Herb (*Ocimum basilicum*), Fresh Leaves, Nutritive Value per 100 g

Principle	*Nutrient Value*	*Percentage of RDA*
Energy	23 Kcal	1 per cent
Carbohydrates	2.65 g	2 per cent
Protein	3.15 g	6 per cent
Total Fat	0.64 g	2 per cent
Cholesterol	0 mg	0 per cent
Dietary Fiber	1.60 g	4 per cent
Vitamins		
Folates	68 µg	17 per cent
Niacin	0.902 mg	6 per cent
Pantothenic acid	0.209 mg	4 per cent
Pyridoxine	0.155 mg	12 per cent
Riboflavin	0.076 mg	6 per cent
Thiamin	0.034 mg	2.5 per cent
Vitamin A	5275 IU	175 per cent
Vitamin C	18 mg	30 per cent
Vitamin E	0.80 mg	5 per cent
Vitamin K	414.8 µg	345 per cent
Electrolytes		
Sodium	4 mg	0 per cent
Potassium	295 mg	6 per cent
Minerals		
Calcium	177 mg	18 per cent
Copper	385 mg	43 per cent
Iron	3.17 mg	40 per cent
Magnesium	64 mg	16 per cent
Manganese	1.15 mg	57 per cent
Zinc	0.81 mg	7 per cent
Phyto-nutrients		
Carotene-β	3142 µg	—
Crypto-xanthin-β	46 µg	—
Lutein-zeaxanthin	5650 µg	—

- ☆ The herbs' parts are very low in calories and contain no cholesterol, but are very rich source of many essential nutrients, minerals, and vitamins that are required for optimum health.
- ☆ Basil herb contains exceptionally high levels of *beta-carotene, vitamin A, cryptoxanthin, lutein* and *zeaxanthin*. These compounds help act as protective scavengers against oxygen-derived free radicals and reactive oxygen species (ROS) that play a role in ageing and various disease processes.

- Zeaxanthin, a yellow flavonoid carotenoid compound, is selectively absorbed into the retinal macula lutea where it found to filter harmful UV rays from reaching the retina. Studies suggest that common herbs, fruits, and vegetables that are rich in zeaxanthin anti-oxidant help to protect from *age-related macular disease* (AMRD), especially in the elderly.
- 100 g of fresh herb leaves contain astoundingly 5275 mg or 175 per cent of daily required doses of vitamin A.Vitamin A is known to have antioxidant properties and is essential for vision. It is also required for maintaining healthy mucus membranes and skin. Consumption of natural foods rich in vitamin-A has been found to help the body protect from lung and oral cavity cancers.
- Vitamin K in basil is essential for many coagulant factors in the blood and plays a vital role in the bone strengthening function by helping mineralization process in the bones.
- Basil herb contains a good amount of minerals like potassium, manganese, copper, and magnesium. Potassium is an important component of cell and body fluids, which helps control heart rate and blood pressure. Manganese is used by the body as a co-factor for the antioxidant enzyme, *superoxide dismutase.*
- Basil leaves are an excellent source of iron, contains 3.17 mg/100 g of fresh leaves (about 26 per cent of RDA). Iron, being a component of hemoglobin inside the red blood cells, determines the oxygen-carrying capacity of the blood.

Medicinal Uses of basil Herb

- Basil leaves contain health benefiting essential oils such as eugenol, citronellol, linalool, citral, limonene, and terpineol. These compounds are known to have anti-inflammatory and anti-bacterial properties.
- An important essential oil, eugenol has been found to have anti-inflammatory function by acting against the enzyme *cycloxygenase* (COX), which mediates inflammatory cascade in the body. This enzyme-inhibiting effect of the eugenol in basil makes it an important remedy for symptomatic relief in individuals with inflammatory health problems like rheumatoid arthritis, osteoarthritis, and inflammatory bowel conditions.
- Oil of basil herb has also been found to have anti-infective functions by inhibiting many pathogenic bacteria like *Staphylococcus, Enterococci, Shigella* and *Pseudomonas.*
- Basil tea (basil water-brewed) helps relieve nausea and is thought to have mild anti-septic functions.

142

Borage

Borage or bee plant is one of the chosen culinary herbs used by our grandmothers to make their meals look and taste better, be it to make popular green sauce, to garnish salads or to pamper children with candied flowers! This ancient garden herb exudes a characteristic cucumber-like aroma to the recipes it added to thus making it as one of the most sought-after herbs in the households. Botanically, it belongs to the family of *Borabinaceae*, of the genus; *Borago* and has the scientific name: *Borago officinalis*. The herb is known also as starflower, since it bears five-petal, deep-blue, star-shaped beautiful flowers in clusters. Some of the other common names include *bee bread, burrage, common-bugloss* etc.

Borage is annual hollow stemmed plant with bristly hairs and reaches about 75-90 cm in height. It grows in plentiful all over the wild highlands of Eastern Europe and Asia Minor regions. The plant features broad oval shaped dark green fuzzy leaves. In general, its leaves are harvested from the plant just after the flower buds form but before flowering. Younger leaves are used in salads while older leaves used as greens. However, as the plant gets older, the leaves get tougher, larger, more fuzzy, and bitter in taste.

Health Benefits of Borage

- ☆ Borage is one of very popular culinary herb, especially in Mediterranean countries. The herb contains many notable phyto-nutrients, minerals, and vitamins that are essential for optimum health and wellness.
- ☆ The herb parts contain essential fatty acid gamma-linolenic acid (GLA), typically in concentrations of 17-20 per cent. Linolenic acid is omega-6 fatty acid that plays a vital role in restoration of joint health, immunity, and healthy skin and mucus membranes.
- ☆ Fresh *burrage* herb has high levels of vitamin C (ascorbic acid); provide 35 mg or 60 per cent of RDA per 100 g. Vitamin C is one of the powerful natural

anti-oxidant help remove harmful free radicals from the body. Along with other anti-oxidants, it has immune booster, wound healing and anti-viral effects.

Borage Herb (*Borago officinalis*), Fresh Leaves, Nutritive Value per 100 g

Principle	*Nutrient Value*	*Percentage of RDA*
Energy	21 Kcal	1 per cent
Carbohydrates	3.06 g	2.35 per cent
Protein	1.80 g	3 per cent
Total Fat	0.70 g	2 per cent
Cholesterol	0 mg	0 per cent
Vitamins		
Folates	13 µg	3 per cent
Niacin	0.900 mg	25.5 per cent
Pantothenic acid	0.041 mg	1 per cent
Pyridoxine	0.084 mg	6.5 per cent
Riboflavin	0.150 mg	12 per cent
Thiamin	0.060 mg	5 per cent
Vitamin A	4200 IU	140 per cent
Vitamin C	35 mg	60 per cent
Electrolytes		
Sodium	80 mg	5 per cent
Potassium	470 mg	10 per cent
Minerals		
Calcium	93 mg	9 per cent
Copper	0.130 mg	15 per cent
Iron	3.30 mg	41 per cent
Magnesium	52 mg	13 per cent
Manganese	0.349 mg	15 per cent
Zinc	0.20 mg	2 per cent

- ✰ Burrage herb contains very high levels of vitamin A (140 per cent of RDA) and carotenes. Both these compounds are powerful flavonoid anti-oxidants. Together, they act as protective scavengers against oxygen-derived free radicals and reactive oxygen species (ROS) that play a role in ageing and various disease processes.
- ✰ Vitamin A is known to have antioxidant properties and is essential for vision. It is also required for maintaining healthy mucus membranes and skin. Consumption of natural foods rich in vitamin A and carotenes are known to help the human body protect from lung and oral cavity cancers.

- The herb has a good amount of minerals like iron (41 per cent of RDA), calcium, potassium, manganese, copper, zinc, and magnesium. Potassium is an important component of cell and body fluids, which helps control heart rate and blood pressure. Manganese is used by the body as a co-factor for the antioxidant enzyme, *superoxide dismutase.* Iron is an important co-factor for *cytochrome oxidase enzyme* in the cellular metabolism. In addition, being a component of hemoglobin inside the red blood cells, it determines the oxygen-carrying capacity of the blood.
- Further, the herb is one of the good sources of B-complex vitamins, particularly rich in niacin (vitamin B-3). Niacin helps lower LDL cholesterol levels in the body. In addition, it has riboflavin, thiamin, pyridoxine, and folates in adequate levels. These vitamins function as co-factors in the enzymatic metabolism inside the body.

Medicinal Uses of Borage Herb

- Borage herb parts, especially its seeds contain notable health benefiting essential oil such as gamma-linolenic acid. This omega-6 fatty acid (18:3 fats) has recommended in the treatment of arthritis, dermatitis, pre-menstrual painful conditions etc.
- An infusion of leaves and seeds is used in traditional medicines to increase breast milk production in the nursing mothers.

143

Burdock Root

Burdock root is a taproot of greater burdock plant, used as a vegetable and medicinal herb. The plant is a short biennial, which believed to be native to Northern Europe and Siberia. In Japan, popular as ***gobo***, it is cultivated as a major root herb since ancient times. However, burdock grows as a wild, easy growing hardy plant almost in any parts of the planet. Botanically, burdock belongs to the family of Asteraceae; of the genus of Arctium and known scientifically as *Arctium lappa*. Greater burdock plant grows to about a meter in height. It features broad heart-shaped coarse leaves as in rhubarb that is deep green on the top and light green on the underside. In summer, thick hairy stems, about 5 feet in height, bear reddish purple tubular flowers, which consequently turn into seed-heads or burrs with hooked spines. The deep taproots which may measure about 2-3 feet long are slender, shaped like carrot and parsnip, are brown with white flesh. Of the four species of Arctium, only greater burdock (A. lappa) and lesser burdock (A. minus) are cultivated for their herbal parts. In general, the roots are unearthed in the fall, slow dried and stored for use in winter. The burdock root has the flavour that resembles the sweet taste of Jerusalem artichokes or parsnips, has a crispy texture with gummy consistency. Almost all the parts of the plant are being used either for culinary purpose *or* as a curative remedy for certain medical conditions.

Health Benefits of Burdock Root

- ✰ Burdock roots, young shoots, peeled stalks, and dried seeds contain numerous compounds that are known to have been anti-oxidant, disease preventing, and health promoting properties.
- ✰ The root is very low in calories; provide about 72 calories per 100 g. Burdock is a good source of non-starch polysaccharides such as inulin, glucoside-lappin, mucilage, etc., that help act as a laxative. Additionally, inulin acts as prebiotic and helps reduce blood-sugar level, weight and cholesterol levels in the blood.

Burdock Root (*Arctium lappa*), Raw, Nutrition Value per 100 g

Principle	*Nutrient Value*	*Percentage of RDA*
Energy	72 Kcal	3.5 per cent
Carbohydrates	17.34 g	13 per cent
Protein	1.53 g	3 per cent
Total Fat	0.15 g	<1 per cent
Cholesterol	0 mg	0 per cent
Dietary Fiber	3.3 g	8 per cent
Vitamins		
Folates	23 µg	6 per cent
Niacin	0.300 mg	2 per cent
Pantothenic acid	0.321 mg	6 per cent
Pyridoxine	0.240 mg	18 per cent
Riboflavin	0.30 mg	2 per cent
Thiamin	0.010 mg	1 per cent
Vitamin A	0 IU	0 per cent
Vitamin C	3 mg	5 per cent
Vitamin E	0.38 mg	2.5 per cent
Vitamin K	1.6 µg	1 per cent
Electrolytes		
Sodium	5 mg	<1 per cent
Potassium	308 mg	6.5 per cent
Minerals		
Calcium	41 mg	4 per cent
Iron	0.80 mg	10 per cent
Magnesium	38 mg	9 per cent
Manganese	0.232 mg	10 per cent
Phosphorus	51 mg	7 per cent
Selenium	0.7 mg	1 per cent
Zinc	0.33 mg	3 per cent
Phyto-nutrients		
Carotene-α	0 µg	—
Carotene-β	0 µg	—
Crypto-xanthin-β	0 µg	—
Lutein-zeaxanthin	0 µg	—

☆ Burdock root is especially containing good amounts of electrolyte potassium (308 mg or 6.5 per cent of daily-required levels per 100 g root) and low in sodium. Potassium is an important component of cell and body fluids that helps control heart rate and blood pressure.

- ✰ This herb root contains small quantities of many vital vitamins, including folic acid, riboflavin, pyridoxine, niacin, vitamin-E, and vitamin-C that is essential for optimum health. Both vitamin C and E are powerful natural antioxidants help the human body stave off infections, cancer and neurologic conditions.
- ✰ Furthermore, it also contains some valuable minerals such as iron, manganese, magnesium; and small amounts of zinc, calcium, selenium, and phosphorus.

Medicinal Uses

- ✰ Just like its fellow Asterceae family member dandelion, almost all the parts of burdock herb too found a place in various traditional as well modern medicines.
- ✰ Burdock has been used in many folk remedies as one of the best blood purifiers. It contains certain diuretic principles, which help expel toxic products from the blood through urine.
- ✰ The herb is employed in the treatment of skin problems such as eczema (dermatitis), psoriasis, skin dryness etc. The plant parts have been used as an herbal remedy for liver and gall bladder complaints.
- ✰ Effusion of burdock seeds has been used for throat and chest ailments.
- ✰ Burdock leaves and stems, in addition to their use as a vegetable, have appetite stimulant and are a good remedy for dyspeptic complaints.

144

Celery

Celery herb is known for its strong aromatic flavour it imparts to variety of popular Mediterranean cuisines. It is a small biennial herbaceous plant originated in Europe. The herb grows easily as potherb and in the kitchen gardens for its flavourful leaves, shoots, roots, and seeds. Botanically, the plant belongs to the family of *Apiaceae* of the genus; *Apium* and known scientifically as *Apium graveolens*. The herb is a small plant; reaching about half a meter in height. It requires fertile, moisture rich soil to flourish. Celery herb bears umbelliform flowers at the top of the stalk in the second year. Edible cluster of long, dark green, somewhat hollow stalks grow upright from the crown of the plant. Its leaves feature similar appearance to that of flat-leaf parsley. Several cultivars exist; wild celery is the parent of the cultivated *stalk-celery (A. graveolens* dulce), *leaf-celery (A. graveolens* secalinum), and *celeriac or knob celery (A. graveolens* rapaceum).

Chinese celery or *oriental variety* features thin, hollow, succulent stalks and has more intense flavour than the continental variety.

Celery seeds, used as spice, are similar in appearance to cumin seeds. They feature dark-brown colour, oblong shape with thin vertical ridges; and flavour that is strongly aromatic.

Health Benefits of Celery

- ☆ One of the very low calories herbal plants, celery leaves contain only 16 calories per 100 g weight and lots of non-soluble fiber (roughage) which when combined with other weight loss regimens may help to reduce body weight and blood cholesterol levels.
- ☆ Celery is a functional food. Its leaves are rich source of flavonoid antioxidants such as *zeaxanthin, lutein, and beta-carotene*, which have been anti-oxidant, cancer protective and immune-boosting functions.

Celery, (*Apium graveolens*), Fresh, Nutrient Value per 100 g

Principle	*Nutrient Value*	*Percentage of RDA*
Energy	16 Kcal	<1 per cent
Carbohydrates	3 g	5.5 per cent
Protein	3.46 g	6 per cent
Total Fat	1.12 g	4.5 per cent
Cholesterol	0 mg	0 per cent
Dietary Fiber	2.10 g	5.5 per cent
Vitamins		
Folates	36 μg	9 per cent
Niacin	0.320 mg	2 per cent
Pantothenic acid	0.246 mg	5 per cent
Pyridoxine	0.074 mg	6 per cent
Riboflavin	0.57 mg	4 per cent
Thiamin	0.021 mg	2 per cent
Vitamin A	449 IU	15 per cent
Vitamin C	3.1 mg	5 per cent
Vitamin K	29.3 μg	24 per cent
Vitamin K		
Electrolytes		
Sodium	80 mg	5 per cent
Potassium	260 mg	5.5 per cent
Minerals		
Calcium	40 mg	4 per cent
Copper	0.35 mg	4 per cent
Iron	0.20 mg	2.5 per cent
Magnesium	11 mg	3 per cent
Manganese	0.103 mg	4.5 per cent
Phosphorus	24 mg	3 per cent
Zinc	0.13 mg	1 per cent
Phyto-nutrients		
Carotene-β	270 μg	—
Crypto-xanthin-β	0 μg	—
Lutein-zeaxanthin	283 μg	—

☆ Its leaves are a good source of *vitamin-A*. 100 g fresh celery comprises of 449 IU or 15 per cent of daily-required levels of this vitamin. Vitamin-A and beta-carotene are natural flavonoid antioxidants. Vitamin A is also required for maintaining healthy mucus membranes and skin, and for eye-sight. Consumption of natural foods rich in flavonoids helps the body to protect from lung and oral cavity cancers.

- The herb is also rich in many vital vitamins, including folic acid (provides 9 per cent of RDA), riboflavin, niacin and vitamin-C, which are essential for optimum metabolism.
- Fresh celery is an excellent source of vitamin-K, provides about 25 per cent of DRI. Vitamin-K helps increase bone mass by promoting osteotrophic activity in the bones. It also has established role in Alzheimer's disease patients by limiting neuronal damage in the brain.
- The herb is a very good source of minerals like potassium, sodium, calcium, manganese, and magnesium. Potassium is an important component of cell and body fluids that helps control heart rate and blood pressure.
- Its leaves and seeds contain many essential volatile oils that include terpenes, mostly limonene (75 to 80 per cent), and the sesquiterpenes like β-selinene (10 per cent) and humulene; however, its characteristic fragrance is due to chemical compounds known as *phthalides* (butylphthalid and its dihydro derivate sedanenolid) in them.
- Essential oil obtained from extraction of celery plant has been used in soothing remedies for nervousness, osteoarthritis, and gouty-arthritis conditions. In addition, its seeds, and root have a diuretic (removes excess water from the body through urine), galactogogue (help breast milk secretion), stimulant, and tonic properties.
- Celery plant contains just 16 calories per 100 g leaves and along with numerous health benefiting phyto-nutrients such as flavonoids, folate, vitamin-A, vitamin-K, minerals and other vitamins, has all the characters to consider it as one of the most valuable low calorie or negative calories weight loss functional foods.

Medicinal Uses

- Wild celery has been used in complementary medicines to reduce blood pressure, to relieve indigestion and as an anti-inflammatory agent. It also used as a diuretic to remove excess water from the body.
- The essential oils in seeds, herb, and root have use as carminative, emmenagogue, galactogogue (help breast milk secretion), nervous system ailments such as headache and nervous irritability.
- The herb has also been claimed useful in rheumatism and gouty conditions.

145

Chives

Sweet, mild-onion flavoured chives are fresh top greens of lily family vegetables. The leaves are tubular and hollow similar to that of onions; however, they are smaller in diameter, and somewhat appear like grass from a distance. They should not be confused to green-onions, which are top greens of young, immature onion plants or to scallions, which are top-greens of *Allium fistulosum* (Welsh onion). Along with chervil, parsley, and tarragon, chive makes the perfectly balanced quartet of classic *French fines herbes* (fine herbs). Botanically, their bright green leave tubules belong to the family of *alliaceae,* of the genus: Allium. Scientific name: *Allium schoenoprasum* (common or onion chives).

Chives are small perennial herbs growing in clumps, probably originated from Siberian highlands. The herb grows best in full sun and a well-drained soil. Plants can be grown from seed or divisions of 2 to 3 bulbs. Fully-grown plant reaches about 8-12 inches in height. In the fields, farmers are generally trim off the leaves periodically to keep the plant growth vigorous. In fact, all plantings should be divided every two to three years to prevent over-crowding and root diseases. Unlike onions and garlic, chive's tiny underground bulbs have unpleasant taste and therefore not sought after in cooking. The flower stems, which rise directly from the base, grow slightly taller than leaves and bear small clusters of mauve or purple flower heads.

Some of different cultivars of chives include:

1. Giant Siberian (*Allium ledebourianum*),
2. Chinese or garlic chives (*Allium tuberosum*),
3. Siberian garlic types (*Allium nutans*).

Health Benefits of Chives

- ✰ Chives are very low in calories; 100 g of fresh leaves provide just 30 calories. Nonetheless, they contain many noteworthy flavonoid anti-oxidants, plant fiber, minerals, and vitamins that have proven health benefits.

Chives (*Allium schoenoprasum* L.), Nutrient Value per 100 g

Principle	*Nutrient Value*	*Percentage of RDA*
Energy	30 Kcal	1 per cent
Carbohydrates	4.35 g	3 per cent
Protein	3.27 g	6 per cent
Total Fat	0.73 g	3 per cent
Cholesterol	0 mg	0 per cent
Dietary Fiber	2.5 g	7 per cent
Vitamins		
Folates	105 µg	26 per cent
Niacin	0.647 mg	4 per cent
Pantothenic acid	0.324 mg	6.5 per cent
Pyridoxine	0.138 mg	11 per cent
Riboflavin	0.115 mg	9 per cent
Thiamin	0.078 mg	6.5 per cent
Vitamin A	4353 IU	145 per cent
Vitamin C	58.1 mg	98 per cent
Vitamin E	0.21 mg	1.5 per cent
Vitamin K	212.7 µg	177 per cent
Electrolytes		
Sodium	3 mg	<0.5 per cent
Potassium	296 mg	6 per cent
Minerals		
Calcium	92 mg	9 per cent
Copper	0.157 mg	17 per cent
Iron	1.60 mg	20 per cent
Magnesium	42 mg	10.5 per cent
Manganese	0.373 mg	16 per cent
Phosphorus	58 mg	8 per cent
Selenium	0.9 µg	2 per cent
Zinc	0.56 mg	5 per cent
Phyto-nutrients		
Carotene-β	2612 µg	—
Crypto-xanthin-β	0 µg	—
Lutein-zeaxanthin	323 µg	—

☆ As in scallions, they too contain more plant-derived dietary fiber than fellow allium members such as onions, shallots and leeks etc. 100 g fresh leaves provide 2.5 g or 7 per cent of daily-recommended levels of fiber.

- Like other allium members, they too possess thio-sulfinites anti-oxidants. Thio-sufinites such as *diallyl disulfide, diallyl trisulfide* and *allyl propyl disulfide* convert to allicin by enzymatic reaction when its leaves disrupted (crushing, cutting, etc.). Laboratory studies show that *allicin* reduces cholesterol production by inhibiting the *HMG-CoA reductase* enzyme in the liver cells. Further, it also found to have anti-bacterial, anti-viral, and anti-fungal activities.
- Additionally, allicin also decreases blood vessel stiffness by releasing vasodilator compound, nitric oxide (NO); and, thereby, help brings a reduction in the total blood pressure. It blocks platelet clot formation besides having fibrinolytic (clot-removal) action in the blood vessels, which helps decrease an overall risk of coronary artery disease (CAD), peripheral vascular diseases (PVD), and stroke.
- Chives surprisingly comprise more vitamin A than any other allium family member vegetables. 100 g of fresh leaves contain 4353 IU of vitamin-A or 145 per cent of daily recommended levels. In addition, the green leaves contain other flavonoid-phenolic antioxidants such as carotenes, zeaxanthin, and lutein. Together, these compounds offer the human body protect from lung and oral cavity cancers.
- They also have some other essential vitamins such as vitamin C, and K, in fact; chives are one of the richest sources of vitamin K, comparatively more than that of in scallions. 100 g of fresh greens provide 212.7 µg or about 177 per cent of daily recommended intake of this vitamin. Scientific studies suggest that vitamin K has a potential role in bone health by promoting osteotrophic (bone formation and strengthening) activity. Adequate vitamin-K levels in the diet help limiting neuronal damage in the brain; thus, has established role in the treatment of Alzheimer's disease.
- Fresh chives are rich source of folates. 100 g leaves provide 105 µg or 26 per cent of DRI of folates. Folic acid is essential for DNA synthesis and cell division. Adequate folate levels in the diet during pregnancy may help prevent neural tube defects in the newborn babies.
- Furthermore, the leaves are packed with other B-complex vitamins as well as some essential minerals such as copper, iron, manganese, zinc, and calcium. The leafy greens contain several vital vitamins such as pyridoxine, pantothenic acid, niacin, riboflavin, and thiamin in healthy proportions.

146

Cilantro

Cilantro is a popular Mediterranean herb commonly recognized in Asia as leaf-coriander. It is widely employed in particularly savory dishes all cultures both in modern as well as traditional cuisines. The herb contains many notable plant derived chemical compounds that are known to have disease preventing and health-promoting properties. It is quite similar to dill in utility terms of its leaves and seeds, which can be used as seasoning. Botanically, the herb belongs to the family of *Apiaceae*, of the genus: *Coriandrum*. Scientific name: *Coriandrum sativum.*

Cilantro herb is native to Mediterranean and Asia Minor (Turkey) regions. This perennial plant requires well-draining fertile soil supplemented with warm summer climates to flourish. It grows 1 to 2 feet in height and features dark green, hairless, soft leaves that are variable in shape, broadly lobed at the base of the plant, and slender and feathery higher on near the flowering stems. The leaves and stem have slightly citrus flavour. It bears umbels of small white or light pink flowers in midsummer, followed by round-oval shaped aromatic seeds. Coriander seeds, used as spice, are round-to-oval in shape, yellowish brown in colour with vertical ridges and have a flavour that is aromatic, sweet and citrus, but also slightly peppery.

Health Benefits of Cilantro (Coriander)

- ☆ Cilantro herb contains no cholesterol; however, it is rich in antioxidants, essential oils, vitamins, and dietary fiber, which help reduce LDL or "bad cholesterol" while increasing HDL or "good cholesterol" levels.
- ☆ Its leaves and seeds contain many essential volatile oils such as *borneol, linalool, cineole, cymene, terpineol, dipentene, phellandrene, pinene* and *terpinolene.*
- ☆ The leaves and stem tips are also rich in numerous anti-oxidant polyphenolic flavonoids such as *quercetin, kaempferol, rhamnetin* and *epigenin.*

Cilantro (*Coriandrum sativum*), Fresh, Nutrient Value per 100 g

Principle	*Nutrient Value*	*Percentage of RDA*
Energy	23 Kcal	1 per cent
Carbohydrates	3.67 g	3 per cent
Protein	2.13 g	4 per cent
Total Fat	0.52 g	2 per cent
Cholesterol	0 mg	0 per cent
Dietary Fiber	2.80 g	6.5 per cent
Vitamins		
Folates	62 μg	15.5 per cent
Niacin	1.114 mg	7 per cent
Pantothenic acid	0.570 mg	11 per cent
Pyridoxine	0.149 mg	11 per cent
Riboflavin	0.162 mg	12 per cent
Thiamin	0.067 mg	5.5 per cent
Vitamin A	6748 IU	225 per cent
Vitamin C	27 mg	45 per cent
Vitamin E	2.50 mg	17 per cent
Vitamin K	310 mcg	258 per cent
Electrolytes		
Sodium	46 mg	3 per cent
Potassium	521 mg	11 per cent
Minerals		
Calcium	67 mg	7 per cent
Iron	1.77 mg	22 per cent
Magnesium	26 mg	6.5 per cent
Manganese	0.426 mg	18.5 per cent
Phosphorus	48 mg	7 per cent
Selenium	0.9 mg	2 per cent
Zinc	0.50 mg	4.5 per cent
Phyto-nutrients		
Carotene-α	36 μg	—
Carotene-β	3930 μg	—
Crypto-xanthin-β	202 μg	—
Lutein-zeaxanthin	865 μg	—

☆ The herb is a good source of minerals like potassium, calcium, manganese, iron, and magnesium. Potassium is an important component of cell and body fluids that helps control heart rate and blood pressure. Iron is essential for red blood cell production. Manganese is used by the body as a co-factor for the antioxidant enzyme, *superoxide dismutase.*

- It is also rich in many vital vitamins, including folic-acid, riboflavin, niacin, vitamin-A, beta carotene, vitamin-C that is essential for optimum health. Vitamin-C is a powerful natural antioxidant. 100 g of cilantro leaves provide 30 per cent of daily recommended levels of vitamin-C.
- It provides 6748 IU of vitamin-A per 100 g, about 225 per cent of recommended daily intake. Vitamin-A, an important fat soluble vitamin and anti-oxidant, is also required for maintaining healthy mucus membranes and skin and is also essential for vision. Consumption of natural foods rich in vitamin-A and flavonoids (carotenes) helps body protect from lung and oral cavity cancers.
- Cilantro is one of the richest *herbal sources for vitamin K;* provide about 258 per cent of DRI. Vitamin-K has a potential role in bone mass building by promoting osteotrophic activity in the bones. It also has established role in the treatment of *Alzheimer's disease* patients by limiting neuronal damage in their brain.
- Coriander seed oil has been found application in many traditional medicines as analgesic, aphrodisiac, anti-spasmodic, deodorant, digestive, carminative, fungicidal, lipolytic (weight loss), stimulant and stomachic.
- Wonderful! Cilantro leaves provide only 23 calories/100 g, but their phyto-nutrients profile is no less than any high-calorie food item; be it nuts, pulses or cereals or meat group.

Medicinal Uses

- The herb parts (leaves, root, and stem) of the cilantro (coriander) plant have been found to have antiseptic and carminative properties.
- The herb contains many phytochemical compounds; phenolic flavonoid antioxidants like quercitin and essential oils have found application in many traditional medicines as analgesic, aphrodisiac, antispasmodic, flatus-relieving (carminative), depurative, deodorant, digestive, fungicidal, lipolytic, stimulant and stomachic.

Culinary Uses

Fresh leaves should be washed thoroughly in the water in order to remove sand and dirt and to rid off any residual pesticides. While in the kitchen, it is generally used just before preparing recipes in order to keep the fragrance and aromatic flavour intact.

- Cilantro (coriander) leaves has been used in preparation of many popular dishes in Asian and east European cuisine since ancient times. When added in combination with other household herbs and spices, it enhances flavour and taste of vegetable, chicken, fish and meat dishes.
- The herb has also been used in the preparation of soups, and sauces. Popular Mediterranean cilantro pesto, prepared by using fresh cilantro, red pepper, garlic cloves, olive oil, pumpkin seeds with few drops of lemon

juie, is a great addition on pasta, in sandwiches or as a marinade to fish, poultry etc. dishes.

- ☆ Freshly chopped and sautéed coriander leaves are a great addition to green salad.
- ☆ Coriander seed powder is one of the main ingredients used in the preparation of curry powder.

147

Dandelion

Revered since earlier times, Dandelion herb is one of the most sought-after herbs to enliven our daily meals. Almost all the parts of the plant, leaves, flower tops, and root, are being used either for culinary purpose or as a curative remedy for certain medical conditions. Botanically, it belongs to the family of *Asteraceae;* of the genus of Taraxacum and known scientifically as *Taraxacum officinale*. There are many common names for this herb like *priest's crown, Irish daisy, monk's head, blowball* and *lion's tooth.* Herb-dandelion is believed to be originated in the Central Asian region and become naturalized in many parts of the temperate and semi-tropical regions, including Mediterranean. It is a very hardy plant, grows vigorously everywhere in the fields, lawns and meadows. It features long stout taproots from which long-jagged dark-green leaves rise directly from the ground surface in radiating fashion.

Golden yellow colour flowers arise at the end of hollow-stalks in late spring to early autumn. Its hollow flower stalks are filled with sweet-scented nectar, attracting bees. Flower-stalks rise straight from the root. Fully-grown plant reaches about 45 cms in height. Almost all the plant parts exude milky navajo-white colour latex from the inured site.

Dandelion Root

The root is a stout, fusiform, and fleshy, dark brown externally and white pulp inside somewhat appears like yam. It contains bitter milky latex; more concentrated than in stems and leaves. Roots are generally dug when the plant turns into second year of life. In general, roots are harvested in summer for medicinal purposes or autumn for drying and grinding for coffee.

Health Benefits of Dandelion Herb

- ☆ Fresh dandelion greens, flower tops, and roots contain valuable constituents that are known to have anti-oxidant, disease preventing, and health promoting properties.

Dandelion Herb Greens (*Taraxacum officinale*), Fresh, Nutrition Value per 100 g

Principle	*Nutrient Value*	*Percentage of RDA*
Energy	45 Kcal	2 per cent
Carbohydrates	9.20 g	7 per cent
Protein	2.70 g	5 per cent
Total Fat	0.70 g	3 per cent
Cholesterol	0 mg	0 per cent
Dietary Fiber	3.50 g	9 per cent
Vitamins		
Folates	27 µg	7 per cent
Niacin	0.806 mg	5 per cent
Pantothenic acid	0.084 mg	1.5 per cent
Pyridoxine	0.251 mg	19 per cent
Riboflavin	0.260 mg	20 per cent
Thiamin	0.190 mg	17 per cent
Vitamin A	10161 IU	338 per cent
Vitamin C	35 mg	58 per cent
Vitamin E	3.44 mg	23 per cent
Vitamin K	778.4 µg	649 per cent
Electrolytes		
Sodium	76 mg	5 per cent
Potassium	397 mg	8 per cent
Minerals		
Calcium	187 mg	19 per cent
Iron	3.10 mg	39 per cent
Magnesium	36 mg	9 per cent
Manganese	0.342 mg	15 per cent
Phosphorus	66 mg	9 per cent
Selenium	0.5 mg	1 per cent
Zinc	0.41 mg	4 per cent
Phyto-nutrients		
Carotene-α	363 µg	—
Carotene-β	5854 µg	—
Crypto-xanthin-β	121 µg	—
Lutein-zeaxanthin	13610 µg	—

☆ Fresh leaves are very low in calories; providing just 45 calories per 100 g. It is also good source of dietary fiber (provide about 9 per cent of RDA per 100 g). In addition, its latex is a good laxative. These active principles in the herb help reduce weight and control cholesterol levels in the blood.

- ☆ Dandelion root as well as other plant parts contains bitter crystalline compounds **Taraxacin**, and an acrid resin, **Taraxacerin**. Further, the root also contains *inulin* (not insulin) and *levulin*. Together, these compounds are responsible for various therapeutic properties of the herb.
- ☆ Fresh dandelion herb provides **10161 IU of vitamin-A per 100 g**, about 338 per cent of daily-recommended intake, one of the highest source of vitamin-A among culinary herbs. Vitamin A is an important fat-soluble vitamin and anti-oxidant, required for maintaining healthy mucus membranes and skin and vision.
- ☆ Its leaves are packed with numerous health benefiting flavonoids such as *carotene-β, carotene-α, lutein, crypto-xanthin* and *zea-xanthn*. Consumption of natural foods rich in vitamin-A and flavonoids (carotenes) helps body protect from lung and oral cavity cancers. Zeaxanthin has photo-filtering functions and protects retina from UV rays.
- ☆ The herb is good source of minerals like potassium, calcium, manganese, iron, and magnesium. Potassium is an important component of cell and body fluids, which helps regulate heart rate and blood pressure. Iron is essential for red blood cell production. Manganese is used by the body as a co-factor for the antioxidant enzyme, *superoxide dismutase.*
- ☆ It is also rich in many vital vitamins including folic acid, riboflavin, pyridoxine, niacin, vitamin -E and vitamin-C that are essential for optimum health. Vitamin-C is a powerful natural antioxidant. Dandelion greens provide 58 per cent of daily-recommended levels of vitamin-C.
- ☆ Dandelion is probably the *richest herbal sources* of **vitamin K; provides about 650 per cent of DRI.** Vitamin-K has potential role in bone mass building by promoting osteotrophic activity in the bones. It also has established role in the treatment of Alzheimer's disease patients by limiting neuronal damage in the brain.

Medicinal Uses

Almost all the parts of dandelion herb found place in various traditional as well in modern medicine.

- ☆ The principle compounds in the herb have laxative and diuretic functions.
- ☆ The plant parts have been used as herbal remedy for liver and gall bladder complaints.
- ☆ The herb is also a good tonic, appetite stimulant and is a good remedy for dyspeptic complaints.
- ☆ The inside surface of the flower stems used as a smoothening agent for burns and stings (for example in stinging nettle allergy).

Safety Profile

Although dandelion herb contains some bitter principles, it can be safely used in healthy individuals without any reservations. However, in patients on potassium sparing diuretic therapy, it may aggravate potassium toxicity. Dandelion herb can also induce allergic contact dermatitis in some sensitive individuals.

148

Dill Weed

Dill weed is a unique perennial herbal plant in the sense that both its leaves as well as seeds are used as a seasoning. It is the member of the *Umbelliferae* family, a large group of flowering herbs and spices, which include caraway, parsley, cumin, fennel, etc. Botanically, it belongs within the family of Apiaceae, of the genus: *Anethum* and scientifically named as*Anethum graveolens*. Dill is native to Mediterranean and East European regions. Just like cilantro, the plant requires warm summer climates with well-drained fertile soil to flourish. the plant requires warm summer climates with well-drained fertile soil to flourish. It grows 1 to 2 feet in height and features dark-green leaves (sprigs) that are wispy and fern-like, have a soft texture with rich pleasant aroma and sweet taste. Dill seeds, used as spice, are similar in taste and appearance to "caraway seeds." They feature light brown colour, oval shape with vertical ridges and flavour that is aromatic, sweet, and citrus, but also slightly bitterly.

Health Benefits of Dill

- ☆ Dill weed contains numerous plant derived chemical compounds that are known to have been anti-oxidant, disease preventing, and health promoting properties.
- ☆ This popular herb contains no cholesterol and low in calories. However, it contains many anti-oxidants, vitamins like niacin, pyridoxine, etc., and dietary fibers, which help to control blood cholesterol levels.
- ☆ Dill leaves (sprigs) and seeds contain many essential volatile oils such as *d-carvone, dillapiol, DHC, eugenol, limonene, terpinene* and *myristicin*.
- ☆ The essential oil, Eugenol in the dill has been in therapeutic usage as local-anesthetic and antiseptic. Eugenol has also been found to reduce blood sugar levels in diabetics. (Further detailed studies may be required to establish its role).

Dill Weed (*Anethum graveolens*) Fresh, Nutrient Value per 100 g

Principle	*Nutrient Value*	*Percentage of RDA*
Energy	43 Kcal	2 per cent
Carbohydrates	7 g	5.5 per cent
Protein	3.46 g	6 per cent
Total Fat	1.12 g	4.5 per cent
Cholesterol	0 mg	0 per cent
Dietary Fiber	2.10 g	5.5 per cent
Vitamins		
Folates	150 μg	37.5 per cent
Niacin	1.570 mg	11 per cent
Pantothenic acid	0.397 mg	8 per cent
Pyridoxine	0.185 mg	14 per cent
Riboflavin	0.296 mg	23 per cent
Thiamin	0.058 mg	5 per cent
Vitamin A	7718 IU	257 per cent
Vitamin C	85 mg	140 per cent
Electrolytes		
Sodium	61 mg	4 per cent
Potassium	738 mg	16 per cent
Minerals		
Calcium	208 mg	21 per cent
Copper	0.146 mg	16 per cent
Iron	6.59 mg	82 per cent
Magnesium	55 mg	14 per cent
Manganese	1.264 mg	55 per cent
Phosphorus	66 mg	9.5 per cent
Zinc	0.91 mg	8 per cent

- ✰ Dill oil, extracted from dill seeds has anti-spasmodic, carminative, digestive, disinfectant, galactagogue (helps breast milk secretion), sedative properties.
- ✰ It is also rich in many vital vitamins, including folic acid, riboflavin, niacin, vitamin A, β-carotene, vitamin-C that is essential for optimum metabolism inside the human body.
- ✰ Vitamin-A and beta carotene are natural flavonoid antioxidants. 100 g of dill weed sprigs provide 7718 IU or 257 per cent of recommended-daily levels of this vitamin. Vitamin A is also required for maintaining healthy mucus membranes and skin and is essential for good eye-sight. Consumption of natural foods rich in flavonoids helps the human body to protect from lung and oral cavity cancers.

- Fresh dill herb is an excellent source of antioxidant vitamin; vitamin-C. 100 g contain about 85 mg or 140 per cent of vitamin C. Vitamin-C helps the body develop resistance against infectious agents and scavenge harmful, pro-inflammatory free radicals.
- Dill weed is a good source of minerals like copper, potassium, calcium, manganese, iron, and magnesium. Copper is a cofactor for many vital enzymes, including cytochrome c-oxidase and superoxide dismutase (other minerals function as cofactors for this enzyme are manganese and zinc). Zinc is a co-factor in many enzymes that regulate growth and development, sperm generation, digestion and nucleic acid synthesis. Potassium is an important component of cell and body fluids that helps control heart rate and blood pressure. Manganese is used by the body as a co-factor for the antioxidant enzyme, *superoxide dismutase.*
- Dill herb has all the characters to consider it has one of the most valuable functional foods. 100 g of dill weed provides only 43 calories, but its phyto-nutrients profile is no less than any other high-calorie food source; be it nuts, pulses, cereals, or meat group.

Medicinal Uses

- The sprigs of the dill weed have antiseptic, digestive, and carminative properties.
- The essential oil, eugenol, in the dill has been in therapeutic use as local anesthetic and anti-septic. Eugenol also has been found to reduce blood sugar levels in diabetics, however, further detailed studies required to establish its role.
- Dill oil, extracted from dill seeds, has been used in traditional medicines as anti-spasmodic, carminative, appetizer, digestive, disinfectant, to help breast milk secretion in nursing mothers and to relieve neurological symptoms like headaches and nervous irritability. Along with caraway, fennel, aniseed, the seeds are most important ingredients in gripe water preparation.

Culinary Uses

Fresh dill weed should be washed thoroughly in the water in order to remove sand and dirt and to rid off any residual pesticides. In order to keep the fragrance and aromatic flavour intact, dill is generally added just before preparing recipes.

149

Epazote

Epazote is a popular Central American herb used by native Mexicans since antiquity. Its strong musky flavour gives unique taste to Mexican and other Latin American cuisines. Its young shoots and leaves used like vegetable in soups while mature, pungent leaves used as digestive and carminatives in bean, fish and corn dishes. Binomially the herb plant belongs to the large *Amaranthaceae* family of herbs and vegetables, including amaranth, spinach, quinoa, beets, etc. Scientific name: *Chenopodium ambrosioides*. Some of the common names include *wormseed, Mexican tea, pazote* etc. Epazote is easy growing annual herb, prefers well draining sandy soil and full sunlight. It grows in the fields, along the roadside as weedy invasive plant. It reaches about 60 to 100 cm in height featuring small pointed leaves with serrated margins. Tiny yellow-green flowers appear in a cluster as in amaranth developing to numerous tiny black seeds.

Health Benefits of Epazote

Epazote has largely been viewed as medicinal herb rather than a culinary plant. In general, its leaves used in the cooking to counter indigestion and flatulence effects of beans, high-fiber and protein food. Nonetheless, the herb has many intrinsic plant nutrients which when used optimally would benefit to overall health and wellness.

- ☆ The herb is very low in calories. 100 g leaves just contain 32 calories. Its plain leaves provide a good amount of fiber, 3.8 g per 100g.
- ☆ Its leaves compose of many monoterpene compounds such as ascaridole (60-80 per cent), isoascaridole, p-cymene, limonene, and terpinene. Ascaridole is toxic to several intestinal worms like roundworm, hookworms, pinworm, etc. Native Mayans drank its infusion regularly to keep off from worm infestation.
- ☆ The herb parts, especially young leaves are an excellent source of folic acid, provide 215 μg or 54 per cent of daily recommended values. Folic acid takes

part in the DNA synthesis and cell division. Expectant mothers, however, may be advised to avoid epazote greens in their diet since it causes uterine cramps and possible risk of termination of pregnancy.

Epazote Herb (*Chenopodium ambrosiodes*), Fresh Leaves, Nutritive Value per 100 g

Principle	*Nutrient Value*	*Percentage of RDA*
Energy	32 Kcal	1.5 per cent
Carbohydrates	7.44 g	6 per cent
Protein	0.33 g	<1 per cent
Total Fat	0.52 g	2 per cent
Cholesterol	0 mg	0 per cent
Ditary fiber	3.8 g	10 per cent
Vitamins		
Folates	215 μg	54 per cent
Niacin	0.639 mg	4 per cent
Pantothenic acid	0.179 mg	3.5 per cent
Pyridoxine	0.152 mg	12 per cent
Riboflavin	0.348 mg	27 per cent
Thiamin	0.028 mg	2 per cent
Vitamin A	57 IU	2 per cent
Vitamin C	3.6 mg	6 per cent
Electrolytes		
Sodium	80 mg	5 per cent
Potassium	470 mg	10 per cent
Minerals		
Calcium	275 mg	27.5 per cent
Copper	0.190 mg	21 per cent
Iron	1.88 mg	24.5 per cent
Magnesium	121 mg	30 per cent
Manganese	3.098 mg	135 per cent
Phosphorus	86 mg	12 per cent
Selenium	0.9 μg	1 per cent
Zinc	1.10 mg	10 per cent
Phytonutrients		
Carotene-β	38 μg	—

☆ Epazote has small amounts of vitamin A, and some flavonoid phenolic anti-oxidants such as beta-carotenes. Together, they act as protective scavengers against oxygen-derived free radicals and reactive oxygen species (ROS) that play a role in ageing and in various disease processes.

- ☆ The herb has a good amount of minerals like calcium (27 per cent of RDA), manganese, potassium, iron, copper, zinc, and selenium. Manganese is used by the body as a co-factor for the antioxidant enzyme, *superoxide dismutase.*
- ☆ It has small but adequate levels of other B-complex vitamins, particularly pyridoxine and riboflavin. These vitamins function as co-factors in the enzymatic metabolism inside the body.

Medicinal Uses

- ☆ Epazote has been found in the traditional medicines in many Central and South American cultures. Its infusion is a popular household remedy for helminthic infestation. Usually, half to one cup of a leaf decoction is given each morning before eating for three consecutive days as treatment.
- ☆ The herb is excellent remedy for stomach and intestinal ailments like indigestion, cramps, and ulcers.
- ☆ Its decoction has been found to have some anti-diabetic properties. Further, certain trial studies suggest it hold hope for some liver cirrhosis and cancers.
- ☆ The herb parts should not be included in the nursing and pregnant mothers for its possible toxic effects.

Safety Profile

Epazote (wormseed) should be used in small quantities. Its seed oil rather contains large concentration of ascaridole and other monopterenes. When taken internally, these chemicals in the oil may cause extensive damage to liver, kidney, rhythm disturbances in the heart and nervous systems. For the same reason, wormseed oil is banned by IFRA (International Fragrance Association) for both external and internal use of its products. The herb parts should not be included in the nursing and pregnant mothers for its possible toxic effect.

150

Garlic

Since time immemorial, garlic has been recognized in almost all the cultures for its medicinal as well as culinary properties. This wonderful herbal plant, grown for its underground root or bulb, contains health promoting phyto-nutrient substances that have proven benefits against coronary artery diseases, infections and cancers. This root herb plant belongs within the family of *Alliaceae* of the genusm, *Allium;* and scientifically known as*Allium sativum*. It is believed to be originating in the mountainous Central Asian region from where it has spread all over the temperate and subtropical regions around the world.

Allium sativum is a perennial crop and is grown by methods similar to those used in growing onions. Fully-grown plant reaches about 50 to 60 cm in height and bears underground bulbous root containing about 8-20 bulblets known as *cloves*. The whole bulb is encased by several layers of white or mauve-tinged thin papery coverings. Several cultivar varieties exist from extra-large *elephant garlic* to small sized *solo garlic*. *Allium oleraceum* or field garlic is a wild, tall variety commonly grown in the United Kingdom. Unlike in onion, garlic flowers are sterile and therefore do not produce seeds. New plants generally are grown from implanting the individual sections of the bulb.

Health Benefits of Garlic

- Strong flavoured, garlic cloves contain many noteworthy phyto-nutrients, minerals, vitamins, and antioxidants that have proven health benefits. Total measured antioxidant strength (ORAC value) is 5346 μmol TE/100 g.
- Its bulbs contain organic thio-sulfinite compounds such as *diallyl disulfide, diallyl trisulfide* and *allyl propyl disulfide* that can form allicin by enzymatic reaction, which is activated by disruption of bulb (like crushing, cutting, etc.).

Garlic (*Allium sativum*), Nutrient Value/100 g.
Total-ORAC value 5346 μmol TE/100 g.

Principle	*Nutrient Value*	*Percentage of RDA*
Energy	149 Kcal	7.5 per cent
Carbohydrates	33.06 g	25 per cent
Protein	6.36 g	11 per cent
Total Fat	0.5 g	2 per cent
Cholesterol	0 mg	0 per cent
Dietary Fiber	2.1 g	5.5 per cent
Vitamins		
Folates	3 μg	1 per cent
Niacin	0.700 mg	4 per cent
Pantothenic acid	0.596 mg	12 per cent
Pyridoxine	1.235 mg	95 per cent
Riboflavin	0.110 mg	8 per cent
Thiamin	0.200 mg	17 per cent
Vitamin A	9 IU	<1 per cent
Vitamin C	31.2 mg	52 per cent
Vitamin E	0.08 mg	0.5 per cent
Vitamin K	1.7 μg	1.5 per cent
Electrolytes		
Sodium	153 mg	10 per cent
Potassium	401 mg	8.5 per cent
Minerals		
Calcium	181 mg	18 per cent
Copper	0.299 mg	33 per cent
Iron	1.70 mg	21 per cent
Magnesium	25 mg	6 per cent
Manganese	1.672 mg	73 per cent
Phosphorus	153 mg	22 per cent
Selenium	14.2 μg	26 per cent
Zinc	1.160 mg	10.5 per cent
Phyto-nutrients		
Carotene-β	5 μg	—
Crypto-xanthin-β	0 μg	—
Lutein-zeaxanthin	16 μg	—

☆ Laboratory studies show that *allicin* reduces cholesterol production by inhibiting the *HMG-CoA reductase* enzyme within the liver cells.

- Further, *allicin* also decreases blood vessel stiffness by release of nitric oxide (NO); thereby bring a reduction in the total blood pressure. It blocks platelet clot formation and has fibrinolytic action in the blood vessels, which helps decrease the overall risk from coronary artery disease (CAD), peripheral vascular diseases (PVD) and stroke.
- Research studies also found that consumption of garlic is associated with a possible decrease in the incidence of stomach cancer.
- *Allicin* and other essential volatile compounds also found to have anti-bacterial, anti-viral, and anti-fungal activities.
- Garlic is an excellent source of minerals and vitamins that are essential for optimum health. The bulbs are one of the richest sources of potassium, iron, calcium, magnesium, manganese, zinc, and selenium. Selenium is a heart-healthy mineral, and is an important cofactor for antioxidant enzymes within the body. Manganese is used by the body as a co-factor for the antioxidant enzyme, *superoxide dismutase*. Iron is required for red blood cell formation.
- It contains many flavonoid anti-oxidants like carotene beta, zeaxanthin, and vitamins like vitamin-C. Vitamin C helps the body develop resistance against infectious agents and scavenge harmful, pro-inflammatory free radicals.
- Garlic cloves have amazingly high levels of vitamins and minerals. Just 100 g provides (in per cent of recommended daily allowance).

Medicinal Uses

- This herb has been used since long time in many traditional Indian and Chinese medicines as a remedy for cold, cough, bronchitis, etc.
- Garlic oil has been used as a local applicant for "ring worm" (fungal dermatitis) infection of skin.
- In modern medicine, this exotic herb is advised as health benefiting food for its anti-microbial, anti-cancer, anti-diabetic, and immune boosting and cholesterol-lowering properties.

Culinary Uses

Both cloves as well as tender green tops of garlic plant are used in a variety of recipes, especially as seasoning. In general, leaves are less pungent than the cloves and used in recipes in a similar way like onion tops. Their outer layers are generally peeled away by hand, and internal creamy white smooth bulblet is either chopped using a knife or crushed just before adding to the recipes.

151

Ginger Root

Ginger root is a popular root herb of culinary as well as medicinal importance. The root still finds a special place in many traditional Indian and Chinese medicines for its unique phyto-chemicals that are known to have disease preventing and health promoting properties. The spicy root is actually an underground rhizome of small herb plant belonging to the *Zingiberaceae* family, of the genus: *Zingiber*. Scientific name: *Zingiber officinale*.The ginger herb is thought to originate in the Himalayan foothills of North India. Today, it is widely grown all over the world as a major commercial crop. Ginger plant grows to about a meter in height. Completely grown plant features thin grass like dark-green leaves and small yellow flowers.

Its root features knotty finger-like projections that grow downward from the ground surface. Fresh raw root has a silver gray outer surface. Cut sections feature creamy white, yellow, or red-coloured crunchy flesh depending upon the variety. The root often contains fibrils running through its center, especially in over-matured. Its pungent, spicy and aromatic smell is due to essential oils and phenolic compounds such as gingerols and shogaols. Galangal (*Alpinia galanga*), also known as blue ginger, is closely related herb that is used extensively in East Asian regions, especially in Thailand, Malaysia, and Indonesian cuisine. Galangal has mild, subtle flavour and less pungent than ginger.

Health Benefits of Ginger Root

- ☆ Ginger has been in use since ancient times for its anti-inflammatory, carminative, anti-flatulent, and anti-microbial properties. Total antioxidant strength measured in terms of *oxygen radical absorbance capacity* (ORAC) of ginger root is 14840 µmol TE/100 g.
- ☆ The root contains health benefiting essential oils such as *gingerol, zingerone, shogaol, farnesene*, and small amounts of β-*phelladrene, cineol, and citral*. Gingerols help improve the intestinal motility and have been anti-

inflammatory, painkiller (analgesic), nerve soothing, anti-pyretic as well as anti-bacterial properties. Studies have shown that it may reduce nausea induced by motion sickness or pregnancy and may help relieve migraine headache.

Ginger Root (*Zingiber officinale*), Fresh, Nutrient Value per 100 g

Principle	*Nutrient Value*	*Percentage of RDA*
Energy	80 Kcal	4 per cent
Carbohydrates	17.77 g	13.5 per cent
Protein	1.82 g	3 per cent
Total Fat	0.75 g	3 per cent
Cholesterol	0 mg	0 per cent
Dietary Fiber	2.0 g	5 per cent
Vitamins		
Folates	11 µg	3 per cent
Niacin	0.750 mg	4.5 per cent
Pantothenic acid	0.203 mg	4 per cent
Pyridoxine	0.160 mg	12 per cent
Vitamin A	0 IU	0 per cent
Vitamin C	5 mg	8 per cent
Vitamin E	0.26 mg	1.5 per cent
Vitamin K	0.1 µg	0 per cent
Electrolytes		
Sodium	13 mg	1 per cent
Potassium	415 mg	9 per cent
Minerals		
Calcium	16 mg	1.6 per cent
Copper	0.226 mg	25 per cent
Iron	0.60 mg	7.5 per cent
Magnesium	43 mg	11 per cent
Manganese	0.229 mg	10 per cent
Phosphorus	34 mg	5 per cent
Zinc	0.34 mg	3 per cent

- ☆ Zingerone, a chemical compound which gives pungent character to the ginger root, is effective against*E.coli* induced diarrhea, especially in children.
- ☆ This herb root only has 80 calories per 100 g and contains no cholesterol. Nonetheless, it composes many essential nutrients and vitamins such as pyridoxine (vitamin B-6), pantothenic acid (vitamin B-5) that are required for optimum health.

- Furthermore, the herb also contains a good amount of minerals like potassium, manganese, copper, and magnesium. Potassium is an important component of cell and body fluids that helps controlling heart rate and blood pressure.

Medicinal Uses

- Ginger root slices, boiled in water with added lemon or orange juice, and honey, is a popular herbal drink in ayurvedic medicine to relieve common cold, cough, and sore throat.
- Its extraction is used as a vehicle to mask bitterness and after-taste in traditional ayurvedic preparations.
- Gingerols increase the motility of the gastrointestinal tract and have analgesic, sedative, anti-inflammatory, and antibacterial properties. Studies have shown that it may reduce nausea caused by motion sickness or pregnancy and may relieve migraine.

Culinary Use

Wash fresh ginger root in cold running water or rinse for few minutes to remove any sand, soil or pesticide residues. Fresh-root has pungent flavour and spicy, peppery taste that hit the palate and nostrils. In order to keep the fragrance and flavour intact, it is generally added at the last moment in the cooking recipes, since prolonged cooking results in evaporation of essential oils.

- Fresh or dried Ginger root along With garlic, cilantro, onion, tomato, cumin and mustard-seeds made to a flavourful curry paste which is then added to variety of vegetable, meat and curry/soup preparations.
- The herb root is also used in the preparation of mango, lemon and spondias (*ambara* in India) pickles.
- Fresh root can be used in the preparation of variety of spicy snacks, candies, and ginger bread in food industry.
- Ginger tea is a popular drink in many Asian countries.

152

Lemongrass

Refreshing, citrus-scented lemongrass imparts unique flavour to recipes. Its coarse tufted stems and leaf buds are among the most sought after herbal parts used in an array of cuisines all over South and East Asian regions. Botanically, the herb belongs to grass family of *Poaceae*. Scientific name: *Cymbopogon citratus.* It is native to Southern part of India and Sri Lanka. The herb is one of the popular ingredients in Thailand, Vietnam, Malaysia, Cambodia, and Indonesia and as far as African and American continents for its culinary and medicinal uses.

Cymbopogon grows in dense clumps, erupting from tough bulbous base with a spread of about 1 meter wide and about 3 feet in height. Its leaves are bright green with sharp edges with appearance similar to that of grass. It flourishes well in fertile sandy soils with tropical climates receiving heavy rain. Several cultivars of cymbopogon based upon their origin, culinary, and oil properties are grown around the world at a commercial level. West-Indian lemongrass (*C. citratus*) is an important culinary herb, and spice used extensively in cooking in many East Asian countries. Indian lemon grass (*C. flexuosus*) is dominantly used in the perfume industry since it contains fewer myrcene and, therefore, has a longer shelf-life.

Health Benefits of Lemongrass

- ☆ Lemongrass herb has numerous health benefiting essential oils, chemicals, minerals and vitamins that are known to have anti-oxidant and disease preventing properties.
- ☆ The herb contains 99 calories per 100 g but contains no cholesterol.
- ☆ The primary chemical component in lemongrass herb is citral or *lemonal*, an aldehyde responsible for its unique lemon odor. Citral also has strong anti-microbial and anti-fungal properties.

Lemongrass (*Cymbopogon citratus*), Fresh, Nutritive Value per 100 g

Principle	*Nutrient Value*	*Percentage of RDA*
Energy	99 Kcal	5 per cent
Carbohydrates	25.31 g	19 per cent
Protein	1.82 g	3 per cent
Total Fat	0.49 g	2 per cent
Cholesterol	0 mg	0 per cent
Vitamins		
Folates	75 µg	19 per cent
Niacin	1.101 mg	7 per cent
Pyridoxine	0.080 mg	6 per cent
Riboflavin	0.135 mg	10.5 per cent
Thiamin	0.065 mg	5.5 per cent
Vitamin A	6 mg	<1 per cent
Vitamin C	2.6 mg	4 per cent
Electrolytes		
Sodium	6 mg	<1 per cent
Potassium	723 mg	15 per cent
Minerals		
Calcium	65 mg	6.5 per cent
Copper	0.266 mg	29 per cent
Iron	8.17 mg	102 per cent
Magnesium	60 mg	15 per cent
Manganese	5.244 mg	228 per cent
Selenium	0.7 µg	1 per cent
Zinc	2.23 mg	20 per cent

- ☆ In addition, its herb parts contain other constituents of the essential oils such as *myrcene, citronellol, methyl heptenone, dipentene, geraniol, limonene, geranyl acetate, nerol*, etc. These compounds are known to have counter-irritant, rubefacient, insecticidal, anti-fungal and antiseptic properties.
- ☆ Its leaves and stems are very good in folic acid (100 g leaves and stem provide about 75 µg or 19 per cent of RDA). Folates are important in cell division and DNA synthesis. When given during the peri-conception period can help prevent neural tube defects in the baby.
- ☆ Its herb parts are also rich in many invaluable essential vitamins such as pantothenic acid (vitamin B5), pyridoxine (vitamin B-6) and thiamin (vitamin B-1). These vitamins are essential in the sense that body requires them from external sources to replenish.

- Furthermore, fresh herb contains small amounts of anti-oxidant vitamins such as vitamin-C and vitamin-A.
- Lemon grass herb parts, whether fresh or dried, are rich sources of minerals like potassium, zinc, calcium, iron, manganese, copper, and magnesium. Potassium is an important component of cell and body fluids, which helps control heart rate and blood pressure. Manganese is used by the body as a co-factor for the antioxidant enzyme, *superoxide dismutase.*

Medicinal Uses

- Pharmacologically, *citral* compound has been used in the synthesis of vitamin A.
- Lemongrass is one of the favoured herbs used in herbal teas.
- It is also helpful in relieving colitis, indigestion, and gastro-enteritis ailments.
- Lemongrass oil when used in aromatherapies revitalizes the body and helps relieves the symptoms of headaches, bodyache, nervous exhaustion, and stress-related conditions.
- Its infusions are often made useful in infections such as sore throats, laryngitis, bronchitis, etc.
- Lemongrass oil is used in massage therapy as a muscle and skin-toner.

Safety Profile

Lemongrass oil can cause skin irritation in some individuals when used in perfumes, cosmetics and as massage oil.

153

Oregano

Oregano is a wonderful perennial culinary as well as medicinal herb. It has long been recognized as one of the "functional foods" for its nutritional, anti-oxidants and disease preventing properties. The herb, whose name means "delight of the mountains" in Greek, is native to the Mediterranean region. Botanically, the herb belongs to the mint (*Lamiaceae*) family, of the genus; *Origanum* and is known scientifically as *Origanum vulgare*.

The plant is a small shrub, growing up to 75 cm in height with multi-branched stems covered with small grayish-green oval leaves and small white or pink flowers. Oregano is particularly used widely in Greek and Italian cuisines. Its leaves have a characteristic aromatic, warm, and slightly bitter taste. The intensity varies; however, good-quality oregano is so strong that it almost numbs the tongue. There are many varieties of oregano cultivated across Europe but the influence of climate, season, and soil on the composition of the essential oils is greater than the difference between various species. Origanum heracleoticum is another Greek variety that is having characteristic sharp scent and flavour. Sweet marjoram (*Origanum majorana*) is one of close related species of oregano. Mexican oregano (*Lippia graveolens*) is a different from origanum species but related to the verbena family of herbs. It is used in place of oregano in many Central American regions.

Health Benefits of Oregano

- ☆ Oregano contains an impressive list of plant derived chemical compounds that are known to have disease preventing and health promoting properties.
- ☆ The herb parts contain no cholesterol; however, are rich source of dietary fiber, which helps to control blood cholesterol levels.
- ☆ Oregano contains health benefiting essential oils such as *carvacrol, thymol, limonene, pinene, ocimene, and caryophyllene*. The leaves and flowering stem of the plant are strongly antiseptic, anti-spasmodic, carminative, cholagogue

(help gall bladder secretion), diaphoretic (sweat production), expectorant, stimulant, and mildly tonic properties. Its decoction is taken by mouth for the treatment of colds, influenza, mild fevers, indigestion, stomach upsets, and painful menstruation conditions.

Oregano Herb (*Origanum vulgare*), Dried Leaves, Nutritive Value per 100 g

Principle	*Nutrient Value*	*Percentage of RDA*
Energy	306 Kcal	15 per cent
Carbohydrates	64.43 g	49 per cent
Protein	11 g	19.5 per cent
Total Fat	10.25 g	34 per cent
Cholesterol	0 mg	0 per cent
Dietary Fiber	42.8 g	107 per cent
Vitamins		
Folates	274 µg	69 per cent
Niacin	6.22 mg	39 per cent
Pantothenic acid	0.921 mg	17 per cent
Pyridoxine	1.210 mg	93 per cent
Riboflavin	0.320 mg	24.5 per cent
Thiamin	0.341 mg	28 per cent
Vitamin A	6903 IU	230 per cent
Vitamin C	50 mg	83 per cent
Vitamin E	18.86 mg	126 per cent
Vitamin K	621.7 µg	518 per cent
Electrolytes		
Sodium	15 mg	1 per cent
Potassium	1669 mg	35.5 per cent
Minerals		
Calcium	1576 mg	158 per cent
Copper	0.943 mg	105 per cent
Iron	44 mg	550 per cent
Magnesium	270 mg	67.5 per cent
Manganese	4.667 mg	203 per cent
Zinc	4.43 mg	40 per cent
Phyto-nutrients		
Carotene-β	4112 µg	—
Crypto-xanthin-β	11 µg	—
Lutein-zeaxanthin	862 µg	—

☆ Thymol is also been found to have anti-bacterial, anti-fungal activities.

- ☆ The herb is rich in poly-phenolic flavonoid anti-oxidants (vitamin A, carotenes, lutein, zeaxanthin, and cryptoxanthin) and has been rated as one of the plant sources with highest anti-oxidant activities. These compounds help act as protective scavengers against oxygen-derived free radicals and reactive oxygen species (ROS) that play a role in ageing and various disease processes.
- ☆ The active principles in the herb may improve the gut motility in addition to increase the digestion power by facilitating copious gastro-intestinal enzyme secretions.
- ☆ This marvelous herb is an excellent source of minerals like potassium, calcium, manganese, iron, and magnesium. Potassium is an important component of cell and body fluids that helps control heart rate and blood pressure caused by high sodium. Manganese and copper are used by the body as co-factors for the antioxidant enzyme, *superoxide dismutase.* Iron helps prevent anemia. Magnesium and calcium are important minerals for bone metabolism.
- ☆ In addition, fresh herb is an excellent source of antioxidant vitamin; vitamin-C. Vitamin C helps the body develop resistance against infectious agents and scavenge harmful, pro-inflammatory free radicals.
- ☆ Oregano is a proven super "functional food." This herb is one of the foundation bases of healthy Mediterranean diet in addition to olive oil, fish and lots of greens, herbs and vegetables.

Medicinal Uses

- ☆ The leaves and flowering stem of the oregano plant have antiseptic, antispasmodic, carminative, increase bile secretion, diaphoretic, expectorant, stimulant, and mildly tonic properties.
- ☆ Oregano decoction is taken by mouth for the treatment of colds, influenza, mild fevers, indigestion, stomach upsets, and painful menstruation.
- ☆ Thymol has been found to have antiseptic, anti-bacterial, and anti-fungal activities.

Safety Profile

Oregano oil extracted from the herb parts may cause skin and mucus membrane irritation and hence, avoided in aromatherapy.

154

Parsley

Parsley is a popular culinary as well as medicinal herb, which is recognized as one of the functional food for its unique antioxidants and disease preventing properties. This wonderful, fragrant rich biennial herb is native to the Mediterranean region and belongs to the family of *Apiaceae* in the genus; *Petroselinum*. Its botanical name is *Petroselinum crispum*. The herb is a small plant with dark-green leaves that resemble coriander leaves, especially in the flat-leaf variety. However, it is milder in flavour than coriander. The herb is widely used in Mediterranean, East European, and American cuisine. There exist several cultivars of parsley grown across Europe. Italian or flat leaf-parsley (*Petroselinum crispum neapolitanum*) is popular around Mediterranean countries and has a more intense flavour than Curley leaf parsley.

Health Benefits of Parsley

- ☆ The herb contains no cholesterol; however, it is rich in anti-oxidants, vitamins, minerals and dietary fiber, which help control blood-cholesterol, prevents constipation, protects the human body from free radicals mediated injury and from cancers.
- ☆ Parsley contains health benefiting essential volatile oils that include *myristicin, limonene, eugenol, and alpha-thujene.*
- ☆ The essential oil, Eugenol, present in this herb has been in therapeutic use in dentistry as a local anesthetic and anti-septic agent for teeth and gum diseases. *Eugenol* has also been found to reduce blood sugar levels in diabetics; however, further detailed studies required to establish its role.
- ☆ Parsley is rich in poly-phenolic flavonoid antioxidants, including *apiin, apigenin, crisoeriol,* and *luteolin;* and has been rated as one of the plant sources with quality antioxidant activities. Total ORAC value, which measures the anti-oxidant strength of 100 g of fresh, raw parsley, is 1301 μmol TE (Trolex equivalents).

Parsley (*Petroselinum crispum*), Fresh, Nutrient Value per 100 g

Principle	*Nutrient Value*	*Percentage of RDA*
Energy	36 Kcal	1.5 per cent
Carbohydrates	6.33 g	5 per cent
Protein	2.97 g	5 per cent
Total Fat	0.8 g	3 per cent
Cholesterol	0 mg	0 per cent
Dietary Fiber	3.3 g	8.5 per cent
Vitamins		
Folates	152 µg	38 per cent
Niacin	1.313 mg	8 per cent
Pantothenic acid	0.400 mg	8 per cent
Pyridoxine	0.090 mg	7 per cent
Riboflavin	0.098 mg	7.5 per cent
Thiamin	0.086 mg	7 per cent
Vitamin A	8424 IU	281 per cent
Vitamin C	133 mg	220 per cent
Vitamin E	0.75 mg	5 per cent
Vitamin K	1640 µg	1366 per cent
Electrolytes		
Sodium	56 mg	4 per cent
Potassium	554 mg	12 per cent
Minerals		
Calcium	138 mg	14 per cent
Copper	0.149 mg	16.5 per cent
Iron	6.20 mg	77.5 per cent
Magnesium	50 mg	12.5 per cent
Manganese	0.160 mg	7 per cent
Phosphorus	58 mg	8 per cent
Zinc	1.07 mg	10 per cent
Phyto-nutrients		
Carotene-β	5054 µg	—
Crypto-xanthin-β	0 µg	—
Lutein-zeaxanthin	5561 µg	—

☆ The herb is a good source of minerals like potassium, calcium, manganese, iron, and magnesium. 100 g fresh herb provides 554 mg or 12 per cent of daily-required levels of potassium. Potassium is the chief component of cell and body fluids that helps control heart rate and blood pressure by countering the effects of sodium. Iron is essential for the production of heme, which is an important oxygen-carrying component inside the red blood cells. Manganese is used by the body as a co-factor for the antioxidant enzyme, *superoxide dismutase.*

- ✰ Additionally, the herb is also rich in many antioxidant vitamins, including vitamin-A, beta-carotene, vitamin-C, vitamin-E, *zeaxanthin, lutein, and cryptoxanthin.* The herb is an excellent source of vitamin-K and folates. Zeaxanthin helps prevent age-related macular degeneration (ARMD) in the retina of the eye in the old age population through its anti-oxidant and ultra-violet light filtering functions.
- ✰ Fresh herb leaves are also rich in many essential vitamins such as pantothenic acid (vitamin B-5), riboflavin (vitamin B-2), niacin (vitamin B-3), pyridoxine (vitamin B-6) and thiamin (vitamin B-1). These vitamins play a vital role in carbohydrate, fat and protein metabolism by acting as co-enzymes inside the human body.
- ✰ It is perhaps the richest of the entire herb source for vitamin K; provide 1640 µg or 1366 per cent of recommended daily intake. Vitamin K has been found to have the potential role in bone health by promoting osteotrophic activity in the bones. It has also established role in the treatment of Alzheimer's disease patients by limiting neuronal damage in the brain.
- ✰ Wonderful! Humble parsley has just 36 calories/100 g, but their phyto-nutrients profile is no less than any high-calorie food sources.

Medicinal Uses

- ✰ Its leaves, stems and roots of this herb plant have antiseptic and carminative properties.
- ✰ *Eugenol* also has been found to reduce blood sugar levels in diabetics; however, further detailed studies required to establish its role.
- ✰ The extraction from the herb has been found to have diuretic effects.

Safety Profile

- ✰ Parsley should not be consumed as a drug or supplement by pregnant women. The herb as an oil, root, leaf, or seed could lead to uterine stimulation and preterm labor.
- ✰ This herbal plant is very high in oxalic acid, 1.70 mg per 100 g. Prolong consumption of oxalate rich foods may results in gouty arthritis, kidney stones and mineral-nutrient deficiencies.
- ✰ Parsley oil contains furanocoumarins and psoralens, which may lead to extreme photosensitivity if used orally.

155

Peppermint

Peppermint has been one of the popular herbs known since antiquity for its distinctive aroma and medicinal value. The herb has a characteristic refreshing cool breeze sensation on taste buds, palate and throat when eaten; and on nasal olfaction glands when inhaled. This unique quality of mint is due to the presence of menthol, an essential oil in its fresh leaves and stem. Botanically, the herb belongs to the *Lamiaceae* family of the genus; *Mentha* and botanically named as *Mentha piperita*. It is actually a natural hybrid-cross between *water mint* (*Mentha aquatica*) and *spearmint* (*Mentha spicata*). Mint herb is originally native to Europe, and now cultivated in all the regions of the world. It grows well under shady conditions and feature lance-shaped purple-veined, dark-green leaves with serrated margins and purple colour whirly-flowers. In general, the mint plant is usually sterile; producing no seeds and reproducing only vegetative reproduction, spreading lateral through its underground rhizomes. There exist more than 20 varieties of mint herbs with a wide range of colour, fragrance, and flavour.

Health Benefits of Peppermint

- ☆ Mint contains numerous plant derived chemical compounds that are known to have been anti-oxidant, disease preventing and health promoting properties. Total antioxidant strength (ORAC) of fresh peppermint herb is 13978 μmol TE/100 g.
- ☆ The mint herb contains no cholesterol; however, it is rich in essential oils, vitamins and dietary fiber, which helps to control blood cholesterol and blood pressure inside the human body.
- ☆ The herb parts contain many essential volatile oils like *menthol, menthone, menthol acetate*. These compounds effect on cold-sensitive receptors in the skin, mouth and throat, the property which is responsible for the natural cooling-sensation that it initiates when inhaled, eaten, or applied on the skin.

Peppermint (*Mentha piperita*), Fresh, Nutritive Value per 100 g

Principle	*Nutrient Value*	*Percentage of RDA*
Energy	70 Kcal	3.5 per cent
Carbohydrates	14.79 g	11 per cent
Protein	3.75 g	7 per cent
Total Fat	0.94 g	3 per cent
Cholesterol	0 mg	0 per cent
Dietary Fiber	8 g	20 per cent
Vitamins		
Folates	114 μg	28 per cent
Niacin	1.706 mg	10.5 per cent
Pantothenic acid	0.338 mg	6.5 per cent
Pyridoxine	0.129 mg	10 per cent
Riboflavin	0.266 mg	20 per cent
Thiamin	0.082 mg	7 per cent
Vitamin A	4248 IU	141 per cent
Vitamin C	31.8 mg	53 per cent
Electrolytes		
Sodium	31 mg	2 per cent
Potassium	569 mg	12 per cent
Minerals		
Calcium	243 mg	24 per cent
Copper	329 μg	36 per cent
Iron	5.08 mg	63.5 per cent
Magnesium	80 mg	20 per cent
Manganese	1.176 mg	51 per cent
Zinc	1.11 mg	10 per cent

- ☆ The essential oil, menthol also has been analgesic (painkiller), local anesthetic and counter-irritant properties.
- ☆ Research studies have also been suggested that the compounds in the peppermint relax intestinal wall and sphincter smooth muscles through blocking calcium channel at cell receptor levels. This property of mint has been applied as an anti-spasmodic agent in the treatment of "irritable bowel syndrome" (IBS) and other colic pain disorders.
- ☆ Peppermint-herb is an excellent source of minerals like potassium, calcium, iron, manganese and magnesium. 100 g fresh herb provides 569 mg of potassium. Potassium is an important component of cell and body fluids that helps control heart rate and blood pressure. Manganese and copper works as co-factors for the antioxidant enzyme, *superoxide-dismutase.*

- ☆ Further, it is rich in many antioxidant vitamins, including vitamin A, beta carotene, vitamin-C and vitamin E. The leaves of mint also contain many important B-complex vitamins like folates, riboflavin and pyridoxine (vitamin B-6); and the herb are an excellent source of vitamin-K.

Medicinal Uses

- ☆ As mentioned above, the essential oils in the peppermint act on cold-sensitive receptors in the skin, mouth and throat, the property which is responsible for the well-known cooling sensation that it provokes when inhaled, eaten, or applied to the skin. This property of mint can be applicable in the preparation of cough/cold reliving remedies like syrups, lozenges and nose inhaler.
- ☆ Peppermint oil has analgesic, local anesthetic and counter-irritant properties and has been used in the preparation of topical muscle relaxants and analgesics.
- ☆ It is also being used in oral hygiene products and bad-breath remedies like mouthwash, toothpaste, mouth and tongue-spray, and more generally as a food flavour agent; *e.g.* in chewing-gum, candy.
- ☆ It is also being used in oral hygiene products and bad-breath remedies like mouthwash, toothpaste, mouth and tongue-spray, and more generally as a food flavour agent; for instance, in chewing gums, candy, etc.

Culinary Uses

Peppermint should be washed thoroughly in the water in order to remove sand and dirt and to rid off any residual pesticides. In order to keep the fragrance and aromatic flavour intact, mint is generally used just before preparing recipes.

- ☆ Mint leaves used extensively in the preparation of herbal tea.
- ☆ As a flavouring base in ice cream and other confectionery.
- ☆ Along with parsley and coriander leaves, mint leaves can be used as a garnish.
- ☆ Mint has also been used in the preparation of soups, and sauces.
- ☆ Freshly chopped mint leaves can be a great addition to green salad.

156

Rosemary

Charming rosemary herb is the perfect potherb to have in your kitchen garden. The plant is one of the recognized herbs for its note-worthy health benefiting phyto-nutrients, anti-oxidants, and essential acids. Rosemary belongs within the family of *Lamiaceae,* of the genus, *Rosmarinus.* Its botanical name is *Rosmarinus officinalis.* It is thought to be originated in the Mediterranean region as a wild, strewing evergreen perennial shrub. Today, it is grown in nearly all parts of the Mediterranean region and Asia Minor as an important culinary herb. Rosemary flourishes in well-drained, alkaline soil. It prefers sunny condition and needs protection shelter from gusty winds. The plant reaches about 1.5-3 meters in height. Its bushy stems and downy young shoots are covered with about 1 inch long, narrow, needle-like aromatic leaves; dark green above and grayish underneath. The plant bears short racemes of small sea blue flowers appearing in early summer. The plant parts; flowers and leaves have an odor that is pungently aromatic and somewhat camphoraceous (camphor-like). Apart from culinary and medicinal purpose rosemary shoots, flowers and leaves are used in ceremonies such as weddings and festivals for decorating banquet halls as incense to ward off bad influences.

Health Benefits of Rosemary Herb

- ☆ Rosemary leaves contain certain phyto-chemical (plant derived) compounds that are known to have disease preventing and health promoting properties.
- ☆ The herb parts, especially flower tops contain phenolic anti-oxidant rosmarinic acid as well as numerous health benefiting volatile essential oils such as *cineol, camphene, borneol, bornyl acetate,* α*-pinene, etc.*These compounds are known to have rubefacient (counterirritant), anti-inflammatory, anti-allergic, anti-fungal and antiseptic properties.

Rosemary Herb (*Rosmarinus officinalis*), Fresh Leaves, Nutritive Value per 100 g

Principle	*Nutrient Value*	*Percentage of RDA*
Energy	131 Kcal	6.5 per cent
Carbohydrates	20.70 g	16 per cent
Protein	3.31 g	6 per cent
Total Fat	5.86 g	20 per cent
Cholesterol	0 mg	0 per cent
Dietary Fiber	14.10 g	37 per cent
Vitamins		
Folates	109 µg	27 per cent
Niacin	0.912 mg	6 per cent
Pantothenic acid	0.804 mg	16 per cent
Pyridoxine	0.336 mg	26 per cent
Riboflavin	0.152 mg	12 per cent
Thiamin	0.036 mg	3 per cent
Vitamin A	2924 IU	97 per cent
Vitamin C	21.8 mg	36 per cent
Electrolytes		
Sodium	26 mg	2 per cent
Potassium	668 mg	14 per cent
Minerals		
Calcium	317 mg	32 per cent
Copper	0.301 mg	33 per cent
Iron	6.65 mg	83 per cent
Magnesium	91 mg	23 per cent
Manganese	0.960 mg	42 per cent
Zinc	0.93 mg	8.5 per cent

- ☆ Rosemary leaves provide just 131 calories per 100 g and contain no cholesterol. Apart from nutrients, this humble herb contains many noteworthy non-nutrient components such as dietary fiber (37 per cent of RDA).
- ☆ The herb is exceptionally rich in many B-complex groups of vitamin, such as folic acid, pantothenic acid, pyridoxine, riboflavin. It is one of the herbs contain high levels of folates; providing about 109 µg per 100 g (about 27 per cent of RDA). Folates are important in DNA synthesis and when given during the peri-conception period can help prevent neural tube defects in the newborn babies.
- ☆ Rosemary herb contains very good amounts of vitamin A, 2924 IU per 100 g; about 97 per cent of RDA. A few leaves a day in the diet, would contribute

enough of this vitamin. Vitamin A is known to have antioxidant properties and is essential for vision. It is also required for maintaining healthy mucus membranes and skin. Consumption of natural foods rich in vitamin A is known to help the body protect from lung and oral cavity cancers.

- Fresh rosemary leaves are a good source of antioxidant vitamin; vitamin-C containing about 22 mg per 100 g, about 37 per cent of RDA. The vitamin is required for the collagen synthesis in the body. Collagen is the main structural protein in the body required for maintaining the integrity of blood vessels, skin, organs, and bones. Regular consumption of foods rich in vitamin C helps the body protect from scurvy; develop resistance against infectious agents (boosts immunity) and help scavenge harmful, pro-inflammatory free radicals from the body.
- Rosemary herb parts, whether fresh or dried, are rich source of minerals like potassium, calcium, iron, manganese, copper, and magnesium. Potassium is an important component of cell and body fluids, which helps control heart rate and blood pressure. Manganese is used by the body as a co-factor for the antioxidant enzyme, *superoxide dismutase.*
- This herb is an excellent source of iron, contains 6.65 mg/100 g of fresh leaves (about 83 per cent of RDA). Iron, being a component of hemoglobin inside the red blood cells, determines the oxygen-carrying capacity of the blood.

Medicinal Uses

- Rosmarinic acid, a natural polyphenolic antioxidant found in rosemary, has been found to have anti-bacterial, anti-inflammatory, and anti-oxidant functions. Apart from the rosemary, other popular herbs like Sage, peppermint, oregano, thyme herbs also contain appreciable levels of rosmarinic acid.
- Rosemary oil which is distilled from the flowering tops contains volatile essential oil such as camphene, cineol, borneol, bornyl acetate and other esters. These compounds are known to have tonic, astringent, diaphoretic, and stimulant properties.
- Its herbal oil is also being used externally as a rubefacient to soothe painful ailments in gout, rheumatism and neuralgic conditions.
- Rosemary herb extractions when applied over the scalp known to stimulate the hair-bulbs and help prevent premature baldness. It forms an effectual remedy for the prevention of scurf and dandruff.
- Rosemary tea is a natural remedy for nervous headache, colds, and depression.

Safety Profile

Rosemary herb might cause abortion in pregnant woman when eaten in large amounts. Further, in some rare cases, rosemary oil products may cause allergic skin reactions. In toxic doses, rosemary has been found to cause kidney dysfunction, and might exacerbate existing neurological conditions like epilepsy, neuroses, etc.

157

Sage

Sharply flavoured, sage herb or garden sage is one of the popular pot herbs known since ancient Roman times. This legendary herb with numerous virtues, long held to be the guardian over all other herbs, has been in use in various traditional European and Chinese medicines for its health promoting and disease preventing properties. Herb sage is an evergreen perennial shrub commonly seen all over the Mediterranean and south-eastern Europe (Balkan) regions. Botanically, the plant belongs to the family of *Lamiaceae*, of the genus: *Salvia*. Scientific name: *Salvia officinalis.* The plant flourishes in well-drained alkaline soil under sunny conditions. It grows up to 75 cm height and feature woody, branching stems. Its aromatic leaves are grey-green, soft and pebble-like textured with fine hair-like filaments growing on either side. It bears violet-blue colour bunches of flowers in summer.

Several cultivars of sage are grown either for medicinal or for culinary purposes.

- ☆ Three-lobed sage (*S. fruiticosa* or *S. triloba*): Large perennial with lobed leaves, grown in Mediterranean countries for making popular sage tea.
- ☆ Pineapple sage (*S. rutilans*): Fresh leaves add flavour to desserts and drinks.
- ☆ Clary sage (*S. sclarea*): Strongly aromatic leaves generally used as throat gargle infusions and in perfumeries.
- ☆ Azure sage (*S. azurea*): The plant is large with blue flowers, used in Mexico as an herbal panacea.

Health Benefits of Sage Herb

- ☆ Sage herb parts have many notable plant-derived chemical compounds, essential oils, minerals, vitamins that are known to have disease preventing and health promoting properties.
- ☆ The primary biologically active component of common sage appears to be its essential oil, which chiefly contain ketones; *α-thujone, and β-thujone*. In

addition, sage leaf contains numerous other compounds, including *cineol, borneol, tannic acid;* bitter substances like *cornsole and cornsolic acid;* fumaric, chlorogenic, caffeic and nicotinic acids; nicotinamide; flavones; flavone glycosides and estrogenic substances. These compounds are known to have counter-irritant, rubefacient, anti-inflammatory, anti-allergic, anti-fungal and antiseptic properties.

Sage Herb (*Salvia officinalis*), Dried, Ground, Nutritive Value per 100 g

Principle	*Nutrient Value*	*Percentage of RDA*
Energy	315 Kcal	15.75 per cent
Carbohydrates	60.73 g	47 per cent
Protein	10.63 g	19 per cent
Total Fat	12.75 g	42.5 per cent
Cholesterol	0 mg	0 per cent
Dietary Fiber	40.3 g	106 per cent
Vitamins		
Folates	274 µg	63 per cent
Niacin	5.720 mg	35.75 per cent
Pyridoxine	2.690 mg	206 per cent
Riboflavin	0.336 mg	26 per cent
Thiamin	0.754 mg	68.5 per cent
Vitamin A	5900 mg	196 per cent
Vitamin C	32.4 mg	54 per cent
Vitamin E	7.48 mg	50 per cent
Vitamin K	1714.5 µg	1429 per cent
Electrolytes		
Sodium	11 mg	<1 per cent
Potassium	1070 mg	23 per cent
Minerals		
Calcium	1652 mg	165 per cent
Copper	0.757 mg	84 per cent
Iron	28.12 mg	351 per cent
Magnesium	428 mg	107 per cent
Manganese	3.133 mg	136 per cent
Zinc	4.70 mg	43 per cent
Phyto-nutrients		
Carotene-α	0 µg	—
Carotene-β	3485 µg	—
Crypto-xanthin-β	109 µg	—
Lutein-zeaxanthin	1895 µg	—

- ☆ Thujone is GABA and Serotonin (5-HT3) receptor antagonist. It enhances concentration, attention span and quickens the senses; hence sage infusion has long been recognized as "thinker's tea." Its effects help deal with grief and depression.
- ☆ Three lobe sages (*S. triloba*) composes flavone called salvigenin. Research studies found that vascular relaxant effect of salvigenin may offer benefits in the prevention of cardiovascular diseases.
- ☆ This herb is exceptionally very rich source of several B-complex groups of vitamins, such as folic acid, thiamin, pyridoxine and riboflavin many times higher than the recommended daily levels.
- ☆ The herb contains very good amounts of vitamin A and beta-carotene levels. 100 g dry ground herb provides 5900 IU; about 196 per cent of RDA. Vitamin A is a powerful natural antioxidant and is essential for eye-sight. It is also required for maintaining healthy mucus membranes and skin. Consumption of natural foods rich in vitamin A known to help the body protect from lung and oral cavity cancers.
- ☆ Fresh sage leaves are a good source of antioxidant vitamin, vitamin C; contain 32.4 or 54 per cent of RDA. Vitamin C helps in the synthesis of structural proteins like collagen. Adequate levels in the body help maintain integrity of blood vessels, skin, organs, and bones. Regular consumption of foods rich in vitamin-C helps the body protect from *scurvy*; develop resistance against infectious agents (boosts immunity) and scavenge harmful, pro-inflammatory free radicals from the body.
- ☆ Sage herb parts, whether fresh or dried, are rich sources of minerals like potassium, zinc, calcium, iron, manganese, copper, and magnesium. Potassium is an important component of cell and body fluids, which helps control heart rate and blood pressure. Manganese is used by the body as a co-factor for the antioxidant enzyme, *superoxide dismutase.*

Medicinal Uses

- ☆ The essential oil obtained from sage has been found to have acetylcholinesterase (Ach) enzyme inhibition activities. This help rose Ach levels in the brain. Ach enzyme improves concentration and may play a role in the treatment methods for memory loss associated with the disease like Alzheimer's.
- ☆ Sage oil, distilled from the flowering tops, contains volatile essential oil such as camphene, cineol, borneol, bornyl acetate and other esters. These compounds are known to have tonic, astringent, diaphoretic and stimulant properties.
- ☆ Sage herb oil is also being used externally as a rubefacient to soothe painful ailments like muscle stiffness, rheumatism, and neuralgic conditions.
- ☆ Further, its oil is used as blended massage oil and in the aromatic therapy. It helps to relieve nervousness, anxiety, headaches, stress, and fatigue.

Safety Profile

Sage herb when used in large amounts causes nervous irritation, convulsions and death. Hence its use is prohibited in known epileptic conditions. The herb should not be used in pregnancy as chemical compounds like thujone in it may cause uterine stimulation resulting in abortion.

158

Savory

Delicate, pleasantly aromatic savory (summer) is a very popular culinary herb widely used in eastern European and American cuisines. Botanically, this annual herb belongs within the mint family of *Lamiaceae,* and known scientifically as *Satureja hortensis*. The herb is well tolerant of different climatic conditions and requires good sunlight to flourish. It reaches about 40-60 cm in height. Its sister variety, winter-savory, known as *Satureja montana,* has pungent flavour and hence, less preferred in cooking.

The plant bears dark green coloured smooth leaves and small purple colour flowers. Winter savory has similar growth characters; however, it features more branching; hard and woody stems. Several wild cultivars exist apart from the garden savory. *Satureja douglasii* or yerba buena is one of the herbs has been in use by the indigenous Americans of California and Alaskan regions as an herbal tea. *Satureja thymbra* is another wild herb that grows in Mediterranean and Middle Eastern countries and is known for its high carvacrol or thymol content.

Health Benefits of Savory

- ☆ Savory leaves and tender shoots contain incredibly high-quality chemical compounds that are known to have been anti-oxidant, disease preventing and health promoting properties. In addition, dietary fiber in this herb helps reduce LDL or bad cholesterol while increasing HDL or good cholesterol levels.
- ☆ Savory leaves contain many essential volatile-oil phenols such as thymol and carvacrol, as well as*linalool, camphene, caryophyllene, terpineol, myrcene, and other terpenoids.*
- ☆ Thymol, one of the important essential oils, has scientifically been found to have antiseptic, anti-fungal characteristics.

Savory (*Satureja hortensis*), Dry, Ground, Nutrient Value per 100 g

Principle	*Nutrient Value*	*Percentage of RDA*
Energy	272 Kcal	13.5 per cent
Carbohydrates	68.73 g	53 per cent
Protein	6.73 g	12 per cent
Total Fat	5.91 g	30 per cent
Cholesterol	0 mg	0 per cent
Dietary Fiber	45.7 g	120 per cent
Vitamins		
Niacin	4.080 mg	25.5 per cent
Pyridoxine	1.810 mg	139 per cent
Riboflavin	0.471 mg	36 per cent
Thiamin	0.366 mg	30.5 per cent
Vitamin A	5310 IU	177 per cent
Vitamin C	50 mg	83 per cent
Electrolytes		
Sodium	24 mg	1.5 per cent
Potassium	1051 mg	22 per cent
Minerals		
Calcium	2132 mg	210 per cent
Copper	0.847 mg	94 per cent
Iron	37.88 mg	474 per cent
Magnesium	377 mg	94 per cent
Manganese	6.100 mg	265 per cent
Phosphorus	140 mg	20 per cent
Selenium	4.6 µg	8 per cent
Zinc	4.30 mg	39 per cent

- ✰ In addition, another phenolic compound, carvacrol in savory inhibits the growth of several bacteria strains like *E. coli*, and *Bacillus cereus*. Carvacrol, therefore, has been used as healthy food additive for its anti-bacterial properties, besides it gives pleasant tangy taste and marjoram like a smell to the food.
- ✰ Savory herb is an excellent source of minerals and vitamins that are essential for optimum health. Its leaves and tender shoots are one of the richest sources of potasium, iron, calcium, magnesium, manganese, zinc and selenium. Potassium is an important component of cell and body fluids that helps controlling heart rate and blood pressure. Manganese is used by the body as a co-factor for the antioxidant enzyme, *superoxide dismutase*. Iron is required for red blood cell formation.

- ☆ The herb is also a rich source of many important vitamins such as B-complex group vitamins, vitamin-A, vitamin-C, niacin, thiamin and pyridoxine.
- ☆ Dry savory provides 1.810 mg of vitamin B-6 or pyridoxine; furnishing about 130 per cent of RDA. Pyridoxine keeps up GABA (soothing neurotransmitter) levels in the brain, which has stress buster function.
- ☆ Vitamin C helps the body develop resistance against infectious agents and scavenge harmful, pro-inflammatory free radicals.
- ☆ Vitamin A is a fat-soluble vitamin and antioxidant that are required maintaining healthy mucus membranes and skin and is also essential for good eye-sight. Consumption of natural fruits rich in flavonoids like vitamin A, carotenes helps protect from lung and oral cavity cancers.

Medicinal Uses

- ☆ Savory herb contains many important essential oils, which are found to have anti-bacterial and anti-fungal applications.
- ☆ As in thyme, distilled tepid savory water is used for throat gargling to help relieve sore throat and bronchitis symptoms.
- ☆ The constituents in this herb part, especially the flowering shoots, has been anti-septic, carminative (anti-flatulence), digestive (helps digestion), expectorant (help clear throat), stomachic and anti-rheumatic (relieves joint pain) functions.

159

Spearmint

Spearmint herb or *garden mint* or *common mint* has long been reputed for its characteristic aroma it imparts to the recipes it added to. The least pungent and subtle among the species of mint family, this unique herb is one of chef's main culinary favourites. Mint herb botanically belongs to the family of *Lamiaceae*, of the genus: *Mentha*. Scientific name: *Mentha spicata*. The plant is a branching perennial herb of Mediterranean origin. It is widely used across Europe and in large parts of Asia and Africa in flavoured drinks, salads and confectionary and as a garnish to recipes. The herb grows easily in fertile, moist and loose soil with underground runners. In general, the divisions are planted to propagate. The plant reaches about 75 cm in height, and bears oppositely arranged leaves all along the thick square stem. The leaves are deep green, deeply-veined, oval shape with pointed ends and serrated margins. Slim pointed spikes of mauve flowers appear during late summer. There are at least 20 species of mint and their hybrids exist, most of them difficult to classify because of their variability and readiness to hybridize among each other. Here are some mint herbs apart from the popularpeppermint and water mints;

- ✰ Pineapple mint (*M. suaveolens*),
- ✰ Ginger mint (*M. x gentilis*),
- ✰ Japanese mint (*M. arvensis* var. *piperascens*),
- ✰ Corn mint (*M. arvensis*),
- ✰ Bergamot or horsemint (*M. piperita* var.*citrata*).

Health Benefits of Spearmint

- ✰ Spearmint is pleasantly aromatic herb packed with numerous health benefiting vitamins, antioxidants and phyto-nutrients.
- ✰ The leaves and herb parts contain essential oil, menthol. Unlike in peppermint, spearmint leaves composes only small amounts of menthol,

0.5 per cent compared to the 40 per cent in peppermint. Less menthol content would make this herb least pungent and subtly fragrant herb in the mint family.

Spearmint (*Mentha spicata*), Fresh, Nutritional Value per 100 g

Principle	*Nutrient Value*	*Percentage of RDA*
Energy	44 Kcal	2 per cent
Carbohydrates	8.41 g	6.5 per cent
Protein	3.29 g	6 per cent
Total Fat	0.73 g	3 per cent
Cholesterol	0 mg	0 per cent
Dietary Fiber	6.8 g	18 per cent
Vitamins		
Folates	105 µg	26 per cent
Niacin	0.948 mg	6 per cent
Pyridoxine	0.158 mg	12 per cent
Riboflavin	0.175 mg	13.5 per cent
Thiamin	0.078 mg	6.5 per cent
Vitamin A	4054 IU	135 per cent
Vitamin C	13.3 mg	22 per cent
Electrolytes		
Sodium	30 mg	2 per cent
Potassium	458 mg	64 per cent
Minerals		
Calcium	199 mg	20 per cent
Copper	0.240 mg	75 per cent
Iron	11.87 mg	148 per cent
Magnesium	63 mg	16 per cent
Manganese	1.118 mg	48.5 per cent
Zinc	1.09 mg	10 per cent

- ☆ The herb has low calories (about 43 calories per 100 g) and contains zero cholesterol.
- ☆ The chief essential oil in spearmint is menthol. Other important chemical components of spearmint are *α-pinene, β-pinene, carvone, cineole, linalool, limonene, myrcene* and *caryophyllene*. These compounds in mint help relieve fatigue and stress.
- ☆ The herb parts are also very good in minerals like potassium, calcium, manganese, iron (148 per cent of RDA), and magnesium. Iron is required for enzymes in cellular metabolism and synthesis of hemoglobin. Potassium is an important component of cell and body fluids that helps control heart

rate and blood pressure. Manganese is used by the body as a co-factor for the antioxidant enzyme superoxide dismutase.

- ☆ Further, the herb is also rich in many antioxidant vitamins, including vitamin A (provides 4054 IU or 135 per cent of RDA), beta-carotene, vitamin C, folates (26 per cent of RDA), vitamin B-6 (pyridoxine), riboflavin and thiamin.

Medicinal Uses

Almost all parts of spear mint herb found a place in various traditional as well in modern medicine.

- ☆ The herb decoction is an excellent remedy for minor ailments such as headaches, nervous strain, fatigue and stress, as well as for the respiratory problems; helping with asthma, bronchitis and catarrh.
- ☆ It is very useful to deal with digestive problems, including nausea, flatulence and hiccups as it relaxes the stomach muscles.
- ☆ The essential oil, menthol, has analgesic, local anaesthetic and counterirritant properties. Menthol also found application in the preparation of toothpaste and mouth refreshers.
- ☆ On the skin, when used as cream or lotion, it may help relieve the itching of pruritis, dermatitis, and hives.
- ☆ Spearmint oil is used as blended massage oil and in the aromatic therapy to help relieve headaches, stress, fatigue, and nervous conditions and to relieve itching.
- ☆ Spearmint tea can be used safely in pregnancy. In women, it helps reduce unwanted hairs through its anti-androgenic properties.

160

Stevia

Stevia plant is a small, sweet-leaf herb of South American origin used by native Guarani tribes of Paraguay since centuries. Is stevia safe alternative to common sugar in carbohydrate-controlled diets? Does it have healthy nutrient profile safe enough for human consumption as in other herbs such as basil, mint, thyme etc? Recent scientific trials firmly establish that this sweet-leaf herb has, in fact, many health benefiting plant-derived phyto-chemical compounds that help control blood sugar, cholesterol, blood pressure in addition to its use as natural sweetener. Together with the rise in demand for low-calorie food alternatives, stevia has drawn the attention of health conscious fitness lovers all over the planet.

Stevia is a small perennial herb belongs to the *Asteraceae* family of the genus *Stevia*. Its scientific name is *Stevia rebaudiana.* Some commonly known names are *honey leaf plant, sweet chrysanthemum, sweetleaf stevia, sugarleaf* etc. Stevia plant grows 2-4 feet in height with slender, branched stems and flourishes well all over temperate and some parts of tropical regions. It is being cultivated as commercial crop in Japan, China, Thailand, Paraguay, and Brazil. Today, China is the leading exporter of stevia products. Almost all the parts of the plant taste sweet; however, the sweet glycosides are typically concentrated in its dark-green serrated leaves. The Guarani Indians of Paraguay have used its leaves to sweeten drinks and foods and in various traditional medicines. In the modern times, it is widely used in foods in many parts India, East Asian region and South American countries.

Health Benefits of Stevia

- ☆ Stevia herb parts are very low in calories. Parts by parts its dry leaves are roughly 40 times sweeter than sugar. The sweetness of sugar in stevia is due to several glycoside compounds including stevioside, steviolbioside, rebaudiosides A-E, and dulcoside.

- ☆ *Stevioside* is non-carbohydrate glycoside compound. Hence, it lack of properties that sucrose and other carbohydrates have. Stevia extracts, like rebaudioside-A, are found to be 300 times sweeter than sugar. In contrast to sugar, however, stevia extracts have several unique properties such as long shelf life, high temperature tolerance, non-fermentative; but contain near-zero calories.
- ☆ In addition, stevia plant has many sterols and antioxidant compounds like *triterpenes, flavonoids* and *andtannins*. Some of flavonoid polyphenolic anti-oxidant phyto-chemicals present in stevia is *kaempferol, quercetin, chlorogenic acid, caffeic acid, isoquercitrin, isosteviol*...etc. Studies found that kaempferol can reduce risk of pancreatic cancer by 23 per cent (American journal of epidemiology).
- ☆ Chlorgenic acid reduces enzymatic conversion of glycogen to glucose in addition to decreasing absorption of glucose in the gut. Thus, it helps reduce blood sugar levels. The lab studies also confirm a reduction in blood glucose levels and an increase in the liver concentrations of glucose-6-phosphate and of glycogen.
- ☆ Certain glycosides in stevia extract have been found to dilate blood vessels, increase sodium excretion, and urine output. In effect, stevia, at slightly higher doses than as sweetener, can help lower blood pressure.
- ☆ Being a non-carbohydrate sweetener, stevia would not favour the growth of *Streptococcus mutans* bacteria in the mouth which is attributed to be a causative agent for dental caries and tooth cavities. On the other hand, certain compounds in stevia are rather found to inhibit caries causing bacteria in the mouth.
- ☆ In addition, being a herb, stevia contain many vitals minerals, vitamins that are selectively absent in the artificial sweeteners.

Stevia Plant Uses in Traditional Medicine

- ☆ Stevia extract has been in use by native South Americans to reduce weight; to treat wound infections, inflammatory conditions, swelling in the legs and as a tonic to treat depression.

Safety Profile

Stevia plant and its processed products are not being used in the European Union countries for their suspected mutagenic effects. In the USA stevia leaf and its extract is limited to use legally as dietary supplements. However, rebaudioside-A, a processed stevia glycoside can be safely used in many of these countries.

However, in 2006, the World Health Organization (WHO) observed the following statement regarding the use of stevia plant: "stevioside and rebaudioside A are not genotoxic *in-vitro* or *in-vivo* and that the genotoxicity of steviol and some of its oxidative derivatives *in vitro* is not expressed *in vivo*." In conclusion, stevia and its products are being used in some advanced countries like Japan and for centuries by Guarani tribes of South America and found to be harmless for human use even in pregnancy.

161

Sweet Marjoram

Sweet marjoram, also known as knotted marjoram, is one of the most popular Mediterranean herbs used for its culinary and medicinal values since ancient times. The herb has the delicate sweet-flavour and mildly pungent (spicy) taste comparatively milder than oregano (*Origanum vulgare*). Botanically marjoram is a member of *Labiatae* family of the genus: *Origanum*. Its scientific name: *Origanum majorana*. The Labatiacea family also includes some of commonly known herbs and spices such as dill, anise,fennel, cumin, etc.

Sweet marjoram is native to Asia Minor (Turkey); however, it grows in plentiful all over the Mediterranean and southern parts of the Europe. This herb plant is a small woody-stemmed shrub grows proficiently in well-drained, lime soil. It reaches about 75 cm in height; feature a hairy stem with soft oval shaped dark-green leaves, and tiny pinkish-white flowers. In general, its leaves are harvested from the plant just after the flower buds form but before flowering. To blanch, hang the harvested stems in a dark, dry room for about 7-10 days, then strip the leaves from the stems and store in air tight container. Several related species of marjoram grown either for culinary or therapeutic purposes. Pot marjoram (*O. onites*) has white or pink flowers. Winter marjoram (*O. heracleoticum*) has the strong flavour similar to that of wild marjoram (*O.vulgare*).

Health Benefits of Marjoram

- ☆ Sweet marjoram is one of very popular herb, especially in Mediterranean region. It was used in ancient times by the Greeks to make wreaths and garlands and was believed as a symbol of peace, harmony and happiness.
- ☆ The herb contains many notable phyto-nutrients, minerals and vitamins that are essential for optimum health and wellness.
- ☆ The herb parts contain certain chemical constituents like *eugenol sabinene*, α *-terpinene, cymene, terpinolene, linalool, cis-sabinene hydrate, linalyl acetate,*

terpinen-4-ol and *terpineol*. These compounds are known to have anti-inflammatory and anti-bacterial properties.

Sweet Marjoram Herb (*Origanum majorana*), Dried Leaves, Nutritive Value per 100 g

Principle	*Nutrient Value*	*Percentage of RDA*
Energy	271 Kcal	13.5 per cent
Carbohydrates	60.56 g	46.5 per cent
Protein	12.66 g	22 per cent
Total Fat	7.04 g	23 per cent
Cholesterol	0 mg	0 per cent
Dietary Fiber	40.3 g	106 per cent
Vitamins		
Folates	274 μg	68.5 per cent
Niacin	0.902 mg	25.75 per cent
Pantothenic acid	0.209 mg	4 per cent
Pyridoxine	1.190 mg	91.5 per cent
Riboflavin	0.316 mg	24 per cent
Thiamin	0.289 mg	24 per cent
Vitamin A	8068 IU	269 per cent
Vitamin C	51.4 mg	86 per cent
Vitamin E	1.69 mg	11 per cent
Vitamin K	621.7 μg	518 per cent
Electrolytes		
Sodium	77 mg	5 per cent
Potassium	1522 mg	32 per cent
Minerals		
Calcium	1990 mg	200 per cent
Copper	1.133 mg	126 per cent
Iron	82.71 mg	1034 per cent
Magnesium	346 mg	86.5 per cent
Manganese	5.433 mg	236 per cent
Zinc	3.60 mg	33 per cent
Phyto-nutrients		
Carotene-β	4806 μg	—
Cryptoxanthin-β	70 μg	—
Lutein-zeaxanthin	1895 μg	—

☆ Fresh herb has high levels of vitamin C (ascorbic acid); provide 51.4 μg or 86 per cent of RDA per 100 g. Vitamin C is one of the powerful natural anti-oxidant help remove harmful free radicals from the body. Ascorbic acid also has immune booster, wound healing and anti-viral effects.

- Marjoram herb contains exceptionally high levels of *beta-carotene, vitamin A, cryptoxanthin, lutein* and *zeaxanthin*. Carotenes, xanthins and lutein are powerful flavonoid anti-oxidants. Together, these compounds help act as protective scavengers against oxygen-derived free radicals and reactive oxygen species (ROS) that play a role in ageing and various disease processes.
- Vitamin A is known to have antioxidant properties and is essential for healthy eye-sight. It is also required for maintaining healthy mucus membranes and skin. Consumption of natural foods rich in vitamin A and carotenes are known to help the body protect from lung and oral cavity cancers.
- Zeaxanthin, an important dietary carotenoid, selectively absorbed into the retinal macula lutea in the eyes where it is thought to provide antioxidant and protective light-filtering functions. It has proven beneficial action against age-related macular disease (ARMD) especially in the elderly.
- Sweet marjoram is one of the richest herbal sources for vitamin K; provide about 518 per cent of recommended daily intake. Vitamin-K has a potential role in bone mass building by promoting osteotrophic activity in the bones. It also has established role in the treatment of Alzheimer's disease by limiting neuronal damage in the brain.
- Marjoram herb has a good amount of minerals like iron, calcium, potassium, manganese, copper, zinc and magnesium. Potassium is an important component of cell and body fluids, which helps control heart rate and blood pressure. Manganese is used by the body as a co-factor for the antioxidant enzyme, *superoxide dismutase.*
- Furthermore, its leaves are an excellent source of iron, contains 82.71 mg/ 100 g (about 1034 per cent of RDA). Iron is an important co-factor for *cytochrome oxidase enzyme* in the cellular metabolism. In addition, being a component of hemoglobin inside the red blood cells, it determines the oxygen-carrying capacity of the blood.

Medicinal Uses

- Marjoram herb parts contain many health benefiting compounds in essential oil such as terpenes, linalool, terpineol, carvocrol, linalyl acetate, ocimene, cadinene, geranyl cetate, citral, eugenol, etc. These compounds are known to have anti-inflammatory and anti-bacterial properties. Marjoram oil has been used to treat minor ailments like flatulence, muscular pain and rheumatism.
- Eugenol in sweet marjoram (10 per cent of eugenol) has been found to have anti-inflammatory function by acting against the enzyme *cyclooxygenase* (COX), which mediates inflammatory cascade in the body. This enzyme-inhibiting effect of the eugenol makes it an important remedy for symptomatic relief in individuals with inflammatory health problems like rheumatoid arthritis, osteoarthritis and inflammatory bowel conditions.

- When taken in moderation, it helps relieve menstruation pain and leucorrhoea. Since it increases menstruation flow the herb should not be taken during pregnancy and in menorrhagia conditions.
- Essential oils in marjoram herb have also been found to have anti-fungal and anti-bacterial functions by inhibiting many pathogenic bacteria like *Staphylococcus, E.coli, Shigella, Proteus,* and *Pseudomonas.*
- Sweet marjoram tea helps relieve nausea, flatulence and is thought to have mild antiseptic functions.

162

Tarragon

Tarragon herb or *dragon wort* is a very popular culinary herb used as flavouring agent, especially in the Mediterranean cuisines. This aromatic perennial growing plant is rich in phytonutrients as well antioxidants that help promote health and prevent diseases. Botanically, tarragon belongs within the family of *Asteraceae* of the genus: *Artemisia* and known scientifically as *Artemisia dracunculus*. The herb is thought to be originated in Central Asia, probably in Siberia. Russian tarragon (*A. dracunculoides*) is a more robust; closely related species. It, however, is quite inferior in flavour to its Mediterranean counterpart and hence, less preferred in cooking. This herb is small shrub featuring slim woody branching stems that reach up to a meter in height. It grows well in rich sandy soil with adequate sunlight. Its leaves are smooth, dark green with pointed ends.

Health Benefits of Tarragon Herb

- This exquisite herb is rich in numerous health benefiting phyto-nutrients that are indispensable for optimum health.
- The main essential oils in tarragon are estragole (methyl chavicol), cineol, ocimene and phellandrene.
- Tarragon has been used as a traditional remedy to stimulate appetite and alleviate anorexic symptoms.
- Scientific studies suggest that poly-phenolic compounds in this herb help lower blood-glucose levels.
- Fresh tarragon herb is one of the highest antioxidant value food sources among the common herbs. Its total measured ORAC (Oxygen radical absorbance capacity) value is 15,542 trolex equivalents (TE) per 100 g.
- Laboratory studies on tarragon extract shows certain compounds in them inhibit platelet activation, preventing platelet aggregation and adhesion to

the blood vessel wall. It thus helps prevent clot formation inside tiny blood vessels of heart and brain protecting from heart attack and stroke.

Tarragon Herb (*Artemisia dracunculus*), Dried, Nutritional Value per 100 g

Principle	*Nutrient Value*	*Percentage of RDA*
Energy	295 Kcal	15 per cent
Carbohydrates	50.22 g	38 per cent
Protein	22.77 g	40 per cent
Total Fat	7.24 g	24 per cent
Cholesterol	0 mg	0 per cent
Dietary Fiber	7.4 g	19 per cent
Vitamins		
Folates	274 µg	68.5 per cent
Niacin	8.950 mg	56 per cent
Pyridoxine	2.410 mg	185 per cent
Riboflavin	1.339 mg	103 per cent
Thiamin	0.251 mg	21 per cent
Vitamin A	4200 IU	140 per cent
Vitamin C	50.0 mg	83 per cent
Electrolytes		
Sodium	62 mg	4 per cent
Potassium	3020 mg	64 per cent
Minerals		
Calcium	1139 mg	114 per cent
Copper	0.677 mg	75 per cent
Iron	32.30 mg	403 per cent
Magnesium	347 mg	87 per cent
Manganese	7.967 mg	346 per cent
Zinc	3.90 mg	35 per cent

- ✰ The herb is very rich source of vitamins such as vitamin-C, vitamin-A as well as B-complex group of vitamins such as folates, pyridoxine, niacin, riboflavin, etc., that function as antioxidant as well as co-factors in metabolism.
- ✰ Tarragon is a notably excellent source of minerals like calcium, manganese, iron, magnesium, copper, potassium, and zinc. Manganese is used by the body as a co-factor for the antioxidant enzyme, *superoxide dismutase*. Iron is essential for cellular respiration (co-factors for cytochrome-oxidase enzyme) and blood cell production.

Medicinal Uses

- ☆ Tarragon herb has been used in various traditional medicines for stimulating the appetite and as a remedy for anorexia, dyspepsia, flatulence and hiccups.
- ☆ The essential oil, eugenol in the herb has been in therapeutic use in dentistry as a local-anesthetic and antiseptic for toothache complaints.
- ☆ Tarragon tea may help cure insomnia.

Safety Profile

- ☆ Tarragon essential oil, estragole, may cause skin irritation in some individuals.
- ☆ In addition, according to the "European commission scientific committee on food" estragole and its metabolite 1'-hydroxy estragole may induce hepatic tumor in laboratory animal. Therefore, reductions in tarragon exposure and restrictions in use levels are thus recommended

163

Thyme Herb

Thyme herb is packed with numerous health benefiting phyto-nutrients (plant derived compounds), minerals and vitamins that are essential for wellness. One of the popular among culinary herb plants, thyme originally is native to southern Europe and Mediterranean regions. Botanically, it belongs to the family of *Limiaceae* of the genus of *thymus,* which includes many subspecies among which the most popular variety is *Thymus vulgaris* or *French thyme.* The thyme plant is a perennial shrub with thin woody base, and square stems. It reaches about 15 to 30 cm in length, featuring very small, light green coloured, paler underneath, slightly curved aromatic leaves. Tiny, fragrant rich, lilac or white colour flowers appear in summer. The other commonly grown varieties of thyme are lemon thyme (*T.x citriodora*), caraway thyme (*T. herba barona*) and wild thyme (*T. septyllum*). Either leaves as well as flowering tips, fresh or dried used for culinary purposes.

Health Benefits of Thyme Herb

- ☆ Thyme contains many active principles that are found to have disease preventing and health promoting properties.
- ☆ Thyme herb contains thymol, one of the important essential oils, which scientifically has been found to have antiseptic, anti-fungal characteristics. The other volatile oils in thyme include *carvacolo, borneol*and *geraniol.*
- ☆ Thyme contains many flavonoid phenolic antioxidants like *zeaxanthin, lutein, pigenin, naringenin, luteolin,* and *thymonin*. Fresh thyme herb has one of the highest antioxidant levels among herbs, a total ORAC (Oxygen Radical Absorbance Capacity) value of 27426-µmol TE/100 g.
- ☆ Thyme is packed with minerals and vitamins that are essential for optimum health. Its leaves are one of the richest sources of potassium, iron, calcium, manganese, magnesium, and selenium. Potassium is an important component of cell and body fluids that helps controlling heart rate and

blood pressure. Manganese is used by the body as a co-factor for the antioxidant enzyme, *superoxide dismutase*. Iron is required for red blood cell formation.

Thyme Herb (*Thymus vulgaris*), Fresh Leaves, Nutritive Value per 100 g. ORAC value 27426

Principle	*Nutrient Value*	*Percentage of RDA*
Energy	101 Kcal	5 per cent
Carbohydrates	24.45 g	18 per cent
Protein	5.56 g	10 per cent
Total Fat	1.68 g	8.4 per cent
Cholesterol	0 mg	0 per cent
Dietary Fiber	14.0 g	37 per cent
Vitamins		
Folates	45 µg	11 per cent
Niacin	1.824 mg	11 per cent
Pantothenic acid	0.409 mg	8 per cent
Pyridoxine	0.348 mg	27 per cent
Riboflavin	0.471 mg	36 per cent
Thiamin	0.48 mg	4 per cent
Vitamin-A	4751 IU	158 per cent
Vitamin-C	160.1 mg	266 per cent
Electrolytes		
Sodium	9 mg	0.5 per cent
Potassium	609 mg	13 per cent
Minerals		
Calcium	405 mg	40.5 per cent
Iron	17.45 mg	218 per cent
Magnesium	160 mg	40 per cent
Manganese	1.719 mg	75 per cent
Manganese	106 mg	15 per cent
Zinc	1.81 mg	16.5 per cent
Phyto-nutrients		
Carotene-β	2851 µg	—

- ✰ The herb is also a rich source of many important vitamins such as B-complex vitamins, beta carotene, vitamin A, vitamin K, vitamin E, vitamin C and folic acid.
- ✰ Thyme provides 0.35 mg of vitamin B-6 or pyridoxine; furnishing about 27 per cent of daily recommended intake. Pyridoxine keeps up GABA

(beneficial neurotransmitter in the brain) levels in the brain, which has stress buster function.

- ✰ Vitamin C helps the human body develop resistance against infectious agents and scavenge harmful, pro-inflammatory free radicals.
- ✰ Vitamin A is a fat-soluble vitamin and antioxidant that is required maintaining healthy mucus membranes and skin and is also essential for vision. Consumption of natural foods rich in flavonoids like vitamin A and beta-carotene helps protect from lung and oral cavity cancers.

Medicinal Uses

- ✰ Thyme contains many important essential oils, which are found to have antiseptic and anti-fungal applications.
- ✰ Throat gargling with tepid thyme water or drinking few sips of thyme tea may help relieve coughs, sore throat, and bronchitis symptoms.
- ✰ In the pharmacy, thyme based formulations has been used as an anti-septic mouthwash in the treatment of caries and gingivitis.

164

Turmeric

Never mind if you have to stain your clothes turmeric stained, but make sure you add this exotic root-herb in the recipes! The herb part is actually an underground rhizome (root), which compose of unique chemical compounds that impart such an intense flavour, colour and distinctive fragrance to it. Binomially, this popular herb belongs to the ginger or *Zingiberaceae* family of root herbs and the genus; *Curcuma*. Scientific name: *Curcuma longa.* Its roots as well as leaves have long been used in traditional Indian and Chinese medicines for their demonstrated anti-inflammatory (painkiller), anti-oxidant, and anti-cancer properties.

Turmeric is native to sub-Himalayan mountain region and now grown widely in many parts of the tropical and subtropical regions as an important commercial crop. The plant grows to a meter in height and features aromatic miniature plantain-like leaves. Turmeric root features dark brown skin on the exterior and deep orange-yellow flesh internally. Its leaves as well as rhizome features unique flavour and fragrance. Taste is described as mild peppery to warm and bitter while its fragrance is sweet and pleasant, slightly reminiscent of a mix of orange-zest and ginger to which it is related. Once harvested, the root is boiled, dried, and ground to make its distinctive bright yellow spice powder. Turmeric plants produce no seeds, and only reproduce via its spreading rhizomes.

Health Benefits of Turmeric

- ✰ The rhizome has been in use since antiquity for its anti-inflammatory (painkiller), carminative, anti-flatulent and anti-microbial properties.
- ✰ The herb contains health benefiting essential oils such as *termerone, curlone, curumene, cineole,* and *p-cymene.*
- ✰ Curcumin, a poly-phenolic compound, is the principal pigment that imparts deep orange colour to the turmeric. In vitro and animal studies have

suggested the *curcumin* may have anti-tumor, antioxidant, anti-arthritic, anti-amyloid, anti-ischemic, and anti-inflammatory properties.

Turmeric (*Curcuma longa*), Nutritive Value per 100 g

Principle	*Nutrient Value*	*Percentage of RDA*
Energy	354 Kcal	17 per cent
Carbohydrates	64.9 g	50 per cent
Protein	7.83 g	14 per cent
Total Fat	9.88 g	33 per cent
Cholesterol	0 mg	0 per cent
Dietary Fiber	21 g	52.5 per cent
Vitamins		
Folates	39 µg	10 per cent
Niacin	5.140 mg	32 per cent
Pyridoxine	1.80 mg	138 per cent
Riboflavin	0.233 mg	18 per cent
Vitamin A	0 IU	0 per cent
Vitamin C	25.9 mg	43 per cent
Vitamin E	3.10 mg	21 per cent
Vitamin K	13.4 µg	11 per cent
Electrolytes		
Sodium	38 mg	2.5 per cent
Potassium	2525 mg	54 per cent
Minerals		
Calcium	183 mg	18 per cent
Copper	603 µg	67 per cent
Iron	41.42 mg	517 per cent
Magnesium	193 mg	48 per cent
Manganese	7.83 mg	340 per cent
Phosphorus	268 mg	38 per cent
Zinc	4.35 mg	39.5 per cent

- ☆ This popular herb contains no cholesterol; however, it is rich in anti-oxidants and dietary fiber, which helps to control blood LDL or "bad cholesterol" levels.
- ☆ It is very rich source of many essential vitamins such as pyridoxine (vitamin B6), choline, niacin, and riboflavin, etc. 100 g herb provides 1.80 mg or 138 per cent of daily-recommended levels of pyridoxine. Pyridoxine is used in the treatment of homocystinuria, sideroblastic anemia and radiation sickness. Niacin helps prevent "pellagra" or dermatitis.

- ☆ Fresh root contains very good levels of vitamin-C. 100 of root compose of 23.9 mg of this vitamin. Vitamin C is a water-soluble vitamin and a powerful natural anti-oxidant, which helps the body develop immunity against infectious agents, and remove harmful free oxygen radicals.
- ☆ Turmeric contains very good amounts of minerals like calcium, iron, potassium, manganese, copper, zinc, and magnesium. Potassium is an important component of cell and body fluids that helps controlling heart rate and blood pressure. Manganese is used by the body as a co-factor for the antioxidant enzyme, superoxide dismutase. Iron is an important co-factor for cytochrome oxidase enzymes at cellular level metabolisms and required for red blood cell (RBC's) productions.
- ☆ Turmeric is one of the readily available, cheap herbs that contain notable phyto-nutrients profile. At 159277 µmol TE/100 g, its total-ORAC value or anti-oxidant strength is one of the highest among known herb and spice species.

Medicinal Uses

- ☆ Research studies have suggested that *Curcumin*, a poly-phenolic compound, found in this herb may inhibit the multiplication of tumor cells, including multiple myeloma, pancreatic and colon cancer.
- ☆ It contains health benefiting essential oils such as *termerone, curlone, curumene, cineole, and p-cymene.* These compounds have applications in cosmetic industry.
- ☆ *Curcumin*, along with other antioxidants, has been found to have anti-amyloid and anti-inflammatory properties. Thus; it is effective in preventing or at least delaying the onset of Alzheimer's disease.
- ☆ The root herb contains no cholesterol; however, it is rich in anti-oxidants, and dietary fiber. Together, they help to control blood cholesterol levels, offer protection from coronary artery disease and stroke risk.
- ☆ Early laboratory studies have been suggestive that turmeric is liver protective, anti-depressant, anti-retroviral effects.
- ☆ It has been in use since a very long ago as an important ingredient in traditional Chinese and ayurvedic medicines for its anti-microbial, anti-inflammatory, carminative, and anti-flatulent properties.

Bibliography

Anderson, R. A. (1962). A note on wet-milling of high amylose corn containing 75 per cent-amylose starch, *Cereal Chem.*, 39, 406-408.

Anderson, R. A. (1963). Wet-milling Properties of grains: Bench- Scale Study, *Cereal Sci. Today*, 8, No.6, 190-92, 195, 221.

Anon, 1964, Joint United States-Canadian tables of feed composition, *Natl. A cad. Sci.*, Nath. Res. Council Publication 1232.

Anonymous, (1982). Prevention of Food Adulteration Act, Government of India Publication, New Delhi.

Arya, P.R. and Rastogi, P.P. (1984). Fruit-Vegetable Preservation at Home (In Hindi). Directorate of Translation and Publication, G.B. Pant University of Agriculture and Technology, Pantnagar, Nainital.

Athalye (1992). Plastics in Packaging. Tata McGraw-Hill Publishing Company Limited, New Delhi.

Atkins, T. A., and Geddes, W. F. (1939). The relationship between protein content and strength of flours in gluten enriched flours, *Cereal Chem.*, 16, 223-31.

Autrey, H. S. (1953). Effect of variables upon milling yields, *Rice J. Annual.*

Bailey, A. E. (1948). Cotton Seed and Cotton Products; Inter-science Publishers, Inc., New York.

Bailey, A. E. (1951). Oil and Fat Products. Inter-science Publishers, Inc., New York.

Baird, P. D., Mac Masters, M. M., and Rist, C. E. (1950). Studies on a rapid test for the viability of corn for industrial use, *Cereal Chem.;* 27, 508-13.

Banerjee, S. N. (1975). Report on the evaluation of the modern dal mill, Department of Food, Govt. of India.

Banwart, G.J. (1979). Basic Food Microbiology. The AVI Publishing Company, Inc., Westport, Connecticut.

Bender, A. E. (1973). Composition of infant foods, In Nutritional and Dietetic Foods, International Textbook Co. Ltd., London, U. K., p. 64.

Bhat, C.M., Sharma, R.N. and Sehgal, S. (1982). A Manual on Food Preservation at Home. Directorate of Publication, Haryana Agricultural University, Hissar.

Bishop, W., and Dustin, J. (1951). Topnotch wet com mill house, *Food Ind.*, 23, No.3, 121-24, 211-12.

Board, P.W. (1988). Quality Control in Fruit and Vegetable Processing. FAO Food and Nutrition Paper 39. Food and Agriculture Organisation of the United Nations, Rome.

Bose, T K. and M. K. Mitra, Eds. (1986). Fruits of India Tropical and Subtropical. Naya Prokash, Culcutta.

Bose, T K. and M.G. Som., Eds. (1986). Vegetable Crops in India, Naya Prokash, Culcutta.

Boundy, J. A., Woychik, J. H., Dimler, R. J., and Wall, J. S. (1967)., Protein composition of dent, waxy and high-amylose corns, *Cereal Chem.*, 44, 160-169.

Bradbury, D., Wolf, M. J., and Dimler, R. J. (1962). The hilar layer of white corns, *Cereal Chem.*, 39, 72-78.

Brautlecht, CiA, (1953). *Starch, Its sources, Production and Uses,* Reinhold Publishing Corp., New York.

Brekke, O. L. (1967). Corn dry milling: Pretempering low-moisture com, *Cereal Chem.*, 44, 521-31.

Brekke, O. L. (1968). Corn dry-milling: Stress crack formation in tempering low-moisture corn, and effect on degerminator performance, *Cereal Chem.*, 45, 291-303.

Brekke, O. L. (1970). Dry milling artificially dried corn: Roller milling of degerminator stock at various moistures, *Cereal Science Today*, 15, 37-42.

Brekke, O. L., and Kwolek, W. F. (1969). Corn dry milling: Cold tempering and degermination of corn of various initial moisture contents, *Cereal Chern.* 46, 545-49.

Brekke, O. L., and Weinecke, L. A. (1964). Corn dry-milling, A comparative evaluation of commercial degermer samples, *Cereal Chern.*, 41, 321-28.

Brekke, O. L., Weinecke, L. A., Boyd, J. N., and Griffin, E. L., Jr. (1963). Corn dry milling: Effects of first-temper moisture, Screen perforations, and rotor speed on Beall egerminator throughput and products, *Cereal Chem.*, *40*, 423-29.

Burton, W.G. (1989). Specific gravity as a guide to the content of dry matter and of starch in potato tubers. In 'The Potato'. 3rd Ed. Longman Scientific and Technical. Essex, England.

C.F.T.R.I. (1981). Home-Scale Processing and Preservation of Fruits and Vegetables. Central Food Technological Research Institute, Mysore.

Chadha, K. L. and G. Kalloo, Eds. (1993). Advances in Horticulture. Vol. 6-7, Malhotra Pub. Co. New Delhi.

Chadha, K. L. and O. P. Parekh, Eds. (1993). Advances in Horticulture. Vol. 1-5, Malhotra Pub. Co. New Delhi.

Chakraverty, A. (2007). Post harvest technology of cereals, pulses and oilseeds. Oxford and IBH Publishing Co. Pvt. Ltd. New Delhi.

Chandra, P. K. (1972). Methodology and Procedures for evaluating performance of commercial rice mills, Unpublished M. Tech. Thesis, L I. T., Kharagpur.

Chavan, U.D. and Chavan, J.K. (1992). Utilization of groundnut flour in preparation of sorghum bhakari. J. Maharashtra Agric. Univ., 17 (20): 346-347.

Chavan, U.D., Chavan, J.K. and Kadam, S.S. (1988). Effect of fermentation on soluble proteins and *in-vitro* protein digestibility of sorghum, green gram and sorghum + green gram blends. J. Food Sci. 53: 1574-1575.

Chevassua-Agnes, S., Favier, J. C. and Joseph, A. (1976). Traditional technology and nutritive value of sorghum beer from camaoon (in French) Cah. Nutr. Diet. 11: 89-104.

Chopra, R. N.; Nayar, S. L and Chopra, I. C. (1956). Glossary of Indian Medicinal Plants. Pubd. by C.S.I.R., New Delhi.

CSIR, (1980). The Wealth of India: Raw Materials, Councill of Scientific and Industrial Research, New Delhi.

Dandekar Machinery Works, Catalogue for rice and pulse milling machinery, *Mis.* G. G. Dandekar Machine Works Ltd., Bhiwandi, Thana, Maharashtra, India.

Dastur. J. F. (1970). Medicinal Plants of India and Pakistan. Pubd. by D. B. Taraporewala Sons and Co. Pvt. Ltd., Mumbai.

Dauthy, M.E. (1995). Fruit and Vegetable Processing. FAO Agricultural Services Bulletin 119. Food and Agriculture Organisation of the United Nations, Rome.

Desrosier, N.W. and Desrosier, J.N. (1987). The Technology of Food Preservation. CBS Publishers and Distributors, New Delhi.

Desrosier, N.W. and Tressler, D.K. (1977). Fundamentals of Food Freezing. The AVI Publishing Company, Inc., Westport, Connecticut.

Earle, F.,R., Curtis, J. J., and Hubbard, J. E. (1946). Composition of the component parts U1 the corn kernel, *Cereal Chem.*, 23, 504-11.

Fennema, O.R. Ed.(1976). Principles of Food Science, Part-l. Food Chemistry, Marcel Dekker Inc, New York.

Frazier, W.e. and Westhoff, D.C. (1983). Food Microbiology. Tata McGraw-Hill Publishing Company Limited, New Delhi.

Gauri Shanker (1984). Practical Manual in Horticulture. Kitabistan, Allahabad.

Girdhari Lal, Siddappa, G.S. arid Tondon, G.L. (1986). Preservation of Fruits and Vegetables. Publications and Information Division, Indian Council of Agricultural Research, New Delhi.

Goel, A. K., Kumar, R., and Mann, S. S. (2007). Post-harvest management and value addition. Daya Publishing House, Delhi.

Gopalan, C, Rama Sastri, B.V. and Bala Subramanian, S.C. (1982). Nutritive Value of Indian Foods. National Institute of Nutrition, R.C.M.R., Hyderabad.

Grist, D. H. (1959). Rice (3rd ed), Longmans Green and Co. Ltd., London.

Haard, N. F. and DK Salunkhe, Eds. (1975). Post-harvest Biology and Handling of Fruits and Vegetables, AVI, Westport, CT.

Hardenburg, R. E., A.E. Watada and C. Y. Wang, (1986). The Commercial Storage of Fruits, Vegetables and Florist and Nursery Stocks, Agricultural Handbook 66, US Dept. Agric. Washington D.C.

Harris, RS. and Karmas, E. (1975). Nutritional Evaluation of Food Processing. 2nd Edition. AVI Publishing Co. Westport, Connecticut.

Haung, M. T, O. Toshihiko, C. T Ho and R.T. Rosen, (1994). Food Phytochemicals for Cancer Prevention, I. Fruits and Vegetables ACS Symposium Series 546, American Chemical Soc. Washington, D. C.

Hofman, V., et al. (1979). Sunflower Oil as a Fuel Alternative. North Dakota Ext. Bul. 13 AENG-735. 4 pp.

Hofman, V., et al. (1981). Sunflower for Power: North Dakota Ext. Circ. AE-735. 12 pp

Houston, D. F. (1973). Western Marketing and Nutrition Research Division, Agricultural Research Service, U. S. Department of Agriculture, Berkeley, California.

Hulme A.C. (1978). Biochemistry of Fruits and Their Products, Vol. I and. II. Academic Press, London.

Ilyas, S. M. (2003). Importance of post-harvest processing and value- addition for additional income and employment generation. Presented at National Symposium on Resource Management for Eco-friendly Crop Production, held at CSAUA and T, Kanpur, February, 26-28.

Ilyas, S. M., Wanjari, O. D. and Goyal, R. K. (2002). Prospects of income and employment generation through agro-processing and value-addition. Presented at Conference on Emerging Opportunities of Agro and Food Processing Industries in Jharkhand held at Indian Institute of Coal Management, Ranchi, during October, 8-9, 2002.

Ilyas, S.M. and Coyal, R.K. (2002). Quality assurance in food processing-impact of WTO, In: Proc. National Seminar on Development of Food Processing Industries in U.P., held at Lucknow during January 16-17, 2002.

Jacob, T. (1987). Poisons in our Food. Publications Division, Ministry of Information and Broadcasting, Government of India, New Delhi.

Javaraman. J. (1981). Laboratory Manual in Biochemistry. Wiley Eastern Limited, New Delhi.

Kader A. A. Ed. (1992). Post-harvest Technology of Horticulture Crops. University of California, Publication No. 3311.

Kaljee, Linda M.; Thiem, Vu Dinh; von Seidlein, Lorenz; Genberg, Becky L.; Canh, Do Gia; Tho, Le Huu; Minh, Truong Tan; Thoa, Le Thi Kim; Clemens, John D. and Trach, Dang Duc (2004): Healthcare Use for Diarrhoea and Dysentery in Actual and Hypothetical Cases, Nha Trang, Viet Nam. *Journal of Health, Population and Nutrition* 22(2): 139-149.

Karel, M., O.R. Fennema and D.S. Lund, Eds. (1975). Principles of Food Science Part II. Physical Principles of Food Preservation. Marcel Dekker, Inc. New York.

Kewdar, A. A. (1986). Biochemical and physiological basis for effects of controlled and modified atmospheres on fruits and vegetables. Food Technol. 40: 99-104.

Kordylas, J.M. (1991). Processing and Preservation of Tropical and Subtropical Foods. Macmillan Education Ltd., Houndmills, Basingstoke, Hampshire.

Kuprits, Y., ed. (1967). Technology of grain processing and provender milling, Tekhnologiya perer atbotki zernai Kombikormovoe, Izd. 'Kolos': Moscow (1 per cent 5). (Translation by Israel Program for Scientific Translation, Jerusalem, Israel)

Lal, G., G.S. Siddappa and G. L. Tandon, (1986). Preservation of Fruits and Vegetables, Indian Council of Agricultural Research, New Delhi.

Ludivina S. de Padua; Gregorio C. Luged and Juan V. Pancho (1977). Hand book on Philippine Medicinal Plants (Vol-1).Pubd. by Documentation and Information Section, Univ. of the Philippines at Los Banos.

Macrae, R., R. K. Robinson and M. J. Sadler, Eds, (1993). Encyclopaedia of Food Science, Food Technology and Nutrition, Academic Press, London.

Mahattanatawee, K.; Manthey, J.A.; Luzio, G.; Talcott, S.T.; Goodner, K. and Baldwin, E.A. (2006): Total antioxidant activity and fiber content of select Florida-grown tropical fruits. *Journal of Agricultural and Food Chemistry* 54(19): 7355–7363. doi:10.1021/jf060566s

Mahfuzul Hoque, M.D.; Bari, M.L.; Inatsu, Y.; Juneja, V.K. and Kawamoto, S. (2007): Antibacterial activity of guava (*Psidium guajava* L.) and Neem (*Azadirachta indica* A. Juss.) extracts against foodborne pathogens and spoilage bacteria. *Foodborne Pathogens and Disease* 4(4): 481–488. doi:10.1089/fpd.2007.0040

Manosroi, J.; Dhumtanom, P. and Manosroi, A. (2006): Anti-proliferative activity of essential oil extracted from Thai medicinal plants on KB and P388 cell lines. *Cancer Letters* 235(1): 114–120.doi:10.1016/j.canlet.2005.04.021 PMID 15979235

Markley, K. S. (1950). Soybean and Soybean Products. Inter-science Publishers, Inc., New York

Meena, RK and Yadav, J.S. (2001). Horticulture Marketing and Post Harvest Management Pointer Publishers Jaipur (Raj.).

Misra, J.B. (1983). A simple arrangement for the accurate determination of the specific gravity of potato tubers. J. *Indian Potato Assoc.* 10: 121-128.

Mudambi, S.R and Rao, S. (1986). Food Science. Wiley Eastern Limited, New Delhi.

Mukhtar, H.M.; Ansari, S.H.; Bhat, Z.A.; Naved, T. and Singh, P. (2006): Antidiabetic activity of an ethanol extract obtained from the stem bark of *Psidium guajava* (Myrtaceae). *Pharmazie* 61(8): 725–727.

Nagy, S. and P.E. Shaw, Eds. (1980). Tropical and Subtropical Fruits AVI, Westport, CT.

Negi, J.P., Singh, B. and Dagar, KS. (2000). Indian Horticulture Database-2000. National Horticulture Board, Ministry of Agriculture, Government of India, Gurgaon (Haryana).

Nissen, M. (1967). The weight of potatoes in water. Further studies on the relation between the dry matter and starch content. *Eur. Potato* J. 10: 85-99.

Odunfa, S. A., Nordsrom, J. and Adeniran, S. A. (1994). Development of starter cultures for nutrient enrichment of ogia West African fermented cereal gruel. Report submitted to HBVC research grants program. USAID, Washington, U.S.A.

Oh, W.K.; Lee, C.H.; Lee, M.S. *et al.* (2005): Antidiabetic effects of extracts from *Psidium guajava.* J. Ethnopharmacol. 96(3): 411–415. doi:10.1016/j.jep.2004.09.041

Pandey, P.H. (1997). Post-harvest Technology of fruits and Vegetables (Principles and Practices). Saroj Prakashan, 646-47, Katra, Allahabad.

Pantastico, Er. B. (1975). Post-harvest Physiology, Handling and Utilization of Tropical and Subtropical Fruits and Vegetables. The AVI Publishing Company, Inc., Westport, Connecticut.

Paulus, K. O. (1984). Modelling in Industrial Cooking. In: *Thermal Processing and Quality of Foods.* Zeuthen *et al.* (eds.), Elsevier Applied Science Publishers, London.

Potter, N. N. (1984). Food Science. 3rd Edition. The AVI Publishing Company, Inc., Westport, Connecticut.

Potter, N.N. (1973). Food Science. The AVI Publishing Company, Inc., Westport Connecticut.

Potter, N.N. and Hotchkiss, J.H. (1996). Food Science. 5th Edition. CBS Publishers and Distributors, New Delhi.

Potty, VH. (1988). Horticultural Industry in India. UNIDO Consultation Meeting, Beijing, Nov. 22-24.

Powar, C.B. and Daginawala, H.F. (1986). General Microbiology Vol II Himalaya Publishing Company, Bombay.

Primo, E., Barber, S., Tortosa, E., Camacho, J., Ulldemolins, J., Jimenez, A, and Vega, R, (1970). Chemical composition of thebyproducts obtained in the different steps of the rice milling process (In Spanish), Rev. Agroquim Teenol, Alimentos 10': 244.

Pryde, E. H. (1981). Vegetable Oil vs. Diesel Fuel: Chemistry and Availability of Vegetable Oils. Alcohol and Vegetable Oil as Alternative Fuels. Proceedings of Regional Workshop: USDA Peoria, IL.

Quick, G. (1980). Developments in use of Vegetable Oils as fuel for -Diesel Engines. ASAE Paper 80-1525r 15 pp.

Quick, Graeme R. (1980). An In-depth Look at Farm Fuel Alternatives. Power Farming Magazine, Feb. 10-17 pp.

Ranganna, S. (1979). Manual of Analysis of Fruit and Vegetable Products. Tata McGraw-Hill Publishing Company Limited, New Delhi.

Ravindran, C. B., Senthi Kumar and Murthy, B. N. S. (2004). Therapeutic Properties of Fruits. Agrobios News Letter, 2 (9): 50-52.

Ryall A. L. and W. T. Pentzer, (1982). Handling Transportation and Storage of Fruits and vegetables, Vol. 1. Vegetables and Melons, AVI, Westport.

Ryall, A.L. and W.T. Pentzer, (1982). Handling, Transportation and Storage of Fruits and vegetables, Vo1.2. Fruits and Tree Nuts AVI, Westport CT.

Salunkhe D. K. and S.S. Kadam, Eds. (1995). Handbook of Vegetable Science and Technology: Production, Composition, Storage and Processing, Marcel Dekker lnc., New York.

Salunkhe, D. K. and B. B. Desai, (1984). Post-harvest Biotechnology of Fruits (Vol. I and II), CRC Press, Boca Raton, Florida.

Salunkhe, D. K. and B. B. Desai, (1984). Post-harvest Blotechnotoqy of Vegetables (Vol. I and II), CRC Press Boca Raton, Florida.

Salunkhe, D. K. and S.S. Kadam, Eds, (1995). Handbook of Fruit Science and Technology: Production, Composition Storage and Processing, Marcel Dekker, Inc. New York.

Salunkhe, D.K., S.S. Kadam and S. J. Jadhav, (1991). Potato: Production, Processing and Products, CRC Press, Soca Raton.

Salunkhe, DK, H. R. Bolin and N. R. Reddy. (1991). Storage, Processing and Nutritional Quality of Fruits and Vegetables 2 nd Ed. Vol. II. Processed Fruits and Vegetables, CRC Press, Boca Raton, Florida.

Schrimshaw, N. S. (1974). Nutritional requirements for high protein foods, Proc. Int. Symp. Protein Foods and Concentrates, Central Food Technol. Res. Institute, Mysore, India, pp. 13-18.

Shakuntala Manay, N. and Shadaksharaswamy, M. (1987). Foods: Facts and Principles. Wiley Eastern Limited, New Delhi.

Shreve, N. R. (1956). The Chemical Precess Industries, McGraw-Hili Book Co. Inc., Tokyo.

Shrivastava, H. C. (1985). Oil Seeds Production Constraints and Opportunities. Oxford and IBH Pub. Co., New Delhi.

Singh, A. (1986). Fruit Physiology and Production. Kalyani Publishers, New Delhi.

Singh, G. (1997). Data book on Mechanization and Agro-Processing Since Independence, CIAE, Bhopal-462038.

Srivastasva, M (1985). Essentials of Food and Nutrition Vol. II (Applied Aspects). The Bangalore Printing and Publishing Company Limited, Bangalore.

Srivastava, R.K and Singh, S. (1987). Utilization of Mushrooms for Preparation of Different Types of Preserved Products. Government Fruit Preservation and Canning Institute,.U.P., Lucknow.

Srivastava, R.P. (1992). Preservation of Fruit and Vegetable Products. Bishen Singh Mahendra Pal Singh, New Connaught Place, Dehra Dun.

Subhadrabandhu, S. S. (1997). Ethnomedical uses of some tropical and subtropical fruits in Thailand. Chronica Horticulturae, 37 (4): 17.

Sukumaran, N.P. and Ramdass, C. (1980). A simple variable load potato hydrometer. *J. Indian Potato Assoc.* 7: 32-37.

Tauro, P., Kapoor, KK and Yadav, KS. (1986). An Introduction to Microbiology. Wiley Eastern Limited, New Delhi.

Thompson, R. A., and r-oster, G. H. (1963). Stress cracks and breakage in artificially dried corn, U. S. Dept. Agr., AMS, 631.

Tressler, D.K. and Joslyn, M.A. (1971). Fruit and Vegetable Juice Processing technology. The AVI Publishing Company, Inc., Westport, Connecticut.

USDA, (2012). United States Department of Agriculture and National Nutrient data base.

Verma, S.c. (1991). Potato Processing in India-An Appraisal. Central Potato Research Institute, Shimla.

Weichmann, J. (1987). Post-harvest Physiology of Vegetables. Marcel Dekker Inc. New York.

Wills, R.HH, T.H. Lee, D. Graham, W.B. Megalasson and E. G. Hall, (1981). Post-harvest: An Introduction to Physiology, Handling of Fruits and Vegetables, New South Wales University Press, Kensington, Australia.

Woodroof, J.G. and Luh, B.S. (1975). Commercial Fruit Processing. The AVI Publishing Company, Inc., Westport, Connecticut.

Woods, A.E. and Aurand, L.W. (1977). Laboratory Manual in Food Chemistry. The AVI Publishing Company, Inc., Westport, Connecticut.

Yamaguchi, M. (1983). World Vegetables: Principles, Production and Nutritive Values, AVI Westport CT.

Index

O

P

A. Photographs of different Types of Fruits

Acai berry

Apple

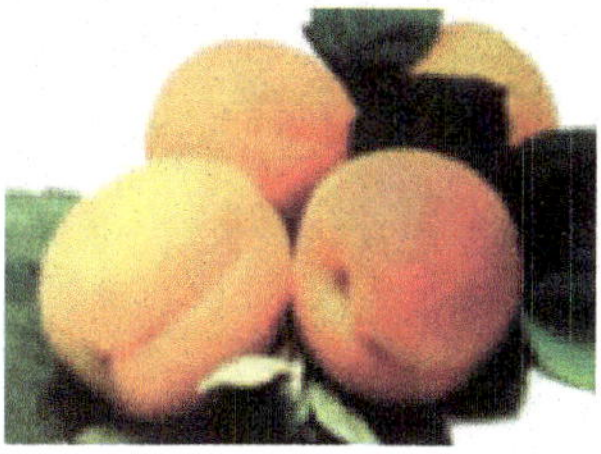

Apricot

Avocado

Banana

Black Berries

Blackcurrents

Blue Berries

Cantaloupe

Cherimoy

Cherry

Chokeberry

Cranberries

Dates

Durian Fruit

Figs

Goose Berries

Grapes

Grapefruit

Guava

Jackfruit

Kiwi Fruit

Kumquat

Lemon

Loquat Fruit

Lychee

Mango

Mangosteen

Mulberries

Olives

Orange

Papaya

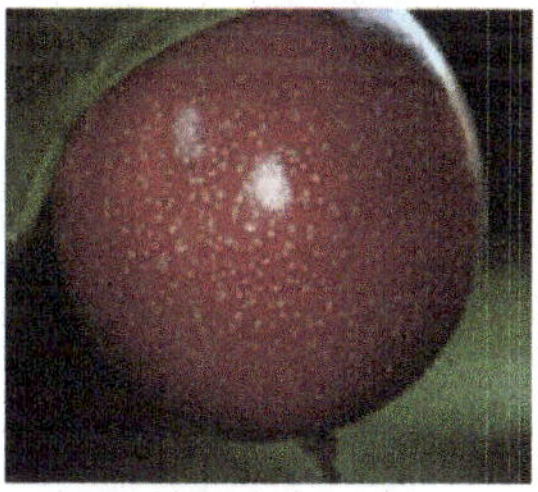

Passion Fruit

Peaches

Pears

Persimmon Fruit

Pineapple

Plums

Pomegranate

Quince Fruit

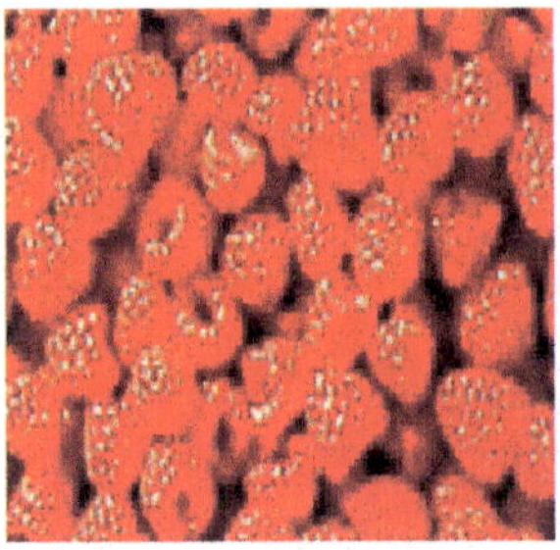

Raspberry

Sapota

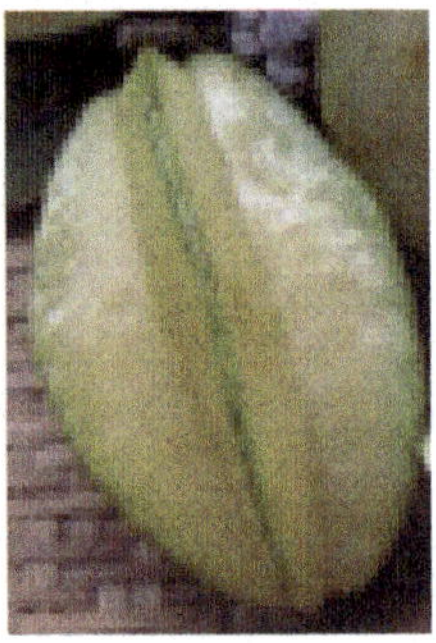

Star Fruit

Strawberry

Tangerine

Watermelon

B. Photographs of different Types of Vegetables

Artichoke

Arugula

Asparagus

Basella

Beets

Bell Pepper

Bitter Gourd

Bok Choy

Brocoli

Brussel Sprouts

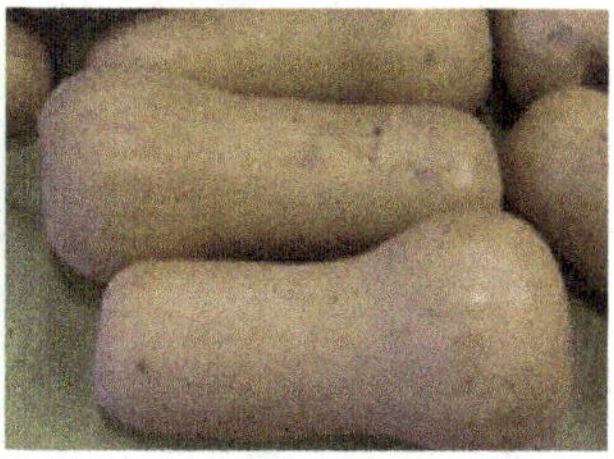

Butternut Squash

Cabbage

Carrots

Cassava

Cauliflower

Crollard Greens

Cucumber

Eggplant/Brinjal

Endive

Fennel

Fiddlehead Ferns

Grren Beans

Cabbage

Carrots

Cassava

Cauliflower

Crollard Greens

Cucumber

Eggplant/Brinjal

Endive

Fennel

Fiddlehead Ferns

Grren Beans

Jicama

Kale

Kohlrabi

Leeks

Lettuce

Lotus Root

Moringa Pods/Drumstick

Mustard Greens

Okra

Onion

Parsnipd

Peas

Potato

Pumpkin

Purslane

Radicchio

Radish

Rhubarb

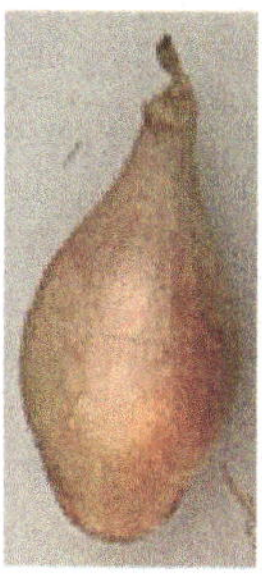

Shallots

Spinach

Sweet Corn

Sweet Potato

Swiss Chard

Tomato

Turnips

Watercress

Yams

Zucchini

C. Phtographs of different Types of Nuts and Seeds

Almonds

Brazil Nuts

Cashew Nut

Chestnuts

Chia Seeds

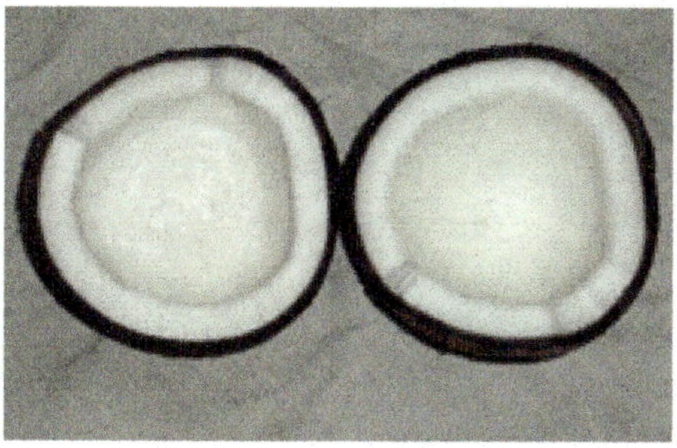

Coconut

Fax Seed/Linseed

Hazelnuts

Macadamia

Peanuts

Pecans

Pinenuts

Pistachio

Poopy Seeds

Pumpkin Seeds

Quinoa

Sesame Seeds

Sunflower Seeds

Walnut

D. Phtographs of different Types of Spices

Allspice

Anise Seed

Bay Leaf

Black Pepper

Capers

Caraway

Cardamom

Carom Seeds

Cayenne Pepper

Chilli Peppers

Cinnamon

Cloves

Coriander Seeds

Cumin

Fennel Seeds

Fenugreek Seeds

Horseradish

Jalapeno Peppers

Mace Spice

Mustard Seeds

Nutmeg

Saffron

Sichuan Peppercorns

Tamarind

Vanilla beans

E. Phtographs of different Types of Herbs

Basil Herb

Borage

Burdock Root

Celery

Chives

Cilantro

Dandelion

Dill Weed

Epazote

Garlic

Ginger

Lemongrass

Oregano

Parsley

Peppermint

Rosemary

Sage

Savory

Spearmint

Stevia

Stevia Dried Leaves

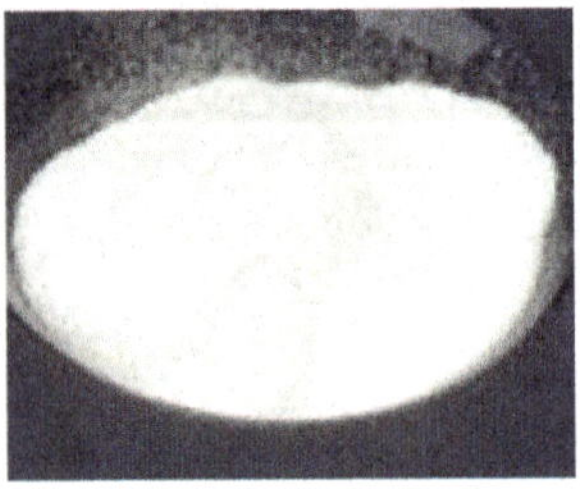

Stevia Sugar

Sweet Marjoram

Tarragon

Thyme

Turmeric

Lightning Source UK Ltd.
Milton Keynes UK
UKOW06n0636131116
287343UK00001B/9/P

9 789351 306153